COLT

COLT

The Making of an American Legend

William Hosley

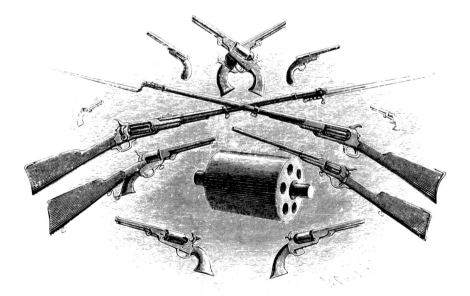

University of Massachusetts Press, Amherst

Published in association with the Wadsworth Atheneum, Hartford

In memory of Eileen (Tony) Learned (1908-1995). She embodied the qualities of self-sovereignty.

To Christine Ermenc, whose companionship, wisdom, and love are my constant inspiration.

*To Hartford, Connecticut—the inner mounting flame of cantankerous old New England.
Its light is inextinguishable so long as memory endures.*

THE ROAD LESS TAKEN IS PAVED WITH LOVE.

LC 96-24139
ISBN 1-55849-042-6 (cloth); 043-4 (pbk.)

Design and composition by *Group* C Inc New Haven/Boston
(BC,CK,MK,FS,CW,EZ)
Set in Adobe Linotype Centennial
Printed and bound by Friesens Corporation

Library of Congress Cataloging-in-Publication Data
Hosley, William N.
Colt : the making of an American legend / William Hosley.
p. cm.
Includes bibliographical references (p.).
ISBN 1-55849-042-6 (cloth : alk. paper). — ISBN 1-55849-043-4 (pbk. : alk. paper)
1. Colt, Samuel, 1814-1862.
2. Gunsmiths—Connecticut—Hartford—Biography.
3. Colt revolver—History.
4. Hartford (Connecticut)—History.
I. Title.
TS533.62.C65H67 1996
683.4'0092'2—dc20
[B] 96-24139
British Library Cataloguing in Publication Data are available.

Page three: Illustration from *Armsmear* depicting a variety of Colt firearms. Engraving by Nathaniel Orr.
Unless otherwise credited, all illustrated items belong to the Wadsworth Atheneum.

CONTENTS

PREFACE

In 1986 the Wadsworth Atheneum's director called my office, anxious for a report on the status of the museum's collection of Colt firearms. The morning news carried a front-page account of an alleged scandal involving the deaccession of Colt guns by a sister institution. The press wanted to know if the Atheneum's Colt collection was intact. Had we sold anything? The fact that I, as the curator of all things American that are not pictures or sculpture, had never paid much attention to this neglected part of the collection was reason enough for concern. I scurried around and was relieved to discover that Sam Colt's personal collection of firearms, bequeathed to us by his widow Elizabeth Hart Colt in 1905, was very much intact and in relatively good order (albeit locked away and inaccessible to our visitors).

Little did I know that this inquiry would set in motion a sequence of discoveries and realizations that would bring the Colts, the Colt legacy, and—most of all—the Colt Collection back to center stage. It was timely. The Atheneum was then reveling in the excitement of having originated its first-ever touring loan show, an exhibition of the collection of the other great name in the pantheon of Atheneum benefactors, J. Pierpont Morgan. The museum's then chief curator, Gregory Hedberg, was so delighted with the success of Morgan that he seriously considered reinstalling much of the museum as a "patronage story," an interesting proposal for an institution connected with patrons, artists, and visionaries like Morgan, Colt, Daniel Wadsworth, E. Everett Austin, Thomas Cole, Wallace Nutting, and Sol Lewitt. The Atheneum was also then hosting a national touring exhibition of American women artists. Hedberg became intrigued with our "other Colt collection," the paintings assembled by Sam's widow Elizabeth. Thrilled to imagine so distinct a claim for the museum as home to America's "first woman collector," Hedberg set out to rehabilitate the Colt collection and to tell the story of our most accomplished female patron.

Changes in administration and priorities, and the realization that both the Colt story and our Colt collection posed challenges of no small magnitude or perplexity, deferred progress. It subsequently became the responsibility of a department little versed in the story and almost entirely ignorant of the character, composition, and context of the collection to make sense of it. It turned out to be a collection twenty times larger than just the guns and of such astonishing diversity and so intensely personal that it had been dismissed years earlier as too bizarre and of no enduring artistic merit. A curatorial colleague recently described the Colt Collection as "memorabilia... of marginal interest," including "lots of cheap nineteenth-century travel souvenirs" that rarely "rise above run-of-the mill Victorian taste," including "European objects" that suggest a level of "taste" that was "uninformed and ordinary."[1] Thus the American Decorative Arts Department became curators of the Colt Collection by default. Aside from the guns and maybe five of the paintings, the collection had fallen into the abyss of forlorn and forgotten agendas from the museum's past. Was it really just too bizarre? Maybe we needed to find a new way of seeing.

Cocurator Karen Blanchfield and I thus set out on a journey into the unknown, and fate being what it is, the snaps, crackles, and pops commenced almost at once. Who could have guessed from the outset that our foray into *this* slice of Victorian life (we'd had others) would force us to confront a wide range of cultural, institutional, and occupational taboos, the whole perplexing legacy of guns in America, the conflict between art and history, and the ongoing challenge of creating collection-based experiences that museum visitors will find entertaining and worthwhile. Early on we discovered the striking parallel (more in kind than degree) between what we were trying to do and what Sam and Elizabeth Colt had done. The Colts sought and occupied many of their generation's frontiers. By confronting the foreign, the exotic, and the unknown and—by faith, conviction, and by hook and by crook—using the knowledge thus gained to pursue the Victorians' cherished quest for "progress," Sam and Elizabeth helped us to better understand the meaning of innovation and the powerful and unyielding legacy of the Victorian age. From the vantage point of a U.S. city at the twilight of the industrial era, the sense of almost living with and in a structure largely created by the Victorians is all too familiar. It is not always a pretty sight.

The Colt project was not lacking in friends or adversaries. Both helped. When the National Endowment for the Humanities withheld its support, declaring it irresponsible to exhibit guns in "minority Hartford," it ratcheted up our determination to redeem, if not the Colts' name, at least the context that gave their work meaning. To study people who overcame mountains of opposition to fulfill their dreams demanded that we attempt to do likewise. Doors close, windows open. We got lucky the day our friends at Fleet Bank, President Richard Higginbotham and his associates Karyn Cordner and Debbie French, stepped forward as our first sponsors. Fleet Bank is today a financial empire no less globally driven than Colt's manufacturing empire was in its day and, while I confess to regretting that none of Hartford's ancient banking dynasties survived the mergers and acquisitions of the 1980s, I couldn't be more delighted that our neighbors in Boston and Providence have secured for New England a position of strength and viability in one of the world's most competitive industries. Fleet has been a faithful ally in rehabilitating one of this community's great stories.

Another window opened when an old friend and kindred spirit, Bruce Fraser, executive director of the Connecticut Humanities Council, accomplished his own dream by creating New England's first large-scale, state-sponsored, merit-based grant-making fund to support programs in the humanities that facilitate cultural tourism. We are all tourists now, especially in the wilderness of our own backyards. What Bruce has done, what Connecticut is doing, is preserving the fragile thread of state and civic memory so that ours and future generations will be able to locate themselves in the continuum of history. That doing so will make those locations more stimulating, meaningful, and entertaining for visitors and residents alike is inevitable.

The State of Connecticut assisted our efforts directly through its Department of Economic Development grants to aid in marketing the state's cultural amenities and tourist attractions. Getting the word out is essential, and more difficult and expensive than ever.

The Hartford Foundation for Public Giving, the jewel of Hartford's charitable trusts, also stepped forward to help as they have in so many tangible, intangible, and often less glamorous ways. The Hartford Foundation is indeed a treasure, and living proof of the frugality and generosity of the hundreds of men, women, and families whose gifts and bequests have earned it a prominent place among this nation's civic foundations. Christopher Hall, director of programs, and James English Jr., the chair of its Distribution Committee, were great advocates for Colt.

Several other individuals and corporations stepped forward to help, including Carol Autorino and her husband, Anthony, president of Shared Technologies, whose stewardship of and passion for the Colt name has made Tony an animated and visionary presence in our work.

A major contribution was also made by my predecessor, the late Henry Maynard, the Wadsworth Atheneum's first professional curator of American arts, whose bequest to the museum was the culmination of his own faithful stewardship of a collection and institution he loved. We are also indebted to the bequest of a friend, the late Robert H. Schutz, whose parents were friends of Mrs. Colt's and who spoke affectionately of Victorian Hartford, almost as if it were in the next room. Thirty years ago "Schutzee," as he was known by his many friends, played a major role in rehabilitating both the Goodspeed Opera House in East Haddam, Connecticut, and the Mark Twain House in Hartford, a legacy of stewardship for our Victorian-age heritage that was years ahead of its time.

The Atheneum's Decorative Arts Council, presided over by Melinda M. Sullivan, supplied badly needed funds to build the cases and environments used to present the collection.

As I write, we are in the midst of negotiations with the owner and CEO of Colt's Manufacturing Company, whose efforts to preserve the integrity of a legend here in Connecticut is inspirational. Together, and in partnership with Connecticut Public Television, we will produce a documentary film to render aspects of the Colts' story in a way that complements our efforts in the media of exhibitry. It is a welcome collaboration.

Without extraordinary financial support, projects like this cannot happen. But we depend no less on people who invest their knowledge, skills, and expertise in the work. I thank the scholar members of the Colt Advisory Committee whose ongoing support and goodwill added so much to the quality and character of our work: Merritt Roe Smith, director of the Program in Science,

Technology, and Society at MIT; Kathleen McCarthy, director of the Center for the Study of Philanthropy at the City University of New York; R.L. Wilson, the godfather of Colt studies; David Barquist, our colleague in the decorative arts at Yale University Art Gallery; and, most of all, Kenneth L. Ames, a guiding light from Winterthur's golden age and one of this generation's most compelling and visionary museologists—to all, our thanks.

To Ellsworth S. Grant, I extend the warmest thanks and appreciation for his pathbreaking work not only on the Colts but also on so many related aspects of Hartford and Connecticut history. Ellsworth is a civic historian in the best tradition, a man whose purposeful and meticulous research and writings are so obviously guided first and foremost by a love of place. It has been an honor to follow his footsteps.

At the Atheneum and among sister institutions in and near Hartford there have been many friends and kindred spirits. I thank Gina Lionette, the Atheneum's marketing director, for her energetic pursuit of excellence; Monique Shira for her steady and persistent hand as publicist and advocate; John Teahan, the Atheneum's librarian and my close confidant and friend; Muriel Fleischmann and Pamela Toma for good cheer and support on the fund-raising front; Zenon Gansziniec for his diligence at cleaning, conserving, and stabilizing the collection; our registrars, Martha Small and Mary Schroeder, for orchestrating loans; Cecil Adams and Edd Russo for implementing and embellishing the design of the exhibition; Linda Friedlaender, associate curator of education; and, finally, most of all, to acting director Kristin Mortimer and former director Patrick McCaughey, without whose support, enthusiasm, humor, and refusal to take no for an answer, the Colt project would have never taken hold.

Among our colleagues in sister institutions I especially thank Judy Johnson, Gary Waite, Christine Bobbish, and Ruth Blair at the Connecticut Historical Society; Jo Blatti and Elizabeth Newell at the Harriet Beecher Stowe Center; Steven Beatty at the U.S. Springfield Armory National Park museum in Springfield; Roy Jinks, historian at Smith and Wesson in Springfield; Jeff Kaimowitz and Alesandra Schmidt at Trinity College's Watkinson Library; and the staff of the Connecticut State Library and its Museum of Connecticut History, especially the museum's director, Dean Nelson, who surpassed everyone in generosity, thoughtfulness, and insight into the qualities of our endeavor.

Finally, I thank Clark Dougan, senior editor at the University of Massachusetts Press, who, with the able support of Kevin Sweeney of Amherst College, transformed a lengthy manuscript into what we hope will be a readable book.

Enough said. The community of inquiry that revolves around the Colts is large, including friends and advocates in Springfield, Massachusetts, up and down the Connecticut River, and across the country.

Karen Blanchfield could not be a better colleague, could not have done more with less, could not have put our mission ahead of personal needs more emphatically, and could not have been a more creative and uplifting spirit in all that has transpired. In speaking for us, I hope I speak of a union of purpose and of the possibilities of teamwork. She is an unreplaceable presence at the core of what our department does, for which I am most grateful.

In the pages that follow we will journey through the Colts' world and encounter, as they did, the conditions that gave rise to the industrial civilization of the nineteenth century. What the reader will not find is a strict chronology of the Colts' lives. This book frames the Colts' stories in relation to the themes and issues in which they were significant players. It is also structured in an effort to extract meaning from the collection of art, firearms, and artifacts bequeathed to the Wadsworth Atheneum by Elizabeth Hart Colt in 1905. In that sense our effort was archaeological and involved interpreting the fragmentary evidence of a collection that had been neglected, in part, because the contexts that originally gave it meaning had been lost. Finally, this work was designed to reassert the relationship between the Colts and the primary place where their drama unfolded, Hartford, Connecticut. Indeed, *Armsmear: The Home, the Arm, and the Armory of Samuel Colt, a Memorial* (1866), the only authorized Colt biography, is also the only study of Colt that attempts to balance the civic, industrial, and personal aspects of his story.[2] But it ends where Elizabeth Colt's story begins and is, therefore, an incomplete rendering of what was, emphatically, a partnership.

When we first conceived the exhibition that gave rise to this book, we hoped it might tour the country, especially to museums with an interest in the American West. I shall not soon forget a prominent western museum's rejecting our script for adopting what they described as an "eastern theme" despite their claiming to be "interested in almost everything pertinent to the Colt firearm." Why the personal, social, and industrial contexts that gave rise to the "gun that won the West" is not "pertinent" I cannot say. But the implication only fueled our desire to reassert the obvious fact that the Colts were easterners and that the much-admired Colt revolver was a direct outgrowth of their personal histories, place of work, and worldview. I confess that I do not subscribe to the nineteenth-century's cult of genius. Environment may not make all the difference, but it is surely important and, as Colt's story testifies, even the most original of ideas backed by the most irrepressible of temperaments, when planted at the wrong place or time, amounts to little. If the gun is important, then the environment in which it was developed is important. The Colts were shaped by and helped shape the place where they lived out their lives. So, yes, this is an eastern version of the Colts' story. It also aspires to be balanced in the weight given to place, product, process, and personalities. And it is emphatically her story as well as his. Sam and Elizabeth were a partnership no less romantic than the legends of the American West in which the Colt revolver is also, assuredly, a part.

Striking the balance did not come easy. From the outset the basic outline appeared less promising: Man invents gun. Man perseveres against odds to make fortune. Man meets woman. Man dies and leaves woman great fortune. Woman spends forty years applying spot-remover to the stain of a fortune made selling guns. Not surprisingly, the reality was not quite so pat. At each fork in the road the Colts faced difficult choices. They proceeded in ways that were often surprising and rarely easy or inevitable. As for Elizabeth Colt's "good works," if the gun was already a progressive and socially acceptable symbol of the age, then we must look beyond the presumption of gun-money guilt to explain her actions. The answer, I think, lies in the inner fire and strong sense of purpose that, more than guns, or religious piety, or the pursuit of an industrial utopia, are the tie that bound Sam and Elizabeth in "faithful affection," one to the other and each to the impulse to make the world a better place than they found it. Whether you think they succeeded in doing so, few could deny that the Colts were among the most intriguing and progressive characters of the industrial age. That said, I confess that more than once during this project I wished the fortune was based on a product as seemingly innocuous as sewing machines, cameras, or automobiles. It certainly would have made our efforts at fund-raising and booking a tour less difficult. We played the cards we were dealt. That it was guns only sharpens the edges between their world and ours, thus making the story more interesting through the perplexity of its contradictions.

As we come to the end of the twentieth century, Hartford is not what it once was. What older American city is? And yet, who with a memory of what American cities once were or, at their best, still are does not feel that they represent the culmination of our civilization and its highest, noblest, and most vibrant possibilities? I remember the city of my youth, Rochester, New York, as a boundless, optimistic, and inspiring place, a place of history and beauty, visual and cultural richness. No aspect of the Colt project has been more poignant or rewarding than the encounter it has provided with the great city-builders of the Victorian age. In seeking the points of intersection between my work as a museum curator and my devotion to the cause of civic history and memory, I have found much to be joyful about. On workday mornings as I cross Bushnell Park past its glorious Soldiers and Sailors Memorial Arch, I cannot help feeling a sense of awe and responsibility and a sense that we are but temporary stewards of the places we inherit and inhabit. Building communities is the work of generations. Fostering a sense of community is a task museums and institutions devoted to learning and discovery are well suited to do. By careful stewardship the resources of the past cannot fail to build a more trustworthy and secure bridge to the future. Our effort to rehabilitate the Colts' collections and the stories associated with the industrial civilization they helped create is but a plank in that bridge. May it fortify all who care to cross.

MEET SAM AND ELIZABETH

Of his virtues and faults—and he had both—we do not now speak.
HARTFORD EVENING PRESS, 1862

1. *Elizabeth Hart Colt*
   (1826–1905), ca. 1867.
   R.S. DeLamater, Hartford,
   photographer.

On Friday morning, January 10, 1862, Elizabeth Colt watched from her boudoir as workmen quietly and deliberately broke through the frozen ground, lifting shovelfuls of earth from the grave site of her deceased husband. Two infants, namesakes of Sam and Elizabeth, already lay buried in the enclosed plot in a grove of trees near the pond on the grounds below. Unaware that she was again pregnant, and with a clinging three-year-old son and a sickly half-year-old daughter, Elizabeth Colt now faced the future alone. At thirty-five she inherited an industrial empire and fortune worth today's equivalent of about $200 million.

Forty years later, having outlived her younger siblings, her children, and most of her husband's associates and friends, Elizabeth Colt worked out the details of her last memorial, a monument to Sam Colt erected by "his wife in faithful affection" on the spot outside her window where he was buried. There had been countless other memorials along the way. Faithful affection and memory had become as much her calling as perfecting the manufacture of firearms with machines had been his. The fortune, still princely, was much diminished by benefactions that transformed the place she called home as assuredly as had the industrial empire that made her devotions possible. Hartford, Connecticut, would never be the same.

Elizabeth Hart Colt (fig. 1) craved privacy almost as much as her husband craved recognition and fame. Before her own death in 1905, at the age of seventy-nine, having set her affairs in order, she destroyed her private papers. A prolific correspondent and a gifted but private poet, she left a legacy of good works and improvements but few words. "The main spring is broken," she wrote at the time of Sam Colt's death, "the works must run down."[1]

Colt's Armory, which had recently doubled its capacity, running multiple shifts since the outbreak of the Civil War, went silent. Its enormous steam engines, belt shafting, and deafening drop forges were now quiet. What was to become of it all? The snow crackled underfoot in the cold winter air and the anxiety was thick on Saturday morning as the armory's thousand workmen filed into Charter Oak Hall. As company officers eulogized their fallen president, a life cut short at age forty-seven, many wondered how they could go on without the man whose name and reputation were inseparably identified with the business he had created.

Tuesday's afternoon funeral was the last drill for a man widely known as "the Colonel." Despite the martial atmosphere that permeated his company town, Sam Colt never wielded arms in combat, never fired at another human being, and was never more than a Sunday afternoon soldier. His lifelong obsession with the military was an abstraction. Who would have guessed this as the workmen, black crepe around their arms, lined up single file to view the body that lay in state in the reception room at Armsmear, the big house on the hill overlooking the Colts' industrial empire on Hartford's South Meadows?

Sam Colt (fig. 2) planned his exit, plotting the site and approving designs for the tomb that would hold his remains. In death, even in January, his body was adorned with fresh camellias from the family greenhouse, the most expansive and exotic in New England. After a service at St. John's Episcopal Church, the silver-mounted steel casket bearing Sam Colt was carried solemnly to the grave site by his beloved political cronies and a delegation of senior aides from the armory. As the bearers approached the grave site, the Armory Band commenced a dirge on its brass instruments. Captain Lewis and the men of Connecticut's 12th Regiment, Company A, the militia unit Colt raised at the beginning of the Civil War, stood in line facing the Putnam Phalanx, Hartford's gentleman's militia dressed in uniforms "modeled in the style of 1776" after their namesake and hero Gen. Israel Putnam. The band stopped playing while

2. Colonel Sam Colt (1814–62), ca. 1854. John C. Buttre, Ridgefield, N.J., based on a photograph by P. Graff, Berlin. This is the earliest portrait of Sam Colt at the height of his fame and is based on a daguerreotype photograph that was probably taken while he was in Berlin in 1854.

Phalanx officer and pallbearer E.D. Tiffany read a resolution on behalf of their fallen friend: "Hartford may well weep over the early Fall of her honored son, to whom she owes much of her prosperity and glory, and the fruits of whose genius and industry will remain to bless our city through unborn generations.... We will wear... the badge of mourning for one year."[2]

Prosperity and glory. The casket was lowered into the tomb and as its marble lid—smooth, white, polished, and engraved with the words "Kindest husband, father, and friend, adieu"—was sealed in place, a community stood frozen in wonder, struggling to bring order to the disorderly feelings that accompanied the loss of this strange and strong-willed man. Had Colt died too young or not young enough? Even his closest friends must have wondered.

Forty-three years later, as friends, family, and well-wishers filed by Elizabeth Colt's open casket to the low, steady tones of Chopin's Funeral March, the August sun radiant, the casket covered with purple velvet and "piled high" with flowers beneath the Colt Memorial Window in the Colt Memorial Church at the north end of Coltsville, few doubted whether "the widow's God," to whom she had long prayed, had at last reunited her with husband and family in heaven. Even in death, lying beneath the hard light and brittle contrasting colors of the stained glass window bearing her husband's likeness in the Old Testament character of Joseph of Egypt, the softness of countenance was undiminished. "That strong, beautiful face transparent to the strength and beauty of the sanctuary of her soul" was how one admirer phrased it. Survived by five nieces, three bearing her name, Elizabeth Colt knew how to make a funeral memorable and seemed to have stage-managed her own, as she had for so many others before, providing flowers that in winter made light of the seasons, hinting at the hopefulness of perpetual abundance.[3]

Her life was described as one "full of good works." She was a woman who had "overcome the ensnaring temptations which attend wealth and position.... She was nobody's enemy; she was everybody's friend.... What she has done for this community... is incalculable.... [She was] the First Woman of Connecticut, [whose] every business undertaking [was] controlled by... loyalty to the memory of her husband and children."[4] In an age that ascribed to women the role of mother and wife, she was, at last, neither, except in the abstract quality of memory to which her devotions were many, often, and excessive. In the twentieth century, the *Hartford Courant* has run the obituaries of Connecticut women on the front page only three times: for Ella Grasso, Connecticut's first elected woman governor, for Beatrice Fox Auerbach, matriarch of the G. Fox retail empire, and for Elizabeth Hart Colt.

The legend and legacy of Sam and Elizabeth Colt is a story of two lives, two visions, a great fortune created and disposed of, and the transformation of an American city. It is a story caught in the cross fire of changing values and definitions of civic life, aesthetic excellence, the meaning of valor and leadership, and, of course, guns. Hartford, Connecticut, where the Colts' story unfolds, is a city left both weaker and stronger by the vision of two individuals who inspired envy and admiration in life and who, a century after their passing, are still invoked as icons of American character. A century and a half ago the French social critic Alexis de Tocqueville marveled at the "immense number of different types of associations" found in the United States. "Americans," he wrote, "combine to give fetes, found seminaries, build churches, distribute books, and send missionaries to the antipodes. Hospitals, prisons, and schools take shape in that way."[5] At its best, the United States has been a nation of dreamers and doers, philanthropists and donors. Sam and Elizabeth portrayed all those roles.

The Colts' story embodies two central themes in the experience of nineteenth-century America: the development of the institutions and apparatus of an industrial civilization, and the creation of the products and technologies that made that civilization possible. In carrying out their respective programs, Sam and Elizabeth also provided a model of personal leadership and style that

3. Colt's Revolver Model, 1831, pine and ash. This is the "little wooden model" Sam Colt reportedly whittled out of pine while on board the ship Carve en route to Calcutta during 1830 and 1831. Colt used this model to help explain his concept for the revolving cylinder repeater to the Hartford gunsmith Anson Chase, who manufactured the first prototype in 1831 (WA #05.1072).

was self-consciously heroic, but which ultimately masked their inner yearnings and aspirations. This book, along with its companion exhibition—Wadsworth Atheneum, September 8, 1996, through March 9, 1997—takes the reader on a journey through the Colts' world and seeks to probe the inner yearnings that earned them a place among the most-celebrated personalities of their age.

Over time the components of Colt's legend have assumed the classic, almost biblical, structure of America's success mythology, with Sam Colt beginning as "a runaway sailor-boy" who endured rejection and failure, persevering against adversity through hard work and single-minded devotion to a mission, and finally achieving "prosperity... in one limpid, sparkling, and unbroken stream" to emerge with a reputation "among the most remarkable inventors of the world."[6] Colt himself created the basic components of his mythology. His widow embellished it, burnished its prestige, and authenticated its pedigree by revealing its roots in childhood experience and asserting its moral significance. Did he make up these stories? Did she? Are they based in fact? As a witness, by the moral authority of her family's social and religious pedigree, and because of her civic and religious philanthropies, Elizabeth Colt emerged as the foremost apostle of the miracle play that became Colt's legend.

Elizabeth sketched the portrait of the seven-year-old inventor "sitting under a tree in the field, with a pistol taken entirely to pieces, the different parts carefully arranged around him," to "prove how early in life his taste for fire-arms was shown." Was Colt's early life really "one of self-reliant exertion," involving "hardships" and "gigantic obstacles," as she reported? Was the path to Colt's fortune truly marked by "industry, self-denial, perseverance... days and nights of weary toil and waiting"? Elizabeth claimed that Colt's "sagacious, far-seeing mind" and the "magnetism of his presence... inspired all with confidence in his power to lead them," and she quoted an associate who claimed that "had he received a military education, he would have been one of the greatest generals the world has ever seen."[7] However factual these claims may or may not have been, they made great copy and signified the emergence of industrialists as subjects of American mythology.

A version of the story of Colt's invention, written during his lifetime, maintains that it was "while firing for amusement at porpoises and whales, off the Cape of Good Hope... that he first conceived, and wrought out... with a common jack-knife and a little iron rod, the rude model" of his revolver (fig. 3). Another version claims Colt was inspired, during his year at sea, by watching the ship's wheel turn and lock in position, while Colt's competitors maintained, as early as 1851, that he was inspired by seeing the Collier and Co. revolving cylinder repeating firearms then in use by British armed forces stationed in India, the ship's destination.[8]

Colt emphatically maintained his priority and originality, claiming to have only discovered while in England in 1835 that he "was not the first person who had conceived the idea of repeating firearms with a rotating chambered-breech." As the myth took form, the idea that Colt had not benefited from prior knowledge of designs for repeating firearms gave way to the grander claim that a lack of knowledge actually helped Colt by preserving the purity of his inspiration and creative process. The "idea of repeating firearms... was entirely original with Colonel Colt," his allies claimed. "Had he been furnished with the results of those who preceded him, probably he would have too nearly followed in their tracks, and thus have been diverted." Thus Colt appeared to depend "solely on his personal resources," relying on "his own personal ingenuity" in developing his patent revolvers.[9]

Such was the legend that Colt propagated to a willing public hungry for "models worthy of imitation," and emblems of this "favored... nation of self-made men."[10] Colt's greatest invention may, indeed, have been the creation of a personal mythology that rooted his invention in divine inspiration and genius.

Had Colt really "suffered years of penury and toil," as he often claimed? Was it truly "a pasle [sic] of dam[n] fools... styling themselves a board of directors" that brought down his first efforts at manufacturing? Colt undoubtedly believed his own publicists, and it did not hurt a man rapidly amassing one of the first great industrial fortunes to appear the victim. "Poor Sam" had no intention of becoming another casualty along the familiar path where "your genuine genius and originator in any... human achievement... [is robbed of] the just fruit of his labors."[11] As explained by Colt's longtime attorney and friend Edward Dickerson, the inventor is a "poet in wood and steel" who wastes "perhaps, eight or ten years to overcome the prejudices of those who are using old contrivances" and "to beat down the opposition of those interested against him.... He plants the seed, he waters the plant, the fruit is just ripe, he sees the market opening before him... [and is] just about to stretch his hand

to take it, when the infringer steps in between him and the prize."[12]

According to the emerging mythology, those who persevere against adversity gain moral authority and are entitled to rewards. "For seventeen years," Dickerson intoned, "Colonel Colt toiled incessantly... against difficulties and trials which most men would have shrunk from.... For years... Colonel Colt hardly knew where the dinner of to-morrow would come from.... Colt's arm... was a failure in the opinion of every one except its inventor; who cherished it, in poverty and in adversity, with the faith and the hope that no one but the man of genius can feel for the creations of his own mind." Such was the puffery released before a committee of Congress investigating Colt on charges of corruption and bribery, a "poor fellow," with "nothing to protect but his life."[13]

Rags to riches, by luck and by pluck, the image of the gifted and faithful struggling through a "night journey of the soul" or a chosen people wandering through the desert before crossing the River Jordan into a promised land, the structure of the American success mythology is always the same. Sam Colt plucked the chords of civic and religious memory to create a persona that sheltered his enterprise through its final historic ascent.[14]

Born in 1814 to Christopher Colt, a socially aspiring, risk-taking, and ultimately unsuccessful entrepreneur, and Sarah Caldwell Colt, who died just before his seventh birthday, Sam had a deep and lasting memory of childhood loss and downward mobility. Sarah Colt was a patrician whose father, Maj. John Caldwell, was one of Hartford's richest and most prominent men, with a fortune based in banking and the West Indies trade. It appears that the Caldwell-Colt connection was severed when Sarah died in 1821.[15] It had not been an easy or comforting alliance.

Family history, birth-order, and money no doubt played an important role in shaping Sam Colt's character and destiny. He was the second youngest of six children. His oldest sister, Margaret, died young. His next oldest sister, Sarah, committed suicide at the age of twenty-one, and his oldest brother, John, was convicted of murder and died in his jail cell in 1842. Sam Colt became estranged from his two remaining brothers, Christopher, who died of natural causes at age forty-three, and James, an eccentric spendthrift, whose primary means of support after 1856 appears to have been waging lawsuits against Sam.

Although Sam Colt never claimed its influence, and his biographers have largely overlooked it, I cannot help wondering if the legacy of his grandfather Lt. Benjamin Colt of Hadley, Massachusetts, did not inspire him. Although blacksmiths do not create products that have generally interested art historians, the trade was always prestigious in agrarian regions like the Connecticut Valley and arguably the most important in the emergence of industrial technology. Of the many blacksmiths, gunsmiths, and metalworkers in the region during the period, none possessed more or more-diverse tools before 1781 than did Colt's grandfather, described as an "ingenious blacksmith" who specialized in making axes and other edge tools and reportedly "manufactured the first scythe ever made in Massachusetts."[16] Lieutenant Benjamin the blacksmith died on Colt's father's first birthday. Neither Sam nor his father ever knew the man. For some reason, the blacksmith's widow, Lucretia, was unable or unwilling to raise her ten children, and neighbors were appointed as young Christopher's guardian. Years later, after Christopher's financial setbacks, Sam Colt was sent off to live with his father's sister Lucretia Colt Price at her home in Hadley, Massachusetts, where he attended Hopkins Academy and, later, nearby Amherst Academy.[17] Might Lieutenant Benjamin's reputation and blacksmithing shop have lived on? In an agrarian society of such interlocking family relations, it is likely that it did. But no specific record has been found, and sorting out the comings and goings of this family has proven difficult. Both Sam Colt and his father appear to have experienced a certain level of abandonment. With the death of Christopher Colt's patrician wife and the loss of not only

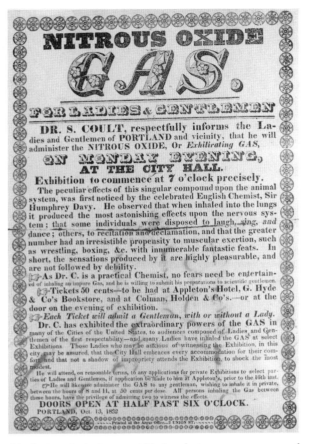

his fortune but also most likely whatever resources and social status she brought to the marriage, Sam's family was ostracized in Hartford, a memory that Sam Colt's wife and friends still recalled half a century later.

In 1829 Sam Colt moved to Ware, Massachusetts, to work in a textile mill where his father was sales agent. The next year he shipped off to sea. It was there that he reportedly conceived and whittled by hand a model of the revolving pistol that would bring him fame and fortune.[18] After commissioning his first prototype revolver in 1831, Colt spent the next two years struggling to pay gunsmiths in Hartford and Albany to work up model long arms based on his revolver concept. Faced with repeated failure, he finally pulled up stakes and moved to Baltimore in 1834.

Why Colt chose Baltimore as a base of operations is unclear. He may have been attracted by its large population of skilled mechanics, or perhaps he thought the city's

4. Advertising Broadside, Portland, Maine, 1832. Printed by N.P. Willis at the Democratic Portland Argus *press.*

For two years (1832–34) Sam Colt supported himself and his firearms design experiments by touring the country giving laughing gas demonstrations as

"Dr. S. Coult, a practical chemist." Newspaper advertisements and broadsides like this were custom printed at local presses where he performed.

proximity to Washington would be an advantage in soliciting government contracts. Whatever the reason, by all accounts Baltimore was the site of Sam Colt's night journey of the soul—a time when "his father, with other friends, opposed him" and he was forced to make it on his own.[19]

With Baltimore as a base of operations, Colt struggled to keep a fledgling gunsmithing operation afloat by performing laughing gas exhibitions on the lyceum circuit as "Dr. Coult," a "practical chemist" from "New York, London and Calcutta" (fig. 4).[20] Having little mechanical training or skills, he rented shop space from gunsmith Arthur Baxter and engaged Baxter's partner, John Pearson, as a technician. On and off for two years, from 1834 to 1836, Pearson worked up experimental designs and prototypes of Colt's repeating firearms. It was no joy. Pearson complained bitterly as Colt's pattern of evasion and broken promises emerged. On the road and out of reach, Colt nonetheless attempted to micromanage the smallest work details, relying on Joseph E. Walker, a fly-by-night showman-musician, as a surrogate project manager until—in the midst of work—Colt vanished to escape creditors.

The surviving letters read like a comedy of horrors.[21] John Pearson never laughed. "I had told him there was no doubt but you would send money to pay him," explained Walker, the erstwhile showman. Indignant at the compounding abuse, Pearson responded to one of Colt's shopping lists of commands by asserting that "your offer of $10 for twelve hours each day I take as an insult.... And I suppose I must find my own candles... and then wait 6 months or [for] your Pleasure to get my money as I have done."[22] After a year, Pearson had grown weary, noting brusquely, "need money, wood, rent is due, shop is cold." Exasperated after two years, Pearson finally declared in a fit of pique that "I worked night and day... so I would not disappoint you, and what have I got for it—why vexation and trouble.... Neither can nor will pay for any more forging.... The manner you are using us in is too Bad.... Come up with some money... In a Devil of a[n] ill humer [sic] and not without a cause."[23]

Compounding Pearson's anger at Colt's broken promises was his resentment of Colt's taking credit for his work. Indeed, fifty years after he quit Colt's employ, Pearson still bitterly claimed to be the true inventor of the Colt revolver.[24] Marred by mistrust as their relationship was, the Colt-Pearson collaboration produced handsomely crafted guns (fig. 5) that succeeded in moving Colt's vision to the next level of performance and credibility.

Prompted by his father's advice to incorporate, Sam Colt had already decided to abandon Baltimore and start again when the last of John Pearson's guns and diatribes came to hand. Having secured British and U.S. patents in

5. *Experimental Colt Revolvers, Hartford, 1833, and Baltimore, 1834–35. Sam Colt's personal collection of firearms was incorporated into the Colt memorial firearms collection, which was mounted and installed at Armsmear by Elizabeth Colt in 1863. The collection includes many early experimental models, including a large concentration of arms cast, forged, and assembled under the direction of John Pearson in Baltimore.*

Top to bottom: rifle, .52 caliber, 6-shot, 8-pointed star and double Colt horse heads inlay, barrel marked "worly" (WA #05.1030); rifle, .30 caliber, 12-shot, with engraved German silver patchbox and stock mountings and double Colt horse heads inlay (WA #05.1028); rifle, *made by Anson Chase, Hartford,* .37 caliber, 9-shot, with engraved German silver patchbox and stock mountings and double Colt horse heads inlay (missing), double trigger

cock-and-fire mechanism; this is the earliest Colt longarm in existence (WA #05.1032); pistol (left), .33 caliber, 6-shot, with extensively engraved mounts, hand-revolving, thumb-cocking mechanism; a rare surviving source model for the original 1836 Colt patent (WA #05.1014); pistol (right), .53 caliber, 6-shot, rare largest-caliber Colt pistol with enclosed recoil shield makes this a veritable hand cannon (WA #05.989);

shotgun, .66 caliber, 6-shot, barrel lug marked "J. Pearson," 8-pointed star and double Colt horse head inlay, engraved "S.Colt./PR" (WA #05.1025); rifle, .53 caliber, 9-shot, ring lever cock-and-fire mechanism (WA #05.1026); rifle, .35 caliber, 10-shot (WA #05.1021); rifle, .36 caliber, 10-shot, "S. Colt P.R." engraved on barrel; forward ring of double ring lever mechanism missing (WA #05.968).

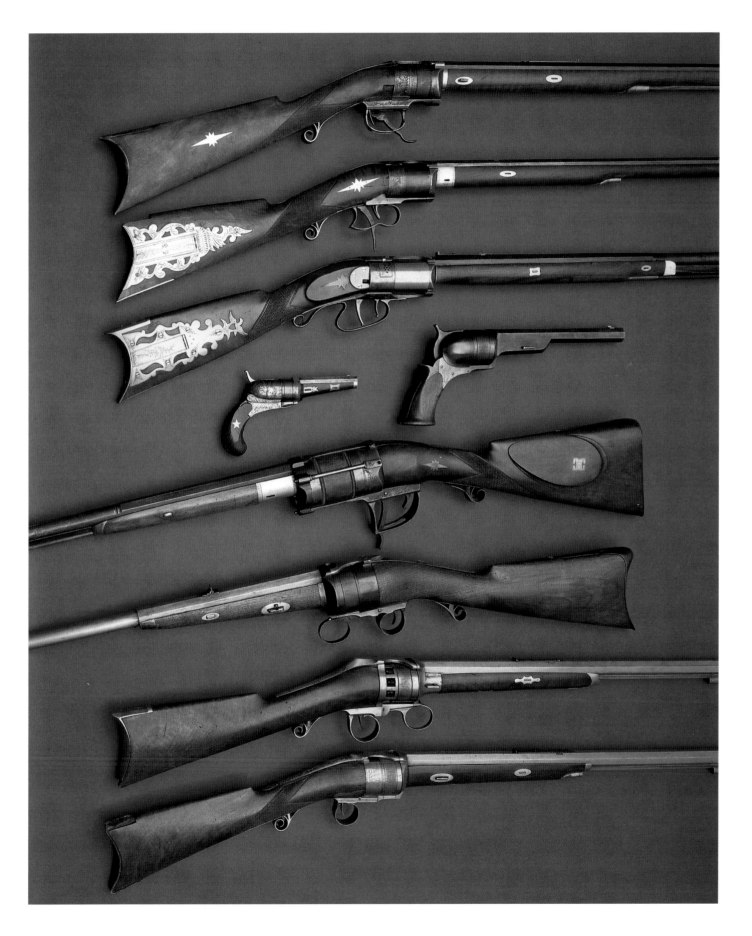

1835 and 1836, Colt, with the support of his extended family, raised enough capital to found the Patent Arms Manufacturing Company in Paterson, New Jersey, which struggled unsuccessfully from 1836 until 1841 to attract enough government contracts and sales to profit from its manufacture.[25]

Paterson was chosen because of its prominence as a manufacturing center, but more importantly because Colt had family there who were rich, well connected, and willing to invest in his dream. Colt's cousin Roswell L. Colt was one of Paterson's most prominent industrialists, and it was in his factory building that the Patent Arms Manufacturing Company's fledgling operations were launched (fig. 6). Colt secured his first U.S. patent (no. 138) in February 1836; the company was chartered in March; and, with his cousin Dudley Selden as treasurer and general manager, the company launched what, in hindsight, seems like an almost inexplicably successful campaign of stock subscription. With capital sufficient to buy machinery and staff a workshop, prospects looked almost promising by the end of 1836. Inquiries from Texas and the Brazilian navy suggested widespread potential demand for a repeating firearm.[26] But competition, the worst financial panic since the one that sent Colt's father into bankruptcy in 1819, dismal results in the gun's first military trials, and the young inventor's insouciance in the face of mounting problems, led to a total collapse of confidence and a straining of relations between Sam Colt and his creditors, even before factory production had begun. It wasn't for lack of trying. Despite later complaints about being betrayed by family members, Colt had two brothers, aunts and uncles, and a still-doting father pulling strings, making loans, marketing, managing, and generally covering up for the centrifugal chaos Sam continued to create.

Despite the evidence of Sam Colt and his father touring state-of-the-art arms factories such as the Springfield Armory and Simeon North's in Middletown, and despite Sam's almost religious devotion to machinery, system, and order the second time around, there is little evidence to support the view that, in 1836, the Colts fully understood what modern arms manufacturing involved, had enough money to carry it out, or the right technical expertise. Pliny Lawton, the machinist from Christopher Colt's Hampshire Manufacturing Company in Ware, Massachusetts, was shop superintendent in charge of production.[27] With his background in woolens manufacturing, he is unlikely to have understood the mechanics of special purpose metal-working machine tools. Looking back on this period, one of Colt's allies laid blame on the fact that it was "so difficult... to invent and construct the machinery necessary to manufacture the article perfectly." Colt's attorney Edward Dickerson acknowledged that the comparable Colt revolver of 1854 was half the price of the 1838 model and "more than twice [its] value."[28] Was the "immense outlay in machinery, tools, and other conveniences" that Dickerson in 1854 regarded as "required" to meet the mass market demand for "cheapness," ever achieved in Paterson? Surely not.[29] With Lawton growing exasperated at the impetuous inventor's moving target of expectations, and with the work force shrinking from a paltry twenty to thirteen, "when we could advantageously employ fifty," and the national economy turning sour, it is surprising that the Patent Arms Manufacturing Company survived as long as it did.[30]

What matters most is that Colt gained critical insight into the problems of marketing, manufacturing, firearms and machinery design, and industrial management by failing at all those things between 1836 when the Patent Arms Manufacturing Company was chartered and his revolver patented and 1842 when the company assets were finally liquidated and its books closed.

The truth about the Paterson experience is that Sam Colt was an abrasive opportunist who lied, cheated, and bluffed his way toward perfecting the "first practical repeating firearm." He failed in New Jersey. And he betrayed the trust of family members and friends who invested, if the record is correct, $230,000 (today equal to $17.7 million) in the flawed invention of an immature and impetuous twenty-three-year-old with limited experience in the system of manufacturing required for success. From beginning to end, Colt was, above all else, a terrific salesman. Product design and manufacturing, however, were challenges unmet until the reincarnation of his vision in Hartford in 1847.

The performance record of the Colt revolvers made by the Patent Arms Manufacturing Company was at best mixed. Indeed, Colt was perhaps as surprised as anyone when, just after the company collapsed, favorable reports of its use in the field began to dribble back from Texas and the frontier. Yet even there, the reputation of "Colt's patent wheel of misfortune" was mixed. Relentless in his pursuit of military endorsements and government contracts—a tactic of mixed value and an ironic obsession from a man famous for boasting of self-reliance—Colt provided damaging testimony when, in an 1840 appeal for yet another military review, he acknowledged: "The arms submitted in 1837 to examination,... the first... ever made on my principle... [were] got up in a great hurry... [and were] very imperfect." This time, he proclaimed with an air of impending reform, "I have... a higher object... pride of making improvements worthy of Government patronage."[31]

After Colt's first prototype exploded on discharge, he began a long process of working out the problems in its design, all the while trumpeting the revolver's virtues to ordnance officials, congressmen, military officers, the press, and investors. Of the many criticisms leveled against Colt's Baltimore- and Paterson–era firearms, some remained unsolved well into the Hartford years. Among their widely cited problems were the tendency for multiple discharge caused by the spread of lateral flame during ignition; too many movable parts, gradually reduced from thirty-six to twenty-eight to nineteen and eventually down to seven; the hindering of aim by the double action cock-and-fire mechanism; the potential for unintended discharge if accidentally bumped or dropped; the risk of its cast-iron cylinder exploding if overcharged (soon rectified by shortening the cylinder and substituting steel); the jamming of the revolving cylinder fouled up by an accumulation of black powder residue on the capping nipples; the difficulties of capping the cylinders in cold weather; and finally, the difficulty of loading cartridges and especially of removing the cylinders, which, until the development of the solid frame in 1855, required that the barrel be disassembled, a slow and unwieldy maneuver.[32] The bottom line, repeated over and over again, is that Colt's early guns were too complicated, too easily fouled up in use, and potentially lethal to the user. And that was just the pistols.

From beginning to end, Sam Colt was committed to manufacturing long arms. They were never as successful as Colt's pistols and were eventually superseded by the tube-fed cartridge mechanisms used in Spencer, Henry, and Winchester rifles. Given the potential for multiple dis-

6. *View of Patent Arms Manufacturing Co., Paterson, New Jersey, ca. 1865. Paterson was America's first planned industrial city. Between 1836 and 1841 Colt manufactured revolvers in a factory complex belonging to his cousin Roswell Colt.*

A ring lever rifle-shaped weather vane was mounted on the tower of the building. (Courtesy of Colt's Patent Fire-Arms Manufacturing Co. Records, Connecticut State Library.)

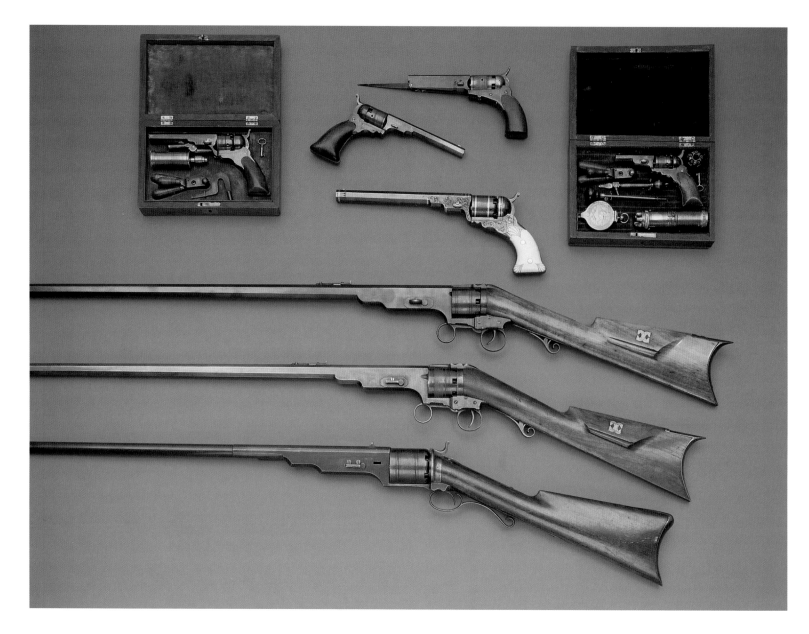

7. Paterson-era Production Colt Revolvers, Paterson, New Jersey, 1836-40. Sam Colt's personal collection of firearms features the most comprehensive representation of Paterson-era Colts known. Most are decorated with the Colt double horse head trademark and engraved cylinder scene featuring prancing centaurs. Pistols have concealed triggers that drop down when hammer is cocked.

Top to bottom, left to right: no. 1, or pocket model pistol, with accoutrements and case, .34 caliber, 5-shot, serial #22 with bulletmold, combination tool, and powder flask (WA #05.1060); no. 2, or belt model, experimental pistol, .29 caliber, 6-shot, with folding bayonet, only known 6-shot from Paterson era (WA #05.1013); no. 3, or belt model pistol, .34 caliber, 5-shot, serial #186, centaur cylinder scene; one of about 500

manufactured; identical to model used by General Jessup in the Seminole War in Florida in 1838 (WA #05.1005); no. 5, or holster model pistol, .36 caliber, 5-shot, serial #984, stagecoach holdup cylinder scene; a deluxe, custom-decorated gun intended for presentation, with silver inlaid band, ivory grips, and extensive engraving including eagle head profile on the sides of the

hammer; this model without ornamentation was made famous by the Texas navy and Texas Rangers; one of about 1,000 manufactured (WA #05.1009); no. 1, or pocket model pistol with accoutrements and case, .28 caliber, 5-shot, serial #7, centaur cylinder scene, with bulletmold, combination tool, powder flask, cleaning rod, extra cylinder, percussion cap dispenser with stamped trademark (WA #05.1059);

no. 1 model ring lever rifle, .36 caliber, 8-shot, serial #125 (WA #05.981); no. 1 model ring lever rifle, .34 caliber, 8-shot, serial #189, centaur and huntsman cylinder scene (WA #05.969); model 1839 smoothbore carbine, .52 1/2 caliber, 6-shot, serial #189, lion, navy, and battle cylinder scene signed by W.L. Ormsby; side-mounted loading lever, model for U.S. patent #1304, the most successful Paterson-era Colt product (WA #05.969).

8. The Last Experiment of Mr. Colt's Submarine Battery in Washington City, 1844. *A. Gibert (1806–75), signed and dated. (Courtesy of the Greg Martin Collection.) Gibert had previously painted* Mr. Colt Testing a Submarine Battery in New York Harbor, near Staten Island, 1842. *The later painting may have been the one exhibited by Gibert at the American Institute Fair of 1844. The same year, Colt won a gold medal at the Fair for his firearms.*

charge, with capped, loaded, open-faced cylinders just inches behind the shooter's steadying hand, Colt's long arms were not unlike the cheap imitations of his pistol that his advocates described as "more dangerous at the breech than at the muzzle."[33] On the other hand, the gun described in 1837 as "incredible" and "invaluable" for its capacity to fire eighteen charges in fifty-eight seconds was bound to find friends as well as foes.[34] Hence, Colt was able to maintain the mirage of rapidly approaching perfection despite repeated setbacks. It helped to keep moving.

By midyear 1837 Lawton, who had "grown as thin as a ghost," was ready to quit, complaining of Selden's "continually hurrying him to get guns on the market and at the same time causing him to dismiss his hands for fear he will not have money to pay them." Selden shot back, accusing Colt of failing to reveal "defects in the original plans" and an unwillingness to cease tinkering and "experimenting" with the patterns.[35] Aside from the lack

of available machinery and manpower, cost-efficient production never stood a chance so long as the inventor continued to make design changes in the midst of production. The prolific and subtle model changes in Paterson-era Colt revolvers have kept firearms collectors busy for years (fig. 7). Before closing operations, the Patent Arms Manufacturing Company turned out between 2,700 and 3,200 pistols in various models and sizes, and between 1,300 and 1,850 rifles, shotguns, and carbines, at a unit cost of between $45.54 and $57.33 (today, more than $3,000), enough, perhaps, to inspire moments of hope, but far too expensive to attract a mass market or sustain operations.[36] By the time the doors closed in 1841, Sam Colt was already on the run, headed for the next campaign.

Sam Colt settled in New York City where, between 1841 and 1846, he pursued other inventions and enterprises, including a secret technique for detonating underwater explosives, a failed telegraph communications company, and an award-winning and marginally profitable design for waterproof powder cartridges.[37] Indeed, Colt's submarine battery plan for harbor defense is one of the most colorful episodes in the interregnum between the first and second incarnations of Colt's work manufacturing firearms.

Colt's biographers described "aquatic pyrotechnics" as his "first love."[38] It was with no small hint of the bombast and bravado for which he became famous that Colt addressed President John Tyler in 1841 with a proposal for "effectually protecting our seacoast." By connecting Robert Fulton's unsuccessful experiments with submarine explosives with Samuel F. B. Morse's experiments with telemagnetic cable, Colt perfected a system for detonating strategically placed underwater mines with an electrical charge. He sought a commission of twenty thousand dollars to perform a public exhibition of harbor defense. A first trial, conducted in New York Harbor opposite Castle Garden on the Fourth of July 1842, was sponsored by a theater tycoon. The first government test was commissioned in Washington, D.C., the following August. In October a third test was sponsored by the American Institute in New York. Colt's grand finale was blowing up a five-hundred-ton schooner moving under sail down the Potomac River in April 1844 (fig. 8) before an audience of thousands.

Accounts of the day's event, which was witnessed by the president and members of Congress who adjourned for the occasion, suggest the theatrics of P. T. Barnum as much as an earnest inventor concerned with national defense. The schooner was tugged into view to the sound of a seventeen-gun salute, the shores of the eastern branch of the Potomac crowded with spectators eager for a view. The mine that sank the boat was preceded by two decoy explosions, "intended as a sort of prelude" to give the exhibition an element of suspense. "Ah! what a pity, it is a failure," the crowd roared after the first mine missed its mark. "Oh, he has missed her! but it was very near," after the second. In the instant that followed the ship was "shattered to atoms" by a third mine. Its sails, surrounded by black smoke, collapsed into the river as the crowd thundered its applause.[39]

Colt's faith in the value of submarine mines was dramatically proven in the Civil War and all subsequent wars, but in 1844 politicians and War Department officials were not sufficiently convinced of their utility to adopt them or commission further experiments.[40] Colt was compensated for each of his four pyrotechnic exhibitions. Yet his failure to secure steady patronage or to see his system adopted for harbor defense was one of his greatest personal disappointments. The principles on which Colt's system of submarine mines was based went to the grave with its inventor in 1862.

Sam Colt's six years in New York were described by friends as a time when he "hardly knew where the dinner of to-morrow would come from," his firearms being considered "a failure in the opinion of everyone." Years later when Colt's critics described him as having "never... given time to mechanical study" being "brought up a gentleman's son, having travelled in Europe," they were half wrong and probably half right.[41] Although documentation is scant, Colt clearly used his time in New York, where he maintained a studio among fellow inventors, scientists, and artists at New York University, to improve himself and read up on applied chemistry and mechanical engineering. During these years Colt was a regular participant and occasional award winner at the annual exhibitions of the American Institute, then one of the nation's leading agents for the promotion of mechanical skill.[42] Colt was also nominated for membership in the National Institute for the Promotion of Science, and while membership and volunteer duties bolstered his sagging reputation, it also undoubtedly provided access to men such as the pioneering chemist and photographer John W. Draper and the artist-inventor Samuel F. B. Morse, with whom Colt collaborated on experiments, and institutions like the Eucleian Society and the New-York Historical Society, all fellow-tenants at New York University.[43]

New York City was pulsating with life and energy, an

international port of entry and crossroads, where music and art, science and literature, libraries and learned societies, prostitutes and theaters crisscrossed a social landscape crackling with political intensity and opinion. New York must have been an utterly transformative experience for even well-traveled provincials like Sam Colt. From Connecticut, a state that was predominately agricultural during most of Colt's life, he found himself in the largest, most cosmopolitan and rapidly growing city in the nation, the power center in an age of American progress and empire. A young man could sample all the vices, virtues, and amenities of life in New York, where only a thin membrane separated its many and varied cultures, subcultures, and subterranean entertainments. A lifelong drinker and carouser, Sam Colt was probably only a few dollars short of having the time of his life during his lengthy and creative sojourn in New York City. From New York, Colt continued to hustle (mostly without success) government contracts for his firearms and his submarine battery system of marine explosives. He dabbled in telemagnetic science and actually made some money patenting and manufacturing waterproof bullet cartridges.

With the outbreak of the Mexican War in 1846, ending a thirty-year-era of peace and pacifism, Sam Colt was finally awarded a government contract and a second chance to succeed at manufacturing firearms. By subcontracting most of the components, which he arranged to have assembled at the New Haven, Connecticut, armory of Eli Whitney Jr., Colt was able to complete the contract. Two years later, with armed conflict and revolution breaking out all over Europe and western migration to the American West in full swing, a new age of militarism had begun. This was one wave Sam Colt was determined to ride. His rapid ascent to the top of the pyramid of American industry had begun.

Colt was certainly on the right track when he wrote glumly to his father Christopher Colt in January 1846 that "I regret exceedingly that you cannot lend a helping hand to renew the manufactory of my Patend [sic] armes [sic].... A little money invested in mashinery [sic]... would

be dubled [sic] every year while the war lasts with Mexico."[44] Sadly, for the family who invested in Sam's dream the first time and from whom he eventually became estranged, only Sam's uncle, Elisha Colt, the cashier at Hartford's Exchange Bank, was willing to provide venture capital. It was 1846, and Colt, due mostly to the intervention of the fabled Texas Ranger Capt. Samuel Walker, finally had a U.S. government contract. This time Colt leveraged every available dollar to amass enough machinery to float the enterprise in high waters.

The Walker-Whitneyville-Colt revolver, so named by firearms collectors to credit Captain Walker's design contribution and because it was first manufactured in Eli Whitney Jr.'s machine shop, got off to so shaky a start that, eighteen months after the contract was awarded, anyone would have again been justified in pulling the plug. Even Walker, who ironically was killed bearing the arm he helped design, fell victim to Colt's renewed pattern of broken promises. Cost overruns and production delays piled up like bodies on the battlefield of Vera Cruz as Colt scurried about frantically, pushing, bullying, and pleading with anyone who could help to stop whatever else they had going and get behind this effort. A list of the people Colt petitioned reads like a who's who of future luminaries in the American arms trade. He approached Edwin Wesson about supplying barrels and bullet molds, sought out Eliphalet Remington in Ilion, New York, and finally made arrangements with Eli Whitney Jr., whose chief inspector and shop manager since 1843 was the machinist-technologist and Springfield Armory alumnus Thomas Warner.[45] Colt bought iron for the forging of barrels and cylinders from Naylor and Co. of Boston, hired the foundry of Slate and Brown in Windsor Locks, Connecticut, to forge and finish the barrels, engaged a Mr. Phillips in New Haven to make brass trigger guards and John Matthewman in New Haven to produce powder flasks.[46] As Captain Walker grew increasingly desperate for delivery, Colt wrote with exaggerated assurance of having "about fifty men engaged to work.... Shall put on every man I can hire at any price... paying... as high as three and four dollars a day to entice them away from other establishments.... Some... work as late as 11 & 12 o clock at night."[47] And, in a self-righteous refrain, Colt promised that, because of his devotion to the cause, I "shall not save one dollar out of the contract." Almost before the ink was dry on the contract Colt was offering to buy out (with what money, who could imagine) the machines used in manufacturing, while boasting to Whitney of the profits he would make once the high reputation of his pistols (inevitably forthcoming), created a flood of new orders.[48] Although the guns were finally delivered, Colt exited New Haven in a flurry of lawsuits and acrimony that earned him the lifelong enmity of Whitney and a group of machinists and technicians who eventually rallied behind the rival Smith and Wesson.[49]

Colt's decision to return to Hartford in 1847 was based more on the availability of credit than on the civic pride Colt later claimed as his reason for returning.[50] A month after the first rather blasé newspaper account of Colt's plan to "establish an Armory in Hartford," the inventor's new friends at the *Hartford Daily Times* reported an "invention of vast importance" with "thirty men at work" using "ingenious machinery, which turns, and cuts and drills, to the greatest perfection." Production could not have commenced much earlier than September 1847, for Colt was still wrangling, at the end of August, to acquire the "inventory of patterns, tools & c. made for Eli Whitney, Esq.," including dies for swaging pistols, barrel-drilling lathes, "one annealing furnace," and related small tools. In addition, Colt purchased some unspecified "machinery from... Ames Manufactory" in Chicopee and undoubtedly commenced operations in Hartford with better machinery than was used ten years earlier in Paterson.[51]

Within a few months of Colt's setting up on Pearl Street, Edwin Wesson, patriarch of the Smith and Wesson dynasty of firearms manufacturers and "celebrated manufacturer of 'Wesson's Patent Muzzle Rifle,'" chose a location right across the street where, in 1848, he "erected... a three-story brick building... [with] coal houses... [an] annealing furnace [and]... valuable machinery."[52] Across town on Commerce Street near the Connecticut River, Solomon Porter began work in November 1847 on the four-story, brick, steam-powered

factory complex (fig. 9) that housed Colt's armory from January 1849 until Colt built and occupied Colt's Armory in the fall of 1855.[53] It was there, as a tenant of a man he later sought to destroy, that Colt's dream finally became a reality.

Eighteen forty-nine marked the turning point in Sam Colt's life and career. It was the first year that big ideas and bigger talk finally began to pay off. In 1849 Colt was granted a controversial renewal of the patent protection that enabled him to beat back his American competitors. He would later insist—and there is no reason to doubt it—that 1849 was the first year he made a profit.[54] That year Colt also began manufacturing arms in the Porter Manufacturing Company's Commerce Street factory and developed the legendary 1849 [.31-caliber] pocket revolver, the most successful handgun, with 325,000 sold, marketed by anyone during Sam's lifetime. This was also the year Colt discovered gold in the American West, not the mineral kind, but the sort coined off the explosive market created by the most frenzied mass migration—to the California gold mines—in American history. The year also witnessed Colt's first foray into international marketing and foreign "diplomacy," notably his sales trip to Europe and Turkey and his clandestine assistance in arming Narcisco Lopez, a soldier of fortune, in a failed invasion aimed at conquest of Cuba. But most important of all, 1849 was the year Colt hired Elisha King Root (fig. 10) as superintendent and technologist-in-residence. Almost every important theme in the early history of what became Colt's Patent Fire-Arms Manufacturing Company was set in motion in a year of new beginnings.

Elisha Root is a central figure in the Colt legend. Credited with perfecting the system of manufacturing used at Colt's, Root was also touted as a symbol of the boundless opportunities for advancement, irrespective of class distinctions, in the new industrial age. As a "man of true merit," Root reportedly rose "from the anvil to the general superintendency of the largest Armory in the world [and the]... largest salary in the State." As a descendant of a French Huguenot family of a mechanical disposi-

tion, Root's origins in the rural New England interior of Belchertown, Massachusetts, were relatively humble. And although modest in comparison to Colt's estate, when Root died in 1865, having succeeded briefly to the presidency of Colt's Fire-Arms, his estate was valued at $123,327 (equivalent to $4.74 million today), a tad higher than the ordinary country blacksmith. More than one-fifth of the estate (almost $1 million today) was the value of his own patents, later purchased by Colt's widow, Elizabeth.[55]

Root was described by his contemporaries as "probably the best mechanic in New England... [and the] inventor of many very ingenious machines," but his gift for inspiring fellow technologists was as important a factor in the armory's success as his personal genius. Having gathered "around him some of the best mechanics in the country," and having installed "thousands of machines [that] do every kind of work... and exhibit every feature of moving known to machinery," Root transformed Colt's abstract obsession with machines into a practical system.[56] So long as the boss was able, through his own gifts, to enlarge markets and market share, and match cash flow with a program of aggressive research and development, Colt's Fire-Arms was able to ride the wave to become, within ten years, "the largest Armory in the world." Did Sam Colt fully imagine such an outcome as he packed up to move from Pearl Street to the Porter Manufacturing Company's Commerce Street factory in the winter of 1849?

According to legend, it was Root who sheltered the fifteen-year-old Sam Colt from pillorying by his Ware, Massachusetts, neighbors after a Fourth of July demonstration in pyrotechnics ended up spraying a crowd of observers with water and mud.[57] Whether true or not, Root was apprenticed in a machine shop in Ware, about ten miles from where he grew up, before heading out on the journeyman's tramping circuit to gain an education as a machinist in Stafford, Connecticut, and Chicopee, Massachusetts, towns within the orbit of Springfield Armory. In 1832 Root joined Samuel Collins in the manufacture of axes in Collinsville, Connecticut. Made superin-

10. Elisha King Root, ca. 1860. (Courtesy of the Connecticut Historical Society.)

9. Colt's Commerce Street Factory, Hartford, 1855. Detail from a View of Hartford to the East, *oil on panel, Joseph Ropes;* *part of a panoramic series of four panel paintings based on photographs taken from the tower of Center Church during the Great Flood of 1854 (WA #05.57).*

tendent there in 1845, Root was credited with developing a system that automated ax manufacturing, by introducing milling and drop-forging machinery.[58] Benefiting from the same westward migration and gold fever that made Colt's revolvers indispensable, demand for the Collins Axe Company's timber axes, pick-axes, and machetes also flourished during this period, with production rising to twenty-five hundred axes per day.[59]

Elisha Root's reputation as a mechanical prodigy and expert in developing manufacturing systems was well established when Colt offered to "pay you again as much as you are now receiving, and if that is not satisfactory, you may yourself fix such compensation as you think fair and reasonable. The important thing is that you come to me at the earliest possible moment."[60] Having been offered the position as "master armorer" at Springfield Armory in 1845, Root was, together with Thomas Warner (then with Whitney and later Wesson), Richard Lawrence, Frederick Howe, and Henry Stone, a member of the small fraternity of master machinists then qualified to carry out the assignment Colt had in mind.

Root's 1853 contract, providing Colt with "exclusive license to use" any improvements "he shall invent," and company publicists anxious to boast of paying the highest wages in the industry verify that Root was eventually paid $5,000 per year (today equal to about $275,000). Before 1861, however, records show payments of about two-thirds that amount, with Root included in the small inner circle that participated as shareholders in the company.[61] Contemporaries described Elisha Root as independent-minded and a humble presence who was relentless in his pursuit of a "safer, shorter, more economical, and more efficient way of accomplishing" manufacturing tasks. In his role as superintendent at the Collins Axe Company, Root demonstrated a remarkable capacity to adapt machinery to unusual tasks and to devise patentable improvements.[62] At Colt's, Root developed an integrated system of special-purpose machines and devised patented improvements in machinery for drop-forging, cylinder boring, rifling, and the slide lathe.[63] His greatest contri-

bution lies in the application of drop-forging machinery to firearms components previously requiring heavy and more labor-intensive grinding and milling.

Astutely aware of the value of a timely story or a well-turned phrase, Colt invited the press to tour his new facility as soon as it was operational. Visiting in the winter of 1849, the Hartford Daily Courant crooned at seeing "a shop 150 feet by 50 feet... [with] long lines of shafting and machinery... in rows" and spoke of the "intense curiosity... [at] witnessing the operations... [the] machines [being] objects of far greater curiosity than the pistol itself, performing difficult work and shaping irregular and intricate forms of solid steel, as though it were soft lead." The new armory, they concluded, "is a museum of curious machinery.... Each part of the pistol passes through an almost endless variety of hands.... Were we ever so skilled in the use of tools, we would not engage to make by hand, during our natural life, one of these frames in the perfection which they are here made in a few minutes by this wonderfully perfect machinery."[64] Production estimates for 1849 vary, but it is doubtful that output much exceeded one hundred revolvers per week, the work of "70 hands."[65] Yet at $25 each, and with workmen paid on average not even $15 per week, Colt probably ended the year with about $75,000 (today about $4.4 million) to plough back into the business. Small wonder that R.G. Dun and Co., in its first review of Colt's credit-worthiness, reported in August 1849, "making money, large business, employs 100 hands, good credit."[66]

Eighteen forty-nine was just a warm up. Two years later, with Colt again "about to enlarge his works and add 200 more" and already employing 300 men, the output soared to about 40,000 revolvers per year, or half that if you prefer R.G. Dun's estimate to newspaper reports. By 1854 Colt was reportedly up to 500 men and 50,000 revolvers per year, though half those numbers would probably be more accurate and no less extraordinary.[67] Production continued to rise: 24,053 pistols in 1856; 39,164 pistols and perhaps 600 workmen in 1857, up to 136,579 pistols at the height of the Civil War in 1863.[68]

Astonished by hyperactive demand on the eve of the Civil War, Colt's father-in-law noted that "orders for Arms are so numerous that his men work till midnight," producing 300 to 400 arms per day.[69] A year later, Colt's Armory, now doubled in size and five times the scale of his 1851 operations, reportedly employed 1,100, a number not surpassed in the half century that followed.[70] If you reckon Colt's Civil War output at half a million arms of various sorts, with an average profit of about $3 per gun, Colt's investment in machinery shows a respectable payoff based on an estimate of about $1.5 million in profits (today about $72 million) on sales of about $9 million.[71]

In 1849 Sam Colt made the first of several sales junkets to Europe and the Middle East. There, more than in the United States, the reputation and market for his invention grew rapidly. In 1851 Sam Colt exhibited his firearms at the Crystal Palace Exhibition in London; he was lionized by the press. "None were more astonished than the English to find themselves so far surpassed in an art which they had practiced and studied for centuries."[72] The following year, hoping to circumvent competition from British manufacturers, he established a state-of-the-art arms manufactory in London.

In the fall of 1851 Colt electrified an audience of military officers and engineers at Britain's prestigious Institution of Civil Engineers when he gave a lecture entitled "On the Application of Machinery to the Manufacture of Rotating Chambered-Breech Fire-Arms, and the Peculiarities of Those Arms."[73] After baiting his hook with the proposition of a lecture on machinery, Colt focused almost exclusively on the virtues of his firearms and only under cross-examination did he explain the process the British described as the "American system of manufacturing."

Lionized by British technologists and eager to plant his flag on British soil, Colt built an arms factory in London, which began manufacturing in the spring of 1853. In recognition of his achievements, Colt became the first American elected to the Institution of Civil Engineers and was awarded the institution's prestigious Telford Medal.

Colt's London "pistol factory" failed in 1856, and although thousands of guns were made and sold there, it never equaled the Hartford facilities in profitability.

With his London armory up, running, and enjoying a constant stream of newspaper coverage and curious visitors, including the Duke of Wellington, Prince Albert, and Charles Dickens, Colt was asked to testify in March 1854 before the Royal Small Arms Committee appointed by Britain's House of Commons. Britain was then on the brink of war with Russia, giving new urgency to matters of armaments and military preparedness. Meanwhile, the technologists and national ordnance experts were in a face-off with the Birmingham small arms industry over plans to adopt machinery and build a national arms factory in Enfield.[74] Not only were costs and the pace of production an issue, but the mid-1850s was also a period of rapid improvement in the design and technology of gunnery. Colt thus faced an audience predisposed to his message. He unloaded both barrels of his politically tinged, messianic vision of an industrial future. Asked where he was born, Colt proudly replied, "Connecticut." "Is it an enterprising State?"—"Yes, it embodies more enterprise than is contained in Great Britain and France combined."[75]

Testifying before the House of Commons, Colt was asked, "Do you consider that you make your pistols better by machinery than you could by hand labor?"—"Most certainly." "And cheaper, also?"—"Much cheaper." In describing the character of English work and workmen, Colt acknowledged problems developing a work force in London:

To start with... I brought over... some Americans... to lead off as my master workmen, but the climate and the habits of the people did not agree with them. [I then began] employing the highest-priced men that I could find to do different things, but I had to remove the whole of those high-priced men. Then I tried the cheapest I could find, and the more ignorant a man was, the more brains he had for my purpose.... The result was... I have men now in my employ that I started with at two shillings a day, and in one short year's time I cannot spare them for eight shillings a day.[76]

Ever ready to cut and close a deal, Colt boldly offered to assume responsibility for British military rearmament. "So confident am I that this system of manufacturing fire-arms is correct... that with one hundred thousand pounds expended in machinery, tools, & c., one million... muskets can be produced at an expense of thirty shillings each... none... inferior to the best that can now be found in her Majesty's service."[77] It was an audacious performance that aimed not only to secure contracts but also to signal that the days of handicraft in arms manufacturing were numbered. Colt's grandstanding played well at home, where local newspapers gloated that "we should not be surprised if he should issue a proposition to England and France to whip Russia into submission on a contract involving a fixed sum of money."[78]

By 1855, with an expanding catalog of products and constant diligence in perfecting both the product design and the means of manufacturing it, Colt's Patent Fire-Arms Manufacturing Company was incorporated in conjunction with the founding of Colt's Armory on

11. Elizabeth Hart Jarvis,
daguerreotype ca. 1852.
(Private collection.)

Hartford's South Meadows. For the first time, the revolver king could confidently shift his attention to his personal life.

Sam Colt had already met his future bride when, in August 1852, he wrote a friend about looking forward to seeing Elizabeth and her sister Hetty on the eve of his departure for the "merry season" at Newport.[79] Elizabeth Hart Jarvis, twelve years younger than Sam, was born in 1826, the first child of a prominent Episcopal minister, William Jarvis, and his wife, Elizabeth Miller Hart Jarvis. Unlike her entrepreneurial husband-to-be, Elizabeth was raised in affluence and enjoyed considerable social status as a descendant of a distinguished line of religious, military, and political leaders. Elizabeth's mother, for whom she was named, had inherited a considerable fortune based in land speculation and the Connecticut Valley's prominent role in the West Indies trade, and she could count several of Rhode Island's royal governors and military leaders among her ancestors.[80] Both sides of the family included many ministers, including her father's uncle, who was the second Episcopal bishop of Connecticut.

Beyond the facts of her wealth and social status, little is known about the early life of Elizabeth Hart Colt. Without reading too much into it, it may be significant that she was the oldest child in a direct line of three women named Elizabeth, each of whom had independent money and, in the case of her mother, was the dominant source of the family's social standing. We also know that from an early age Elizabeth, though raised in privilege, was imbued with a deep sense of piety and responsibility. While her religious faith no doubt owed much to the influence of a family steeped in the practice of Christianity, the tragic loss of half her siblings—all of whom died, not as infants, but as children and playmates in the bloom of life when she was in her early teens—must have tested her convictions severely.

In 1838 Elizabeth's grandmother Elizabeth Bull Miller Hart and her maiden aunt, Hetty Buckingham Hart, moved to Hartford, where Elizabeth attended school during the 1840s. Her parents and younger siblings remained in Portland, Connecticut, where the Reverend Mr. Jarvis served as rector of Trinity Church. In the summer the family moved to Newport, Rhode Island, New England's fabled resort community, where Sam and Elizabeth most likely first met.

The Newport of Sam and Elizabeth's courtship was an international playground ideally suited to making connections with the kinds of political and military leaders Colt cozied up to throughout his career. The first summer Sam and Elizabeth are known to have courted in Newport was "to be an important one" in which "several gentlemen who are to be prominent candidates for the Presidency in 1852... will gather here for the purpose of interchanging sentiments." All the "leading politicians," boasted the Newport press, "will... vacation here." In the summer of 1851 Newport's prominent guests included First Lady "Mrs. [Millard] Fillmore... the family of Gen. [Winfield] Scott... [and] Count Gurowski."[81] Accounts of "pistols... popping at all hours, the bowling alleys... thundering," throngs "pour[ing] through the broadway of the Ocean House [Hotel],... boats and yachts," theatrical performances of "Shakespeare and Dickens," the "air full of the music" of the popular "Germania Musical Society,... fast horses... fast men [and]... fast women,... desperate damsels anxiously searching for rich husbands, and rakish men seeking... wealthy wives" suggest how far Newport had evolved from the purse-lipped puritanism of its New England neighbors.[82]

For Sam Colt, the "whirl of fashionable equipages... dancing, music, scandal, flirtation, [and] serenades" of Newport, was a continuation of the celebrity and high living he had grown accustomed to during his summer of glory as America's industrial poster boy and enfant terrible at London's Crystal Palace exhibition in 1851.[83] Like Newport, Colt also had grown away from the purse-lipped puritanism of his youth. Stunning, poised, well-connected, and then only twenty-five years old, Elizabeth Hart Jarvis (fig. 11) was obviously swept away by the thirty-eight-year-old industrialist, not so much by his wealth—she had

plenty of her own—but by his audacious charm.

Years later, Elizabeth wrote an uncharacteristically revealing account of Sam Colt's character and their relationship. She described her husband as a "quick, high-tempered, impulsive man," who "distrusted all men until they were proved true," and "fairly awed me" with the "magnetism of his presence.... More truly than any other, he filled my ideal of a noble manhood, a princely nature, an honest, true, warm-hearted man." She contrasted his "industry, self-denial, [and] perseverance," and the "majesty in his forbearance," with "the gentleness and tenderness of his nature," citing the "tender memories he ever cherished of the mother and sisters, whose love had blessed his earliest years," as "to me one of the most... beautiful traits in his strong character."

Although Newport was an important focal point for the Colts' romance and social aspirations, few details of their life there have survived. Were the Colts among the twelve hundred guests invited to the "fete champetre" given by William S. Wetmore for the historic homecom-

12. Samuel Colt's Turkish costume, bought in Constantinople, and worn at the Ocean House in Newport, R.I., in 1860.
(WA #05.1585).

ing of expatriate financial legend George Peabody, the "event of the summer" of 1857, attended by "foreign legations, officers of the army and yacht squadron in uniform… statesmen, scholars, [and] divines"? In 1860 the Colts attended a Newport society ball at the famed Ocean House, in which the Colonel, dressed in the Turkish costume acquired on his visit to Constantinople in 1849 (fig. 12), danced and caroused with guests from "Richmond, New Haven, Nashville, Boston, Philadelphia, South Carolina, New Orleans, Cincinnati, Brooklyn, [and] Virginia."[84] With the sound of the waves "breaking through the fog and dust" of the evening air and the "play and hop" of the dashing Germanians in the parlor, the "world beyond the piazza… a vast white opacity," few could have imagined this scene as the end of an era, when a year later, North and South faced each other at war.[85]

On a balmy summer afternoon in June 1856, Sam and Elizabeth were joined in matrimony by Connecticut's Episcopal bishop in the parlor of her family's home in Middletown. Following a glorified bachelor's party, in which Colt, his closest friends, and several hundred armory workmen drank, feasted, and danced into the night, the Colts' wedding day began with Sam and his friends boarding the steamboat Washington Irving. With

the dock and armory office draped in bunting and flags, "a grand salute of rifles was fired from the cupola of the Armory" as Colt and such distinguished guests as Commodore Matthew Perry from the famed Japan Expedition stepped aboard for the twenty-five-mile ride down the Connecticut River.[86]

Colt furnished the bride with a dress and jewelry rumored to cost eight thousand dollars (today equivalent to more than four hundred thousand dollars). Following the ceremony, the newlyweds (figs. 13 and 14) and guests feasted on a cake reportedly "six feet high… trimmed with pistols and rifles made of sugar" with a "young colt on top of it."[87] After the reception the Colts, accompanied by the bridal party and the Armory Band, "took the evening express for New York," where they reportedly "hired the whole of one of the largest Hotels," the St. Nicholas, for a "large reception party" before setting off the next morning for Europe, accompanied by the bride's brother Richard, sister Hetty, and a multilingual translator.[88]

The bridal party arrived in London in mid-June, where they remained for almost a month before setting off for the continent.[89] The highlight of their stay in London was a lavish Fourth of July dinner hosted by the London-based American financier and renowned patriot George Peabody, head of the international banking

15. Richard Jarvis in
Diplomatic Uniform,
1856. From a carte
de visite *photograph.*
ca. 1867.

dynasty of Hartford native J. Pierpont Morgan.[90] The arms collection in the Tower of London, St. Paul's Cathedral, Westminster Abbey, the British Museum, and Hyde Park were among the attractions outlined in Elizabeth's copy of *Saunterings in and about London* (1853), whose well-worn pages survive in the library collection at the Colts' home. No doubt the bridal party also visited Colt's London Armory, although their mood must have changed markedly since Sam's last visit. Under intense competitive pressure from British manufacturers, the London armory was barely limping along; by the end of the year it would be forced to shut its doors permanently.

After London the Colts visited Belgium, and from there they went to Holland, Bavaria, Vienna, and the Tyrolean Alps. The climax of the tour came in Russia, where Sam and Elizabeth spent six weeks and achieved national fame as guests at the coronation of Czar Alexander II. It was an event described by their Hartford friend, former Gov. Thomas Seymour, then U.S. minister to Russia, as one of the "most magnificent pageants ever witnessed in the world," in which "the display of diamonds and costly attire representing the wealth of Europe, were never surpassed."[91]

By securing Colt a temporary appointment to the embassy in St. Petersburg, Seymour assured the bridal party a place at one of the nineteenth-century's most ostentatious international celebrations. Colt had been received at the Winter Palace of the Czar's father, Nicholas I, in 1854, an occasion he regarded as one of the highlights of his life. But this was even more extraordinary. In attendance were princes and princesses from throughout Europe and the top-ranking diplomats and military officers from England, Austria, Belgium, Prussia, Portugal, Brazil, Greece, Sweden, Sardinia, Denmark, Bavaria, and Turkey. Colt was so concerned not to appear rough-hewn before European society that he employed attendants and carriages, and outfitted the entire delegation to the Czar's coronation in martial splendor (fig 15).[92] At the head of the U.S. delegation was Governor Seymour, followed by "Mr. Kolt (et Madame

16. *Gifts from the Russian Czars, 1854–59. Czar Nicholas I presented Sam Colt with a diamond ring (right) in appreciation for Colt's assistance in equipping the Imperial Armory at Tula in 1854. Nicholas died in 1855 and was succeeded by his son, Alexander II, whose coronation Sam and Elizabeth attended in 1856. Alexander and his brothers received deluxe presentation Colt revolvers from Sam Colt in 1859.*

In return Alexander instructed his minister of the household, Count M. de Aldeburg, to present Colt with a diamond-adorned, gilt and enamelled snuffbox, bearing the czar's cipher and an inscription to Samuel Colt inside the lid (WA #05.1535 and accompanying letter). Additional tokens of appreciation from the czar included a bronze medal with Alexander's likeness in bas-relief (WA #05.1099) and another diamond ring (WA #05.1537).

17. Bolshoi Theater,
Moscow, 1856. From
deluxe commemorative
book, Description
du Sacre et du
Couronnement de leurs
Majestés Imperiales
l'Empereur Alexandre II
et l'Impératrice Marie
Alexandresna, 1856,
printed in 1863.
(Courtesy of the New
York Public Library.)

Kolt), Mr. Gervis (et madame sa soeur)."[93] Years later, the commemorative medal, gold and enamelled snuff box encrusted with diamonds, and diamond ring presented to Colt at the coronation, along with a second diamond ring from his reception by the Czar's father in 1854 (fig. 16), remained the most valued treasures owned by the Colts—permanent testimony to their moment on the grand stage of European society.

During their stay in Russia, the Colts and Elizabeth's siblings, Hetty and Richard, toured St. Basil's Cathedral in Moscow, attended royal receptions, an imperial banquet in the Great Hall of the Palace of Facettes, a lunch for the diplomatic corps in the Gold Hall, an opera at the new (1856) Bolshoi Theater (fig. 17), and a beguiling array of balls and banquets hosted by various European ambassadors in honor of the Czar and Czarina. Colt later complained of having been "driven to death... in the whirlwind of sightseeing, coronation, fetes, balls, parades," and dinners, through their "whole stay in Russia."[94] Glorious and glamorous as it may have been, it was all a bit much, and by December, following a final shopping spree in Paris, Sam and Elizabeth were happy to head home.

After their return to Hartford, the Colts quickly settled into what seems to have been a very happy marriage. Indeed, for Sam, who had traveled incessantly since 1829, the five years following the bridal tour were the most stable and prosperous of his life. For Elizabeth, life with the "impulsive man" who embodied her "ideal of a noble manhood" seems to have been no less fulfilling. Yet for all their prosperity and contentment, the Colts suffered more than their share of personal tragedy. During their five years of marriage they lost two children; Elizabeth was pregnant with a fifth child destined to be stillborn when Sam died in 1862; and the two remaining children would predecease Elizabeth. It is impossible to gauge the cumulative weight of those losses, but Elizabeth's account of Sam's reaction to the death of their first daughter, Lizzie (fig. 18), suggests the depth of their sorrow. "For weeks [Sam] did not leave his room," she wrote, "sitting with the portrait of our baby before him, and convulsed with such grief as one seldom sees."[95]

The winter of 1862 was a dark and lonely time for Elizabeth Colt. As in the past, however, she maintained the stoic composure and resignation of the devout. "To human sight," she wrote of her deceased husband, "it seemed that there was much in the future for his guiding hand to do; but 'He who doeth all things well,' and who smiteth in love, judged otherwise."[96] There was, indeed, much in the future to do. If Sam could no longer do what needed to be done, Elizabeth would. The empire her husband had built would endure, and his legacy would not be forgotten.

In the pages that follow I have tried to re-create a sense of the environment in which the Colts lived their lives and accomplished their deeds. My hope is that doing so may provide a richer sense of their place and time, and of the possibilities of individual initiative and vision. Sam and Elizabeth were extraordinary people who lived in extraordinary times. Together, they built an empire on a foundation of art, religion, guns, and personal memories. Their story evokes many of the key issues of their times, from the struggle to create an industrial economy after the decline of New England agriculture and maritime trade, to the quest by American women to define the meaning of personal sovereignty and to play an important role in public affairs, to their generation's achievement in creating the modern American city. Building the cultural, educational, and humanitarian apparatus of an industrial civilization required means. For a generation that had imbibed so deeply the waters of puritan theology and Christian duty, creating that apparatus became an end in itself. Toward that end, Sam and Elizabeth Colt made a contribution no less distinguished than it was distinctive to their place in time.

18. *"Baby Lizzie," Elizabeth Jarvis Colt, 1860. Pastel on cardboard by Matthew Wilson, American, 1815–92 (WA #97.1936). Wilson was well known for his pastel portraits of children; all three of the Colt children who died in infancy were memorialized in Wilson portraits.*

PRACTICALLY PERFECT
SAM COLT AND THE REVOLUTION IN MACHINE-BASED MANUFACTURING

There is nothing that cannot be produced by machinery.

SAM COLT, 1854

That world of complex machinery... must have required more brains to invent all these things

than would serve to stock fifty Senates like ours.

MARK TWAIN, 1868

The Connecticut River Valley's fifteen minutes of fame—not just wealth and prosperity, but real attention-grabbing, culture-altering achievement—occurred at the intersection of machine-based manufacturing and guns, and peaked during the 1850s, a decade of growth and transformation more convulsive than any since the region was first colonized by English Puritans. No other claim to regional distinctiveness compares with the coalescence of creativity and capital that made the river towns of Connecticut, Massachusetts, Vermont, and New Hampshire what California's Silicon Valley is today, the vanguard of an internationally significant, technology-based transformation that changed the world of work. Despite its association with guns, this revolution was never, in either cause or consequence, primarily about guns. Then and now, gun-making was a relatively small industry. But the particular needs of gun-making in that place and time caused a convergence of capital and technologies that launched the revolution in machine-based manufacturing that changed the way everything from typewriters and cameras to sewing machines and bicycles would be made. The perfection of systems of integrated, special-purpose machines was a transformative milestone that eventually earned the United States a position as the dominant industrial power on earth.

Colt's Armory was neither the first nor the only center of arms manufacturing in the region, and the region, although dominant, never monopolized the industry. Colt was the most prominent of a phalanx of fellow technologists, the high priests in the new religion of machine-based manufacturing. Those high priests—men like Colt's Elisha K. Root; Springfield's Thomas Blanchard; Windsor,

Vermont's, and Hartford's Richard S. Lawrence; and Springfield's and New Haven's Thomas Warner—were, by and large, compulsive improvers more driven by process than products. Controversy and conflict followed them every step of the way. Their achievement was neither inevitable nor easily gained. Had they been motivated by anything less than a nearly messianic desire for system and uniformity—a desire shared by religious sects, shopkeepers, writers, and artists throughout the period—their system of machine-based production would not have risen so far so fast.

The Connecticut Valley technologists were also driven by nationalism, at a time when the United States craved validation in the eyes of the world and yet was being tested to the breaking point at home. Was the United States still united in its sense of mission? Was the United States manifestly destined to achieve "prosperity and glory"? The technologists and manufacturers, as officers in a domestic holy war of values between North and South, and an international holy war between monarchical and democratic forms of governance, were determined to file a convincing brief. Not content to simply measure up, they sought overwhelming dominance as a way of validating the American way of life.

In 1855, the year Colt's Armory was completed, Connecticut claimed the world's largest manufactories of axes, pins, and clocks, was home to the nation's largest silk factory, and enjoyed considerable export markets for such value-added goods as silverware, rakes, scythes, pianos, coffee grinders, steam engines, pumps, glass bottles, and pewterware.[1] By 1850, the last year in which the census ranked agriculture first in Connecticut's economy, the state and region had endured forty years of waning agricultural productivity and minuscule wealth formation and growth. Ten years later, Hartford, the region's leading manufacturing center, had more than doubled in population, while Connecticut led the nation in patents per capita, absorbed eighty thousand European immigrants, and was second only to Massachusetts in the per capita formation of wealth.[2] Industrialization became the springboard that hurled Hartford into the Gilded Age affluence that earned it a reputation as the richest small city in America.

Dreams of an industrial millennium seem naive in hindsight. But the swaggering, bragging, blustery bravado of the 1850s—a time when Americans willfully boasted that in the mechanical arts "we stand before the world as the most fertile and inventive people that now exist"— gives Sam Colt a context and opens the way to a wider consideration of the challenges and issues of his time.[3]

Today, as New England staggers to regain its economic momentum after forty years of declining competitiveness in the industrial sector, the legacy of industrialization is perhaps as controversial as was its prospect a century and a half ago. New England embraced manufacturing in stages, running the gamut from aversion, to ambivalence, to evangelical enthusiasm. In 1817 the Connecticut Society for the Encouragement of American Manufactures proclaimed, "Manufacturing establishments are the last best hope of Connecticut.... Few countries are so well watered as New England... [a place] excellently calculated for mills and machinery of all kinds. New England must either manufacture or dwindle into insignificance."[4] It was not long before the region's

clergymen provided the necessary religious underpinning.

In 1846 Hartford's most influential Christian thinker and civic visionary, the Reverend Horace Bushnell, launched a campaign on behalf of manufacturing and scientific agriculture as a dress rehearsal for the thunderous endorsement he preached in 1847, a few months before Sam Colt's arrival from New York. Attacking the impression that poor land and harsh weather diminished Connecticut's agricultural prospects, Bushnell challenged an audience at the Hartford County Agricultural Society to seek out new frontiers of creative enterprise. Because "our country abounds in rapid streams and waterfalls.... [We are] destined by nature to be a great manufacturing region." The "cheap land" frontier of "the great West," he prophesied, "will be known... as the great American corn-field, the Poland of the United States," while "New England... will be sprinkled over with beautiful seats and bloom as a cultivated garden.... The true farmer," he chided, "must... have his wits at work... must be a different style of man."[5]

The following year Bushnell issued "Prosperity Our Duty," a blistering call to arms that asserted an immutable correlation between virtue, courage, faith, and prosperity. Although explicitly a prayer of hopefulness delivered just as New England was emerging from a decade-long depression the rest of the country had cast off several years earlier, "Prosperity" represented an endorsement of manufacturing from one of New England's most influential clerics. Cited, twenty years later, in the opening pages of Sam Colt's biography, this sermon unofficially launched an era of unprecedented growth and industrial achievement.

We have come now to a great and final crisis... [and must face the question] whether we are to go on maintaining our growth and numbers, or to sink into decline.... [From] Joseph in Egypt [to]... Hezekiah of Gihon... the Scriptures represent... the fixed connection between virtue and prosperity.... One great mechanic rising into wealth and public note among us would rectify many false impressions and breathe new life and courage into all the mechanic professions.... Hartford can prosper; therefore,... it shall. Go then... all together to the task of... proving to mankind... that growth and progress are the right, under God, of every people that will do their duty.[6]

Tough talk for tough times. Bushnell, who was described by contemporaries as a "first-class man of war... bound to have that prevail which ought to prevail," was no stranger to controversy.[7] Clear enough, from its tone and content, "Prosperity Our Duty" was meant to rock the foundations of civic denial and confront the forces that, well after the rise of Colt's industrial empire, continued to regard manufacturing with disdain and its operatives, evangelists, and beneficiaries as a menacing presence.

Hartford's old-money, merchant-banker elite campaigned against the new-money interests of manufacturers, tapping the deep vein of resentment that shaped civic and political discourse throughout the era. Hartford's antebellum bankers no doubt were as cautious and risk-averse as bankers often are, while Hartford newspapers across the political spectrum were predictably progrowth. The anti-industrialists waged their campaign mostly by indirection, more in action than words, while manufacturing's progrowth agents were open and aggressive. The Hartford press, especially the Democratic *Hartford Daily Times,* lobbied relentlessly on behalf of manufacturing interests. *Times* editor Alfred Burr, who eventually emerged as one of Sam Colt's most enthusiastic advocates, was rabidly progrowth and promanufacturing, consistently proimmigrant, and strongly pro-Union and virtually pro-South in the Civil War. In 1847 the *Times* was the most progressive newspaper in town, a dependable cheerleader and provocateur for manufacturing interests.

The "old fogy element" was ridiculed by the *Times* throughout the industrial era as a remnant of Connecticut's "standing order."[8] Historians have described it as the "self-perpetuating group of ministers, lawyers and merchants from the leading families who controlled the affairs of church and state."[9] "Years ago," the *Times* recalled, "there was a prevailing sentiment among prominent men in Hartford, which retarded the prosperity of the City, and crushed... every project which would have... increased our population and wealth.... These men were either rich or getting rich, and they wanted no change which should give a chance to others. The brawny arms and dusky faces of the honest-hearted mechanics were offensive to them."[10] Manufacturing enterprises, argued the *Times,* were greeted in "a cold and indifferent manner [by] Hartford capitalists... and monied men."[11] Throughout the era, the Democratic press lobbied for more and different kinds of manufacturing and affordable housing.

Hardly immune to criticism and deeply aware of the growing litany of complaints about working conditions and class conflict in the industrial cities of Europe, Connecticut's manufacturers sought stability and continuity as they grafted the apparatus of industrialization onto the structure of New England's traditional agrarian economy. Tapping a reservoir of patriotic mythology, New England's industrialists argued that the American experience would again prove different from Europe. In 1847, with Bushnell preaching the gospel of hard work, courage, and adjustment, and with Sam Colt soon to launch Hartford's greatest manufacturing enterprise, the *Times* struck a surly but patriotic note by sneering at the "Englishman's general idea of the United States" as a society "in a rude state of semibarbarism: with no social refinement, and little intellectual culture; a mere money delving, grave digging, lean, lank, generation of Yankees." In contrast, the United States imagined itself as "centuries in advance of Great Britain," which the *Times* reviled for its prostitutes and the degraded conditions of its factory workers.[12] Ten years later, at the consecration of Charter Oak Hall, a symbolic focal point in Sam Colt's factory town, dedicated to "the grand idea of placing the labor of this valley upon the platform of intelligence," orators spoke of the distinction between Europe and America: "Labor... is a far different and loftier element than it is or ever was in the Old World.... With us... it is emphatically the parent of capital."[13] What astonished Europeans about this rosy capitalist's scenario is that U.S. workers appeared to agree, at least for a while.

Industrialization was the stage on which Americans acted out the drama of passage from national adolescence to adulthood, a drama centered in its relationship with the parent figure of Great Britain. The competitive rhetoric peaked at the Crystal Palace during the Great Exhibition of the World's Industry in London in 1851, as both sides attempted to reconcile expectations with performance. Throughout the summer, amid mixed reviews and the unmistakable sense of being outclassed, Americans exhibited an odd and vulnerable mixture of childlike craving for affirmation and bellicose posturing. But it was at the Crystal Palace that American guns and machines first attracted international attention, beginning an epoch of sustained technological transfer from the New World back to the Old that reversed, for the first time, the pattern of economic dependence. Then and now, technological sophistication was a key indicator of a nation's economic and cultural independence, prestige, and power. The equation *technology equals power* was widely accepted on both sides of the Atlantic, and it was no small measure of the revolution at hand when British technologists spoke of "waging war with our neighbors—not on the battle-field... but in the foundry; engineers being our generals, and founders our admirals."[14] Astonished to at last find themselves in the winner's circle, technology's cheerleaders at *Scientific American* heralded mechanics as the new "Agents of Power" reserving special praise for "mechanical genius in the production of useful machinery."[15]

Sam Colt, as a master of stagecraft and public posturing, became a field general in the transatlantic rivalry between British and American technology. Colt appeared indifferent to the distinction between the practical needs of his enterprise and his personal pleasure in upstaging a rival parent figure. Joy in up-ending authority was one of the most consistent themes in Colt's life. In the months and years immediately following his widely reported triumph at the Crystal Palace, Colt waged war on two fronts: the war to successfully launch and expand his manufacturing facilities and the private patriotic war to dominate, and when possible, humiliate, British and European arms manufacturers, preferably on their own turf.

Curiously, Colt's patriotic war received the full support of British technologists, who were themselves battling the handicraft traditions of the British arms industry. Technologists on both sides of the Atlantic evolved into a transcontinental brotherhood, ultimately more loyal to their shared quest for right principles and technological innovation than to national boundaries. Colt's British peers became his staunchest advocates, confirming that he had, in fact, seized on one of the wedge issues of the age.

Inspired by the Crystal Palace exhibits of America's armorers and machine shops, the British government resolved to create its first state-run royal armory. A representative was dispatched to attend the American Crystal Palace exhibition in New York in the fall of 1853, and the Royal Small Arms Commission conducted hearings in March and April 1854, then appointed a Committee on the Machinery of the United States to tour the United States. The committee was to investigate and report on the revolution in machine-based manufacturing and its potential applications for the new Royal Small Arms Manufactory at Enfield, England. Comprised of three prominent technologists and ordnance experts, the committee spent the spring and summer of 1854 surveying the state of the art of arms and machine manufacturing in the eastern United States. This was an epic prelude to the moment when Americans were able to reverse the flow of technology from Europe and begin the sustained export of value-added manufactured goods and machine tools.

The head of the committee, John Anderson, had visited Colt's London arms factory the previous December as chief inspector of machinery for the British Ordnance Department. He was overwhelmed by a sense of possibilities, noting that it was "impossible to go through the work without coming away a better engineer." Like his peer the inventor and chairman of the Royal Small Arms Commission, James Nasmyth, whose impression of Colt's London armory "was to humble me very considerably," Anderson was concerned that Britain had fallen "very far behind."[16]

The committee arrived in the United States in the winter of 1854 and toured arms factories, machine shops, iron foundries, and related industrial facilities from Harpers Ferry, West Virginia, to Pittsburgh and the industrial belt of the Mohawk Valley in New York. But their primary destination was the Connecticut Valley, where they visited the U.S. Armory at Springfield, four major private arms factories (Colt, Whitney, Sharps, and Robbins and Lawrence), and a handful of machine shops. The British were not alone in their curiosity about machine-based manufacturing. During the summer of 1853 many of the same facilities were visited by the U.S. Armory Commission, appointed by the secretary of war, Jefferson Davis, "to satisfy themselves whether the government can make arms better or more economically than individuals."[17] At home and abroad the impression of a major breakthrough was at hand.

So impressed were the commissioners by the manner in which machinery was adapted to special purposes and deployed to the task of rapid production, that the merits of American firearms and machinery quickly became a secondary concern. As Anderson reported, "Americans... display an originality and common sense in their arrangements which are not to be despised.... The great good which the country at large will derive" from the proposed Royal Small Arms Manufactory will make the actual musket "a secondary consideration." George Wallis, dispatched to New York the year before to report to the Royal Small Arms Commission on the machinery exhibited at New York's Crystal Palace Exhibition, noted plainly that the "originality of... application is one of the most remarkable features in the progress of industry in the United States."[18]

The "Report of the Committee on the Machinery of the United States," presented to the House of Commons in 1855, is a remarkable and understudied text, the industrial and technological companion to Alexis de Tocqueville's legendary and often-cited analysis of American society and politics, *Democracy in America*. This eminently readable travel account describes and analyzes dozens of American factories in a variety of industries, examined firsthand "in the localities in which they are carried on."[19] It provides a snapshot of American industrial technology at the moment when it achieved international distinction as the "American system of manufactures." Americans were as surprised by the revolution as everyone else. George Wallis noted that the "absurd prejudices or vicious cupidity of the [American] mercantile classes often compels manufacturers not only to imitate, but absolutely to brand [his product] with the name of his foreign rival.... [Thus they] repress the rising energies... of their own countrymen."[20]

The visitors were almost giddy with excitement at what they observed, convinced that they were, in part, bearing witness to the "different style of man" Bushnell had prophesied. If the difference between American and British manufacturing capability was more a matter of deployment and attitude than technology or design, what did this say about the effects of democracy and independence on the character and habits of the American workmen? Or was the committee's fawning over American workers meant more to provoke and undermine the authority of British artisans, whose rigid hierarchies and time-worn divisions of labor had proven so intractable to technologists and bureaucrats anxious to introduce a new system? In the report of their journey, Anderson and his colleagues hammered away at issues of values and cultural adaptation, noting that "the American machinist takes every opportunity of becoming acquainted with our inventions and scientific literature while... [we] are comparatively careless." What they found especially astonishing and worth emulating was the "avidity with which any new idea is laid hold of" by American workmen and the fact that American "workmen hail with satisfaction all mechanical improvements... as releasing them from the drudgery of unskilled labour." Where British technologists were painfully familiar with "combinations to resist" the introduction of machines or labor-saving devices at home, in America such combinations were allegedly "unheard of."[21]

In a special report on the American Crystal Palace exhibition, George Whitworth came close to invoking the mythic essence known as "Yankee ingenuity." While observing that the absence of Britain's traditional guilds and apprenticeship systems favored innovation, Whitworth also noted that "this necessity for self-supply... this knowledge of two or three departments of one trade" and the "long and well-directed attention paid to the education of *the whole people* by the public schools... has been the means of originating many ingenious machines." Whitworth and the Committee on Machinery attributed the "inventive spirit" they observed to the "successful violation of the economic law" of the division of labor, the keystone of industrial orthodoxy since Adam Smith's *Wealth of Nations* (1776) codified the theory and practice of industrial manufacturing.[22]

Why this revolution in machine-based manufacturing took place in the Connecticut Valley is a story rooted half a century earlier in a remarkably successful program of government industrial patronage. Today, with government's role in economic development so widely contested, it is perhaps useful to point out that the government has not always failed when it has tried to nurture technology and economic growth. Historian Merritt Roe Smith argues persuasively that if one program in fifty equals the effect that the U.S. Springfield Armory had on American industry, it potentially pays for forty-nine flops. The seemingly incongruous image of the federal government as a bold and visionary innovator and venture capitalist is repeatedly validated in the case of military design and contracting, first and most specifically in the story of the U.S. Springfield Armory.[23] And yet the difficulties faced by the government's corresponding initiative at Harpers Ferry suggests that nature counted as much as nurture in the equation. Hence, government patronage achieved almost miraculous results when it was driven by a sense of mission and where it coincided with a cultural and economic predisposition toward adaptability and change.[24]

The military's quest for uniformity became a powerful economic engine when its cost-saving application to high-volume production became apparent. Initially, cost savings were neither the primary goal nor an inevitable outcome. Historians concede that the United States' advantage over the British in the adoption of machinery owes at least as much to differences in relative costs as to the resistance of British workman to innovation. The United States had a long-standing advantage in raw materials and a disadvantage in the cost of labor. Nathan Rosenberg has argued that although "labor-saving," American gun-stocking machinery was also "resource-extravagant" and initially might "not have been worth adopting in England at prevailing English timber prices."[25] So when British technologists praised the "'go a-head'... habits of thought and determination" among American workmen, they may have been more interested in securing government patronage for the British machine tool industry than in lowering the cost or improving the performance of British firearms.[26]

With enabling legislation sponsored by George Washington and approved by the Second Congress in 1794, the United States established national armories in Springfield, Massachusetts, and Harpers Ferry, Virginia, sites selected for political reasons and because their inland locations were relatively protected from invasion in times of war. By 1802 both federal armories were turning out a regulation smooth-bore (as opposed to rifled) flintlock musket modeled after the Charleville pattern introduced by the French in the 1760s, a traditional arm manufactured by traditional means on a larger-than-usual scale. Gun stocks were shaved and bored by hand, barrels were forged by hammers on anvils, and the locks, a part traditionally imported from Europe, were manufactured one at a time by skilled craftsman who ground, drilled, filed, and custom-fit the locks, stocks, and barrels. Unifying the production of components under one roof was a considerable achievement at a time when private gunsmiths typically finished and assembled components brought together from specialists scattered throughout the country.[27]

The notoriously haphazard and cyclical character of supply and demand was the problem the U.S. armories were created to solve. The War of 1812 brought to a head the problems of inspection and supply, prompting the War Department's newly established Ordnance Department to launch a thirty-year quest for uniformity in the manufacture of military arms. As thousands of arms damaged in war returned to national and regional arsenals, the enormous costs of custom manufacturing became increasingly apparent. Field repair—the ability to substitute damaged parts quickly and without complex tools—had also proven elusive and costly. Under the direction of Col. Decius Wadsworth and Lt. Col. George Bomford, the Ordnance Department launched a single-minded program of reform and innovation aimed at achieving uniformity and interchangeability of parts. For twenty-five years, until the 1840s, when private arms manufactures like Sam Colt seized the vanguard by adapting the innovations created in the quest for uniformity to the rather different concerns of volume production and profits, the government armories and government-sponsored small arms contractors occupied the cutting edge of technological innovation. The Springfield Armory (fig. 19) was the most important factor in spawning the network of interrelated companies, whose work forces and projects became remarkably interwoven. The progress made at Springfield and Harpers Ferry in developing machine-based solutions to the problem of uniformity provided the foundation on which Colt, Winchester, Smith and Wesson, Sharps, Remington, and the other legendary names in American firearms were ultimately built.

Innovation moved through open doors. The most intriguing aspect of the culture of arms manufacturing is the synergy created between the government and private armories, inventors, and machines shops. At the point of its greatest impact, the Springfield Armory became a "classroom for all the private arms companies... which surrounded the armory like satellites" and a "pivotal clearing house for the acquisition and dissemination of technical information."[28] Merritt Roe Smith has noted that the "rapid movement of technical information from shop to shop and from armory to armory was critical to the formation of the... 'American system.'" Springfield's "ability to monitor and assimilate work being done elsewhere" and the Ordnance Department's "open door policy" of insisting that private contractors share their innovations and improvements with the national armories on a royalty-free basis was the key factor in the convergence of talent and skill that led to the perfection of machine-based manufacturing.[29]

When Sam Colt was later questioned about the dynamic character of the unusual public-private partnership among American arms manufacturers, he explained: "I think it is good policy in any person who has to negotiate with another, always to hold within himself the power to produce that which he requires of another.... [The] Government of America... can produce at the national armories double what they now do; they do not do so; they wish to encourage the artisans of the country, and are willing to employ them."[30] More than altruism was at play. With the carrot of patronage, the stick of quality control and inspection, and a mandated open-door policy requiring technology sharing among federal manufacturers and their subcontractors, a public-private partnership of remarkable dynamism was created. With the federal armories seeking appropriations, the private armories seeking contracts, and everyone seeking improved efficiency, innovations were prolific and inevitable.

Contracts fed innovation and innovation spiraled back into the federal armories in a steady line. At each stage more machines were deployed to more-varied tasks

19. Upper Water Shops in 1830, U.S. Springfield Armory, ca. 1830. Watercolor by Lt. Thomas B. Linnard (ca. 1817–ca. 1851). The Springfield Armory was situated in three locations. The stocking, filing, and assembly took place in the hill shops and arsenal on State Street; barrels were turned, bored, welded, and polished at the upper water shops and lower water shops on the Mill River in Springfield. (Courtesy of the Springfield Armory Historic Site.)

to the point where muskets that had once involved several dozen operations carried out by a handful of operatives were now made by a process involving hundreds of discrete, precisely choreographed operations performed by hundreds of machines, each carrying out a single precise function. The status of machine builders soared, and while the American system is best understood as a revolution in choreography and quality control, the machines that performed the tasks were also increasingly experimental and innovative. For the most part, the innovations were homegrown in a seed bed of Yankee ingenuity fertilized by government patronage.

Springfield's strategic advantage was its location near the many small machine shops and private arms manufactures of Worcester County and the Connecticut Valley. The valley and hill towns of interior New England were home to hundreds of blacksmiths, woodworkers, and toolmakers, seasoned by a tradition of adaptability, always searching for an advantage that would buttress their fortunes against ebbs and flows of a volatile and meager economy. The age of the craftsman generalist plying his trade for a neighborhood market was quickly passing by. Those who survived to the eve of industrialization were well suited to an important role in the industrial order, that of the machinist-inventor-engineer.

One of the most celebrated of the machinist-inventors was Thomas Blanchard, whose story is that of the quintessential Yankee improver. As a young mechanic in Worcester County, Blanchard invented a pattern lathe that greatly improved the speed and accuracy by which identical irregular forms could be mass produced. Its introduction was controversial. When Blanchard's barrel-turning lathe was introduced at Springfield in 1818, skilled workmen grumbled about its effect on their livelihood. One of the gun-stockers apparently boasted that his job could never be "spoiled" because it was impossible to turn a gun stock. Not only were the old jobs spoiled, but the skilled craftsmen were replaced by machines manned by workmen painfully aware of how quickly and easily they could be replaced. By 1827 Blanchard had adapted machines to perform more than a dozen discrete operations on the walnut stocks, the fastest and most accurate system of production yet achieved.[31]

Blanchard's patron and the man responsible for carrying out the government's quest for uniformity and interchangeability at Springfield was Roswell Lee, armory superintendent from 1815 until his death in 1833. Lee's tenure marked the peak years in the formation of the armory system. Growth, collegiality, and communication between the Springfield Armory and the surrounding network of private contractors and between Springfield and the U.S. armory at Harpers Ferry created optimal conditions for innovation. With the help of several

gifted machinists, Lee oversaw the system of quality control through gauging and inspection that would prove so essential to interchangeability and, eventually, volume production. Lee also put Springfield's fiscal house in order by adopting a standardized system of accounting, no small matter in an organization that employed more than a hundred workmen and had half a million dollars' worth (today about $35 million) of capital equipment. He encouraged public scrutiny and guaranteed that, as it grew, the Springfield Armory would also rise in public esteem.[32] Springfield became a destination for technologists, tourists, and curiosity seekers whose impressions reveal firsthand the way contemporary society was then struggling to find a balance between agrarian virtue and industrial power. A deeply religious man and a student of science, history, and human behavior, Lee helped facilitate the delicate transition from handicraft to machine-based manufacturing.[33]

One of the most revealing documents of the open-door policy promoted by government patronage is the circuit ride Sam Colt and his father made in 1836 while preparing to launch Colt's first arms factory. Christopher Colt and his twenty-one-year-old son spent a month interviewing key personnel and inspecting iron mines and machinery, beginning, appropriately, at Samuel Collins's ax factory in Collinsville, Connecticut, and ending up at Simeon North's armory in Middletown. Given their circumstances—no track record, little money—it is extraordinary how much information and access these two potentially fly-by-night entrepreneurs received. At Collinsville, Colt "found that the [air] used at the forges was supplide [sic] by the operation of a large [air] pump made from the sillinder [sic] & piston of a steam engine." From there they "met Mr. Lorton [Pliny Lawton] at Ware [Massachusetts, site of the woolen factory Colt's father represented as sales agent], examined the mashinary [sic] there in use & noticed paticually [sic] what kinds... [are] applicable to the manufacture of the Patent Gun." A week later they "had an interview with Mr. Rippley [Springfield Armory superintendent James Ripley] in

relation to the different Iron mines.... [The] best Iron now made [is] in... Salisbury, Ct." They traveled to Salisbury to "see the mines, forges & furnaces" and then returned to "examining the mashinery [sic] & c. at the Springfield U.S. Armory." A few days later they headed South "in company with Mr. Lawton" to inspect "the private armory of Mr. [Simeon] North where we saw exelent [sic] mashinary [sic]." Finally, on May 21 "my father and myself met Mr. [Lawton] at Springfield and agreed with him for servaces [sic] for the P[atent] A[rms] Man[ufacturing] Co." and concluded with a visit to "Mr. Amesies [Nathan P. Ames] swoard [sic] factory [in Chicopee near Springfield] & all parts of the U.S. Armory," followed that evening with "a long interview with Mr. Bates, the inspector of arms. [Showed] him my invention."[34]

The open door swung both ways. By the mid-1850s the U.S. Springfield Armory was surrounded by independent contractors, opportunists, and competitors, who, through a combination of contracts, capitalization, and borrowed ideas, finally reached a point where they could spring free of government patronage. As power and technical know-how shifted from the public to the private sector, the open door began to close, and the communal ethic that had knit technologists, capitalists, and the Army Ordnance Department together in a common quest for uniformity was replaced by a spiral of competitiveness in which only the strongest would survive. Years later Maj. James Wolfe Ripley, superintendent of the Springfield Armory at the time of the shift, became almost irrationally hostile toward inventors and entrepreneurs like Colt who he believed "owe many of their modifications and much of their successful working to the ingenuity and skill of the mechanics in government employ, for which the inventor obtains all the compensation and lays claim to all the credit."[35]

One of the casualties was the greatest of the Connecticut Valley's private machine-building, arms-making technology emporiums, the Robbins and Lawrence Co. of Windsor, Vermont (fig. 20). More than Colt's, Smith and Wesson, or Winchester, Robbins and

Lawrence exemplifies the rise of the private arms factory and its influence on the science and technology of machine building. The quest for uniformity and interchangeability of firearms was the engine that powered the race to perfect the system of machine-based manufacturing. As a product in great demand, composed of complex metal parts, and for which the personal preferences of users were of little significance, military arms emerged as the perfect instrument for exploring the frontiers of machine-based production. Hence, all the important armories, both private and public, divided their efforts and resources almost equally between making arms and making the machines that made the arms. The chicken-and-egg aspect of the business proved dangerous to those who placed too great an emphasis on one side or the other. In the end, Colt might have been better off if he had been less single-minded about guns and more alert to the possibilities of the technology he had mastered. Robbins and Lawrence had the opposite problem. Lacking abundant capital and a controlling interest in a great product, they engaged in a variety of manufacturing enterprises and became overextended. But during the critical fifteen years between the Mexican War and the Civil War, Robbins and Lawrence occupied the vanguard of machine building and technology, and their influence on European and on all subsequent arms and complex metal parts products manufacturers was considerable.

Robbins and Lawrence was founded in 1844, in a successful bid for a military arms contract awarded at the outset of the Mexican War. Under the technical triumvirate of Richard Lawrence, Frederick Howe, and Henry Stone (Samuel Robbins provided venture capital), Robbins and Lawrence launched a period of prolific enterprise and inventiveness. They refined and assembled a network of machinery that included a complete line of drilling machines, milling machines, and planers. They were among the first to use the turret lathe, a revolutionary adaptation of the milling machine that held a cluster of tools on a vertical axis and made it possible to perform a sequence of operations on a piece of work without removing it.[36] They also developed a plain milling machine that became the basis for the famous Lincoln Milling Machine that eventually sold more than one hundred thousand units worldwide.[37] Robbins and Lawrence was one of the two suppliers tapped by the Committee on Machinery to provide machine tools for the Enfield Armory and by so doing played a pioneering role in the development of the American commercial machine tool industry. Machine building was suddenly big business, and it was Richard Lawrence who, looking back on his career, acknowledged that failure to expand their production of machine tools "brought Pratt & Whitney into the business," a company that soon surpassed Robbins and Lawrence in size and importance.[38] Robbins and Lawrence failed in 1856 and spent several years in receivership before final liquidation in the fall of 1859. By that time the regional network of interrelated companies was already in place, with inventors, machine builders, and manufacturers happy to pick away at the carcass of a fallen competitor, whose tools and machinery were auctioned off in Hartford the following spring.[39]

Robbins and Lawrence were not alone in tapping the new market for American machine tools before the Civil War. The most prominent machine shop within the immediate orbit of the Springfield Armory was that organized as a joint stock company in 1834 by Nathan P. Ames Jr. and his brother James T. Ames. The Ames Manufacturing Company in nearby Chicopee were pio-

20. Armory and Car Shops of Robbins and Lawrence, Windsor, Vt., ca. 1849. B.F. Smith, del., E.W. Bouve, Boston, publisher. (Courtesy of the American Precision Museum, Windsor, private collection.)

neers in electroplating, bronze casting, and contract arms manufacturing. Ames, the nation's leading manufacturer of military swords, also manufactured bayonets and cannon. In 1840 Nathan Ames, in company with a commission from the U.S. Ordnance Department, toured Europe to investigate gun-making machinery, in hopes of introducing new machines at the federal armories. One of the first American companies, along with Robbins and Lawrence, to manufacture machine tools commercially, Ames achieved acclaim by supplying milling and gun-stocking machines for Britain's Royal Small Arms Manufactory in 1855.[40] Another high-tech machine tool enterprise in the Springfield orbit was the American Machine Works. Founded in 1848, five years later it employed 150 men making and designing gun-making machinery, with a brisk market in the South.[41]

If increased competition hastened the development of ever more sophisticated tools and machines, it also stimulated the evolution of better and faster guns. Since the invention of firearms in the fifteenth century, various attempts had been made to devise weapons that could be fired many times without reloading.[42] But not until the 1830s did the quest for speed and repetition shift into overdrive, involving technologists and inventors throughout Europe and America. The Space Race of the 1950s and 1960s has known many analogues throughout history. The race to perfect fast guns—breechloaders and repeaters—was among the most urgent and colorful.

"The discovery of a perfect weapon loading at the breech," one nineteenth-century analyst observed, was "second in importance to the invention of gunpowder" because it enabled an "army lying on the ground" to "pour into its antagonist a 'sheet of flame and shower of lead.'"[43] Breechloaders were typically five times faster to load than traditional muzzle loaders and much safer to use, allowing soldiers to reload from a crouching or protected position. Carrying as many as fourteen loads at a time, the repeaters increased the speed of firing to more than ten times that of the traditional flintlocks that had

been the dominant military firearm for centuries.

The earliest U.S. patent for breech-loading guns was assigned to John H. Hall of the U.S. Armory at Harpers Ferry in 1811. A design ahead of its time that was manufactured on a small scale, Hall's rifle ended up collecting dust in state arsenals after war secretary Joel Poinsett concluded in 1840 that "every attempt to increase the rapidity of firing will fail... after involving the government in great expense."[44] Then, as always, the argument was that fast guns, by reducing the time of it took to reload, wasted ammunition.

In 1848 Christian Sharps secured a patent for what came to be regarded as the "first successful Breech Loading Rifle" that was inspired by John Hall's original design. A "true inventor... genuine genius and originator [and] a very decided spiritualist," Sharps was described by contemporaries as "not sufficiently hard, selfish, keen, self-asserting, and 'practical'... to gain and hold the just fruit of his labors." Nevertheless, within three years he incorporated as a holding company of Robbins and Lawrence and began manufacturing his new firearm in Hartford (fig. 21).

Unfortunately, as Maj. Alfred Mordecai, the government's leading ordnance expert, noted in 1856, Sharps's development of a workable breech-loading mechanism was not by itself sufficient to render the invention fully practical—at least not in the view of the U.S. military. "Loading at the breech," he reported, "if it can be accomplished in a perfect manner, offers a complete solution of the question of easy loading and close fitting; it is... remarkable that no method of making a practical application of it has yet been suggested which can command general, or even extensive, approbation; mechanical ingenuity seems to have been thus far incapable of removing all the difficulties of having an opening or joint exposed to the action of the charge of powder."[45]

Mordecai was mistaken. By the mid-1850s a convergence of interlocking innovations in gunnery had eliminated the last obstacles to the creation of an effective breach-loading firearm. In 1848 the French captain

C.E. Minié resolved the search for the perfect self-expanding bullet needed to take advantage of the principle of the rifled gun barrel's difficult-to-load, spiral groves by designing an oblong bullet, the famous Minié ball, composed of a tapered cylinder with a hollowed out base that was easy to load and yet bit into the rifling grooves upon firing. By 1855 the smoothbore musket was obsolete. Even more important was the development of the metallic cartridge. By uniting the components of bullet, ignition system, powder, and casing, it enhanced the uniformity and weather resistance of projectiles, while greatly increasing the speed of loading and simplifying storage. The greatest blunder of Sam Colt's career was his failure to recognize the significance of the metallic cartridge. Colt's two major rivals, Winchester and Smith and Wesson, emerged in part as a result of his failure to take stock of an invention that, ironically, was developed by one of Colt's own workmen, Rollin White.[46]

The parallel development of the percussion cap, which replaced the awkward and weather-sensitive spark and external priming powders of the flintlock system, made the revolver practical in the same way that the metallic cartridge made breech-loading long arms practical. Each validated the other, to the point where the combination of rifled barrels, breech-loading mechanisms, metallic cartridges, and repeating mechanisms achieved by either revolving cylinders (Colt's) or spring-fed magazines, created the ping-pong sequence of innovations that brought firearms into the modern age.

The breathtaking pace of technical innovation that occurred throughout the Connecticut Valley between 1845 and 1860 firmly established the region as the center of the American gun industry (fig. 22). In addition to Colt's revolver and Sharps's rifle, there was Lewis Jennings's invention of the first lever-action rifle with a tubular magazine, patented in 1849 and manufactured by Robbins and Lawrence, as well as Daniel Leavitt's 1849 patent of a revolver developed in collaboration with Edwin Wesson, and later Daniel B. Wesson, and born of a conspiracy to get around Colt's patent.[47] Horace Smith, who had supervised the manufacture of the Jennings Repeating Rifle at Robbins and Lawrence in 1850 and was listed as a gunsmith in Hartford that year, soon entered into partnership with the Wesson faction.[48] In 1852 Horace Smith and Daniel Wesson founded the Volcanic Repeating Arms Company, which manufactured lever-action pistols and carbines that later evolved into Tyler Henry's rifle (fig. 23). The Henry rifle in turn evolved into the Winchester Repeating Arms Company's famous lever-action, tube-fed, rapid fire, sixteen shot, .44-caliber killing machine.[49] As early as 1858 *Scientific American* enthused that the Volcanic rifle was "beyond all competition by the rapidity of its execution. Thirty shots can be fired in less than one minute," which "far outdoes the best revolving firearms yet produced."[50] Imagine facing mounted troops armed with such capacity.

In 1859 Christopher Spencer, a former Colt machinist who was then working as a shop superintendent at

21. *View of Sharps' Rifle Co., Hartford, Conn., 1853. Photograph of a watercolor by C.S. Porter. (Courtesy of the Connecticut Historical Society.)*

22. *Hartford and Connecticut Valley Arms, the Civil War Era, 1856–62.* Top to bottom, left to right: Springfield model 1855 rifle-musket, *ca. 1858, U.S. Springfield Armory; this was the first standard-issue rifled musket made available to U.S. troops; features Edward Maynard's patented trapdoor on lockplate with tape primer; superseded the famous 1841 model Mississippi musket (courtesy of the Museum of Connecticut History);*

Sharps model 1853 shotgun, *1854, by Sharps' Rifle Manufacturing Company, Hartford, .26 caliber, 1-shot, serial #9,377, deluxe grade engraving attributed to Gustave Young (private collection);*

Spencer shotgun, *1868, .52 caliber, 7-shot, patented 1860 by Christopher Spencer, Manchester, Conn., manufactured by the Spencer Repeating Rifle Co., Louis Daniel Nimschke (1832–1904), engraver; the Spencer, the first repeating rifle to use a metallic cartridge, was designed while Spencer was employed at the Cheney Silk Mills in South Manchester (Courtesy of the Connecticut Historical Society);*

Smith and Wesson "volcanic" pistol, *ca. 1854, .41 caliber, 5-shot, Norwich, Conn. (private collection); this was the first cartridge loaded, tube-fed repeating pistol;* Pepperbox, *1849, Robbins and Lawrence Co., Windsor, Vt., .31 caliber, 5-shot (private collection);* Smith and Wesson model 2 army pistol, *1863, Springfield, Mass., .32 caliber, 5-shot, Louis D. Nimschke, engraver (private collection).*

Cheney Brothers Silk Mills, devised a tube-fed, breech-loading rifle of his own that surpassed even the Henry rifle in speed and firepower. Patented in March 1860, the Spencer rifle contained a magazine of seven spring-fed metallic cartridges in a tube with a lever-action trigger guard that fed and ejected cartridges and empty cases at breakneck speed. With the financial backing of Charles Cheney, the Spencer Repeating Rifle Co. was incorporated in January 1862, with manufacturing based in the Chickering and Sons piano factory in Boston.[51]

The Spencer rifle could discharge seven bullets in twelve seconds, or twenty-one per minute and was renowned during the Civil War as the rifle that "was loaded on Sunday and fired all week." Maj. Gen. James H. Wilson field tested the Spencer, which he described as "the best fire-arm yet put into the hands of the soldier.... I have never seen anything else like the confidence inspired by it."[52] The Henry rifle and the Sharps rifle were also widely used during the war, contributing to Hartford's reputation as the center of firearms manufacturing in the United States.

In spite of the deadly effectiveness of the breech-loader and the tube-fed repeater as instruments of warfare, it was not the rifle but the revolver—especially the Colt revolver—that became synonymous in the popular imagination with "the fast gun." Although Colt himself claimed to have invented the revolver, the evidence suggests otherwise. The first U.S. patent revolver, Elisha Collier's revolving flintlock of 1813, was almost certainly known to Colt when he began dabbling in the design of chambered-breech, revolving pistols. Moreover, Colt's own collection of repeating firearms (fig. 24), the first assembled in the United States, contains specimens of

European and American make clustered around and predating the time of his own invention.

Nor was Colt's revolver the best or only design of its type. The British-made Adams and Deane double-action revolver, patented by Robert Adams in 1851, was in many ways a better gun.[53] With the expiration of Colt's patent in 1857, a rash of imitators, notably the Joslyn Arms Company's (Stonington, Connecticut) revolver of 1858, Savage and North's (Middletown, Connecticut) patent of 1859, the Starr Arms Company's (New York City) double-action revolver of 1859, the Remington Arms Company (Ilion, New York) solid-frame revolver patented in 1858, and a near duplicate of Colt's revolver manufactured by Rogers and Spencer in Utica, New York, all entered into competition with Colt's. But Colt created the market and retained a production advantage that earned his gun its status as the standard by which all others were judged. Design considerations aside, the convergence of myth and reality gained Colt's revolver a status that transcended the particulars of design during a period when Colt's design and engineering were never far behind and often ahead of all his direct competitors. By the time the *Hartford Times* reported that a "tailor in London has invented a waistcoat on the principle of Colt's revolver... a garment with four fronts," the association between Colt and the revolver had already entered the realm of mythology.[54]

Sam Colt never acknowledged his indebtedness to his predecessors or his competitors. Nor did he recognize the role of the Springfield Armory as an incubator of technological innovation in the arms industry. Even during his controversial feud with the Hartford Common Council, when he threatened to relocate to New York City,

23. Model 1860 Henry Rifle, 1861, .44 caliber, 15-shot, serial #9, gold mounted and engraved as a presentation to President Abraham Lincoln from the New Haven Arms Co. (Courtesy of the National Museum of American History, Smithsonian Institution.)

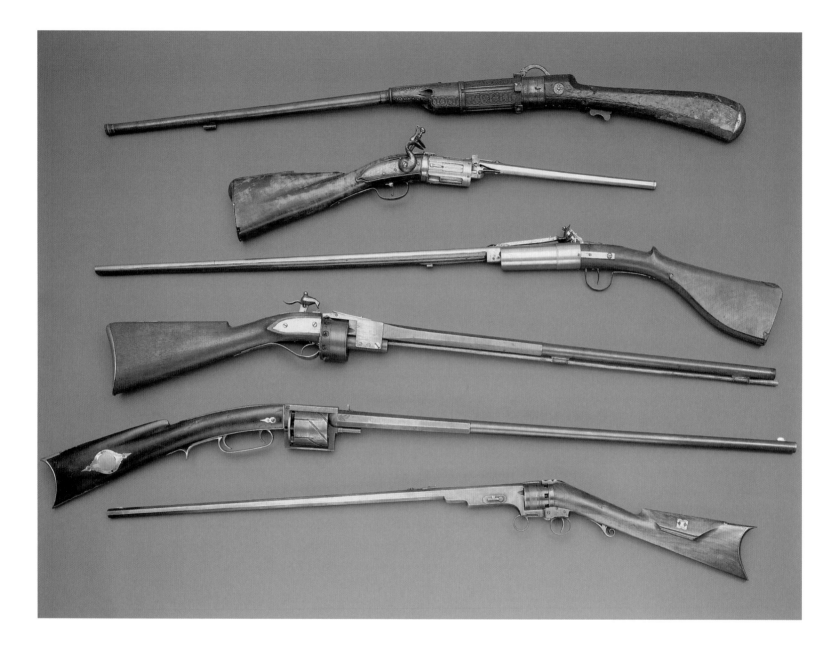

24. Colt's Collection of
    Revolving Cylinder
    Repeating Firearms,
    1650–1837.
    The collection of firearms
    Sam Colt began amassing
    included a survey history
    of the revolving-chamber,
    breech-loaded rifles,
    including samples
    acquired in 1851 during

Colt's stay in England,
examples used in the
patent infringement
lawsuit of 1851, and
examples illustrated
in his article
"On the application of
Machinery to the manu-
facture of... Arms."

Top to bottom:
matchlock revolving
musket, 1750–80,
India,.60 caliber, 5-shot
(WA #05.1024);
snaphaunce revolving
musket, ca. 1650,
John Dafte, London,
.46 caliber, 6-shot
(WA #05.1022);

flintlock revolving musket,
ca. 1650, German
or Dutch, .45 caliber,
6-shot (WA #05.1031);
flintlock revolving musket,
1826, invented by Rufus
Porter, Billerica, Mass.,
made by Joseph H. Center,
Boston, .42 caliber, 9-shot;
Colt purchased the rights
to this design in 1836
(WA #05.1009);

Colt seemed unaware that the government's patronage of arms and machines had created an atmosphere in the Connecticut Valley that greatly increased his likelihood of success. Colt profited from the discovery that the machinery developed in the quest for uniformity—a security and logistics concern—had brought down the unit costs of firearms, a profit and marketing concern. Colt praised both the "uniformity and cheapness" of "the American system." When queried about his costs, Colt explained: "The wages of those who attended... machines was 10 per cent, and those of the best class of workmen engaged in putting together, finishing and ornamenting the weapons, was also 10 per cent," leaving "80 per cent" of the cost, as work "done by the machines."[55]

Having been burned once by introducing his revolvers at a cost beyond what the market would bear, Colt became obsessed with cost control. More importantly, he eventually discovered that cost control was a better defense than patents, lawsuits, and litigation in protecting the value of his creative property. In arguing, unsuccessfully, for a second patent extension, Colt's attorneys claimed that "by the increase and subdivision of machinery," Colt would "be able so to perfect his armory... that he will be able to furnish... a perfect arm at a price which will defy... spurious imitations."[56] Sam Colt eventually pioneered the art of "creative destruction," cannibalizing his own product lines in an effort to stay a few steps ahead of the competition in perfecting the system and lowering the costs of production. "I have made a great many improvements in the machinery," Colt boasted in 1854, "and every day adds an improvement now."[57]

Sam Colt's failure in New Jersey during the 1830s suggests, among other things, the importance of timing. In his classic study of creativity and innovation, George Kubler pointed out how the timing of one's entry in the life cycle of an innovation makes all the difference. Kubler noted that the great variation in success among creative people depends not so much on talent as timing.[58] Luck and timing as much as talent and temperament gave Colt an edge up in the second incarnation of his arms manufactory. Temperamentally well suited to the rough-and-tumble, litigious jockeying for position that coincided with the commercial phase of the firearms manufacturing revolution, Colt thrived on difficult challenges and craved the personal recognition that went along with opening up new markets and beating the drum for his product. While a unique product, brisk demand, innovative marketing, and relentless cost control made Colt rich, it was government-supplied venture capital, in the form of military contracts, that got him started.

Improvements in the design of Colt's revolver piled up during his Hartford years, so that by time of his death in 1862, it really was, if not the first or even the best, surely the most practical and efficient revolver in the world. Considering that his first prototype exploded in testing and that the arms made in Paterson, New Jersey, were dangerously flawed in a number of particulars, it is remarkable how nearly perfect Colt's revolver eventually became. Rusting, breaking, the problem of multiple discharges, and the jamming of percussion caps were among the many problems tackled and overcome in the quest to perfect Colt's arm. Such an outcome was by no means assured when Colt commenced manufacturing the Walker .44-caliber dragoon, or cavalry pistol (fig. 25), so named in recognition of Texas Ranger Samuel Walker's participation in its design. The Walker was a hand cannon, so unwieldy in weight and size, and with an overlong cylinder so prone to exploding from overcharging

revolving shotgun, *ca. 1837, invented and patented by Otis W. Whittier, Enfield, N.H., .70 caliber, 6-shot; Colt later experimented with the use of the "zigzag" cylinder design shown here (WA #05.1033);*

no. 1 model ring lever rifle, *1838, invented by Sam Colt, manufactured by the Patent Arms Manufacturing Co., Paterson, N.J., .34 caliber, 8-shot, serial #189;*

Colt regarded this as the culmination of two centuries of progress in the design of rotating breech firearms (WA #05.969).

25. Walker Model Army Pistol, 1847. Invented by Sam Colt, manufactured by Eli Whitney Jr., Whitneyville, Conn., .44 caliber, 6-shot, serial #1020, faint impression of the ranger and Indian cylinder scene, one of about 1,100 manufactured. Sam Colt acquired this gun in 1860 with a history of having belonged to Captain Samuel Walker during the Mexican War. This is one of the first couple dozen Colt revolvers to arrive on the front of the Mexican War (WA #05.998).

that it is no surprise that most of the contracted first thousand ended up on the scrap pile with burst cylinders.[59] With the 1850 patent for the second model army pistol, Colt made several improvements. Lighter in weight, at four pounds two ounces, it introduced a locking device that held the cylinder in place between chamber openings, thus preventing accidental discharge in the event that the gun was dropped or bumped, a serious flaw in all of Colt's prior designs.[60] In 1850 Colt also introduced both the "1851 model" navy revolver, a smaller and lighter gun of .36-caliber, and the "1849 pocket revolver," smaller and lighter still, at .31-caliber and the least expensive and most popular gun in Colt's line, selling 325,000 from the time of its introduction until it was discontinued in 1873. The 1851 navy revolver became the favorite sidearm of Confederate officers in the Civil War. Its name notwithstanding, it was popular in all branches of the military.

Sam Colt never shook his obsession with producing military long arms (fig. 26). Despite their relatively poor sales, Colt's revolving rifles and carbines were the best repeating long arms available until invention of the metallic cartridge rendered them comparatively obsolete. Although design modifications gradually reduced the problem of lateral fire and multiple discharge, the difficulty was not completely eliminated until Colt adopted the use of metallic cartridges. The idea of steadying the barrel of a rifle with a hand placed in front of a cylinder that might discharge bullets from the sides as well as through the barrel was understandably disquieting. The pistol

was held in one hand, or if two, with both hands behind the breech, eliminating the danger. But Colt's carbines, shotguns, and rifles were never perfectly safe, and as soon as faster, safer guns were available, sales diminished and the models were soon discontinued. Even Colt's own agents complained that they could "sell a great number of the shot guns if [they] had the proper confidence in them."[61] Indeed, most of the revolving rifles with sword bayonets made for Colt's Rifle Regiment remained as unsold inventory at the factory well into the Civil War.[62] Produced in several calibers, these rifles were more impressive to look at than to use and are today a much-sought-after collectors' item and rarity.

The first Hartford-made long arms were actually pistols fitted with detachable shoulder stocks (fig. 27). The earliest of these were produced about 1854. By 1859 Colt had patented several variations of the shoulder stock, including the canteen gun stock, with its built-in liquid container. Part of the problem with Colt's long arm was its open-frame construction with detachable barrel. Ironically, Colt experimented with solid-frame construction during the Paterson-era but rejected it later on as rendering the removal of cylinders for cleaning and maintenance too difficult. With solid-frame construction, a feature found in both the Adams and Deane pistol and in the earliest Remington-made revolvers, the joint between the hammer and cylinder remained snug and secure, thus diminishing the lateral flame that was the primary cause of multiple discharge. The snugger the joint, however, the more difficult it was to rotate the

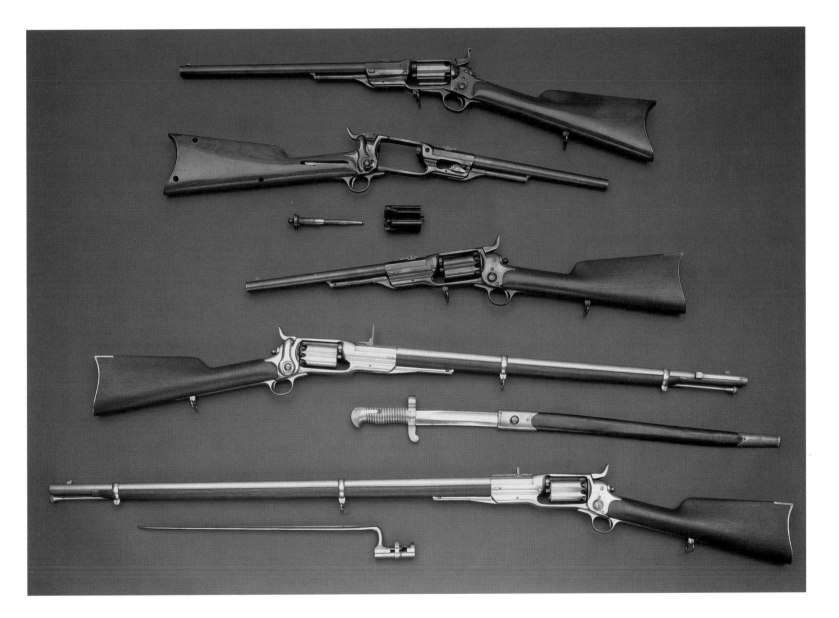

26. *Colt's Military Longarms, 1858–62.* Top to bottom: new model carbine, .50 caliber, 6-shot, serial #7; one of only ten made (WA #05.972);

new model carbine, .56 caliber, 5-shot, serial #326 centerpin, solid frame construction disassembled, parts skeletonized for demonstration purposes (WA #05.983);

new model carbine, .56 caliber, 5-shot, serial #852 military-grade blued steel (WA #05.983);

new model rifle with bayonet and scabbard, .56 caliber, 5-shot, serial #4775; military-grade white steel finish, brass extension for cleaning rod stored inside buttstock (WA #05.974);

new model rifle with bayonet, .56 caliber, 5-shot, serial #7391; military-grade white steel finish, brass extension for cleaning rod stored inside buttstock (WA #05.975).

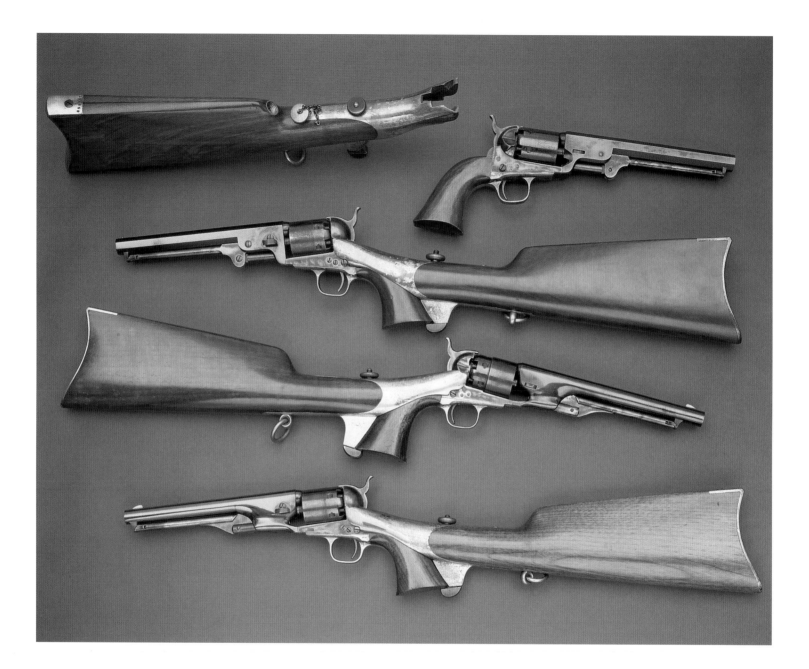

27. *Colt's Production Pistols and Shoulder Stocks. These were gathered into the Colt Memorial Firearms Collection by Elizabeth Colt from new factory inventory in 1862.*

Top to bottom: old model navy or belt pistol (model 1851) with detachable stock, .36 caliber, 6-shot, serial #128,429; almost 250,000 made, 1850–73. *The detachable canteen/ gunstock (U.S. patent #22,627) was introduced in 1859* (WA #05.985);

old model navy or belt pistol (model 1851) *with detachable stock, .36 caliber, 6-shot, serial #128,519 (WA #05.987);* new model army or holster pistol (model 1860) with detachable stock, .44 caliber, 6-shot, serial #1617; about 200,000 made, 1860–73 (WA #05.986);

new model navy or belt pistol (model 1861) with detachable stock, .36 caliber, 6-shot, *serial #11,634/s; almost 40,000 made, 1861–73 (WA #05.984).*

cylinder, thus requiring frequent adjustment, not unlike the rotary blade of a chain saw that loosens with use.

With his patent about to run out in 1857, Colt and chief engineer Elisha K. Root entered into a frenzied campaign to perfect a new model solid-frame revolver. Colt's first Hartford-made solid-frame revolver adopted a zigzag type cylinder, modeled after a design by O.H. Whittier represented in Colt's revolving firearms collection. Colt's only production model solid-frame revolvers accommodated the removal of cylinders by fastening the cylinder with a cylinder pin screwed through the center back in the location formerly reserved for the revolver's hammer. To accommodate the center pin, the hammer was removed to the side, hence the designation of Colt's solid-frame designs as "sidehammer revolvers" (fig. 28). Although almost forty thousand sidehammer revolvers of various (mostly small) calibers were made beginning in 1856, the design was not very successful owing, in part, to the tendency of the cylinder pin to freeze up, thus requiring severe force to loosen it, a grim prospect in the field.[63] So in spite of the fact that the sidehammer Colt's were the first design that did not require removal of the barrel when changing cylinders, it could hardly feel like an improvement when the cylinder pin was stuck.

Thirteen years after establishing his Hartford armory, with sales in the high hundred thousands, Sam Colt introduced the most perfect design of his lifetime in the model 1860 army, navy, and police revolvers (fig. 29). Although less popular with collectors because of the huge volume of their eventual manufacture, the 1860 model revolvers are a beautiful, sleek, elegant design, lighter, cleaner, and more efficient than anything else of its class. Almost from the beginning, Colt's revolvers were distin-

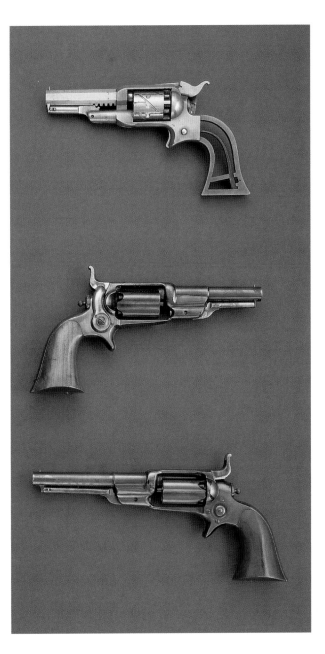

28. Colt's 1855 Model Solid Frame Sidehammer Pocket Pistols.
Top to bottom: sidehammer pistol, experimental prototype, .26½ caliber, 5-shot, serial #5; related to U.S. Patent #13,999 assigned to Elisha Root, 12/25/55; never went into production (WA #05.1001); sidehammer pistol, .31 caliber, 5-shot, serial #6,373; acquired for Memorial Collection in 1863 (WA #05.999); sidehammer pistol, serial #7,1623; a production model based on the Root-Colt patent (U.S. patent #20,144) of 1858 (WA #05.997).

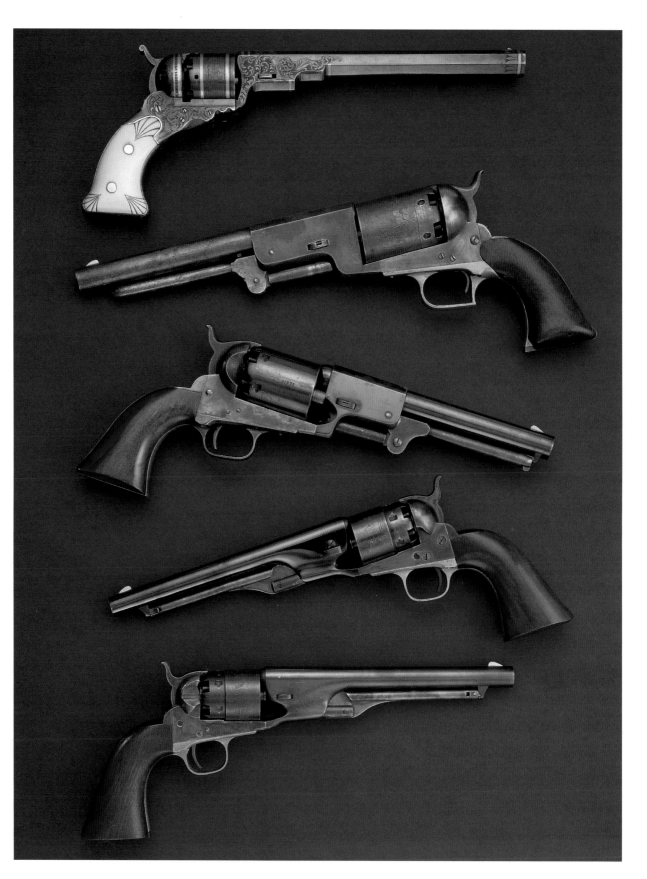

29. *Evolution of Colt's Army Pistol, 1839–60.* Top to bottom: Paterson holster pistol, *.36 caliber, 6-shot, serial #984 (WA# 05.1009);* army or holster pistol, *Walker model, .44 caliber, 6-shot, serial #1020 (WA# 05.988);* old model army or holster pistol, *third model dragoon, .44 caliber, 6-shot, serial #15,016, (WA# 05.922);* new model army or holster pistol, *model 1860, .44 caliber, 6-shot, serial #43,831 (WA# 05.966);* new model army or holster pistol, *model 1860, .44 caliber, 6-shot, serial # 55794 (WA# 05.1010).*

guished for the macroscopically scratch-free surface attained by a technique called bluing, a chemical cooling agent that protects against rust, turning the steel components blue.[64] As steel replaced cast iron in construction, in combination with brass back straps, walnut grips, and decorated cylinder scenes, even the most pedestrian of Colt's revolvers gained an artistic quality. The voluminous writings on Colt's firearms fail to fully account for the changes that were made in the last years of Colt's life. But when the new model army .44-caliber revolver hit the market in 1860, priced to move and weighing in at just two pounds eleven ounces, with its sleek modern styling, streamlined barrel, and fluted cylinders, the world finally had a sidearm even the worst military curmudgeons could love.

Colt introduced the "New Model" with a blitz of hype about a "new" material he described as "Silver Spring Steel," a "gun metal" that possesses "all the qualities of the most refined cutlery steel, without the brittleness," being "three times stronger or tougher than the best cast steel heretofore made," requiring "less weight of metal... for barrels and cylinders," and with a "spring temper" that "propels the ball with from one-quarter to one-third greater force." Citing a report by the Ordnance Board of Practice, Colt described a test in which his "Holster Pistol was fired twelve hundred times, and his Belt Pistol fifteen hundred times, cleaning only once a day," with "neither... injured by the firing," and achieving a "penetration... through seven inches of board... while the highest penetration of the common Dragoon Pistol was only through five inches."[65] Whatever the merit or veracity of the tests, Colt's silver spring steel is metallurgically almost indistinguishable from the steel used in the "Old Model" revolvers.[66] Pure marketing. Pure hype. So how did Colt's get the weight down? Like Dorothy's magic slippers in the Wizard of Oz, the power had been there all along. Numerous design revisions along the way had long since eliminated the cause of burst cylinders. The Walker pistol exploded not because its metal was too light but because the cylinder was half again as long as it

needed to be, a problem rectified by 1848. By streamlining the barrel and frame and cutting flutes in the cylinder, Colt eliminated all extraneous metal. Was the new model really more powerful? It depends on what it was compared with.

By the time of his death in 1862, Sam Colt had made and sold almost one million guns, a first in the history of American manufacturing. The final revolution, after years of incremental progress, was, again, in the mind. The revolver carried by officers and mounted troops in the Civil War was indeed a masterful culmination representing thirty years of hard labor and love. To the end, the nineteenth-century's fascination with the mystery of genius almost required that invention appear to spring forth fully formed, without source and without trouble. For Colt, it had been anything but. Colt's genius was the passion that burned within, the inner flame that earned him the confidence and support of other men. With relentless toil, perseverance, and attention to details, Sam Colt, Elisha Root, and the countless machine builders and tenders who worked at Colt's Armory marched a new American industry to the head of its class, while producing a weapon both perplexing and amazing, even today, for its capacity to attack and defend.

When Sam Colt returned to Hartford in 1847, the city was markedly different from the place he had left fifteen years earlier. With the economy stagnant from recession and the best and the brightest of the city's youth heading west, resistance to manufacturing finally crumbled. No one predicted the turnaround that resulted in a doubling of population between 1850 and 1860, a demographic transformation that absorbed thousands of Irish and German immigrants and resulted in Hartford's emergence as the "largest manufacturer of firearms, of any place in the Union."[67] The railroad arrived in 1839 and was connected with Springfield in 1844. Although Hartford had manufactured woolens and paper on a limited basis since the eighteenth century, its location was ill suited to manufacturing during the age of water power. Not until steam power proved its competitive advantage did Hartford's industrial boom take off.

Several complementary manufacturing concerns emerged alongside Colt's during the late 1840s and 1850s. Prompted by the same Mexican War contracts that launched Colt, Edwin Wesson's Rifle Works, a parent company to Smith and Wesson, settled within spitting distance of Colt's first shop on Pearl Street, where it grew rapidly until Wesson's death in 1849. By the end of 1851, the city boasted of the "talent of our own mechanics," including a wide range of increasingly mechanized traditional handicrafts, and the manufacture and export of such complex, value-added products as church organs, fire engines, pianofortes, railroad cars, and electroplated wares. In 1848 Colt's second landlord and eventual nemesis, Solomon Porter, built the first and largest steam-powered manufacturing facility in the city, a symbolic turning point among the community's merchant-bankers. Located in the city's first manufacturing district on Commerce Street, near the confluence of the Connecticut and Mill Rivers, this four-story, forty-by-fifty-square-foot brick building was the location of Colt's armory when he first achieved international fame and profitability.[68]

Rising on the same wave with Colt were several neighboring manufacturing concerns that shared with firearms an emphasis on sophisticated technology, as well as products and processes that required the use of skilled machinists and engineers. When the *Hartford Daily Times* chided manufacturers for relying on industrial products such as "plain cottons and woolens which require no great skill for their production" and urged "diversifying our employments," they could already see the future in Hartford's fast-growing high-tech manufacturers. As early as 1849, the *Times* reported with astonishment on "the mammoth buildings for manufacturing purposes which have sprung up recently," noting particularly, "Tracy & Fales leviathan car factory" and the boiler works of Woodruff and Beach, which "give a sort of Birmingham aspect here, which seems strange... for

old Hartford." As the roller-coaster ride of industrialization picked up speed, the *Times* cheered from the sidelines, noting in 1853 that "this city is now experiencing a rapid improvement in population, wealth, enterprise, and the ornaments of society.... [The] course and character of its business has wholly changed. From being a mere market town for the county, with half a dozen wholesale stores, it has become a manufacturing place." With manufacturing came new and varied social amenities deemed essential in adjusting to industrialization. The *Times* gloated that a "new park is projected... new church edifices have multiplied," and the Wadsworth Atheneum, occupied by the Historical Society, the Picture Gallery, and the Young Men's Institute, was flourishing, a growing cluster of prized civic ornaments in the new industrial order.[69]

Appropriately, Hartford was home to Woodruff and Beach, one of the nation's first large-scale commercial manufacturers of the steam engines that liberated manufacturing from its reliance on water power. The parent firm to Woodruff and Beach, the foundry established in 1821 by Henry Beach's father-in-law, Truman Hanks, built and installed Hartford's first steam engine and was among a small group of Hartford machine shops that built small pumps and engines before 1840.[70] By 1848 the company, reorganized as Woodruff and Beach in 1846, employed one hundred men and was "doing large business" as "makers of heavy and complicated machinery, especially such as required skill and ingenuity in designing."[71] In 1863 the company employed more than four hundred workmen in the manufacture of marine engines for commercial ships and gunboats, large double-piston pumps, and custom iron castings. With steam derricks, lathes, gearing of all sizes, and an iron planer described as "the largest ever erected in this country... capable of planing a mass of iron twelve feet square by thirty feet long," Woodruff and Beach were a clanking, clamorous marvel of grease, coal, and smoke and a "pride to the citizens of Hartford."[72]

G.S. Lincoln and Company, founded in 1834 and

renamed the Phoenix Iron Works in 1850, contributed to Hartford's international reputation for machines and machine-based manufacturing. Although described by analysts for the R.G. Dun credit service as "smart... capable, energetic & enterprising" and "making money all the time," it is not certain when and how the company expanded from its technologically rudimentary emphasis on architectural and mortuary cast iron and bank vaults into the manufacture of machine tools.[73] The Phoenix Iron Works advanced what historians have described as "one of the most striking... chapters in the history of the machine tool industry" when, in 1855, Francis Pratt, the Phoenix Works supervisor and a former pistol maker and machinist at Colt's Armory, developed an improved commercial version of the plain milling machine made famous by Robbins and Lawrence.[74] The Lincoln Milling Machine became the most successful and widely exported machine tool of its generation. The man who developed the Lincoln miller went on to found Pratt and Whitney, a machine tool industry legend, that specialized in, among other things, the commercial manufacture of machine tools designed for making guns and sewing machines.[75]

During the Civil War and in the decades that followed, Hartford's manufacturers maintained a brisk pace in job formation and growth, with the arms industry leading in the numbers and diversity of workmen employed. Hartford's special advantage was in its transportation amenities and investment capital applied to a "technologically convergent" network of manufactures relying heavily on similar types of machinery.[76] Boasting of having the "largest banking and insurance capital of any city... in proportion to population," once Hartford embraced manufacturing, its development capacity was considerable. Manufacturing employment remained high throughout the depression that followed the crash of 1873, led by 1,000 workers in Cheney Brothers' Hartford and South Manchester silk factories, 700 at Colt's, 300 at Weed Sewing Machine, 400 at Pratt and Whitney's, and 200 at the National Screw Company.[77] When, in 1876, *Scribner's*

Monthly described Hartford as the "richest [small] city in the United States," it honored not only the capitalists and industrialists but also the relatively high wages of the city's high-tech, machine-based manufacturing work force.[78]

In the 1860s and 1870s Hartford experienced a proliferation of new high-tech manufacturing enterprises, many with roots in Colt's Armory. Francis Pratt, Amos Whitney, George Fairfield, Christopher Spencer, Charles Parker, Asa Cook, and Charles Billings, seven of Hartford's most influential technologists, each worked in the arms industry before launching the enterprises that earned them fame and fortune. The Pratt and Whitney Company, once "one of the largest builders of accurate and ingenious tools and machinery in the country," was launched by its founders while moonlighting from day jobs at George Lincoln's Phoenix Iron Works.[79] Pratt's work in developing the Lincoln Milling Machine was a prelude to an illustrious career manufacturing machine tools distinguished for unprecedented standards of gauging and precision. After only two years in partnership, Pratt and Whitney was Hartford's most sophisticated commercial machine tool manufacturer, employing about sixty men. In 1865 they embarked on a major expansion, building a factory on the Park River (formerly Mill River), where they produced patented milling machines, thread spooling and related textile machinery, gun machinery, turret lathes, tools used in making sewing machines, and patented lawn mowers, chucks, machine guns, and vices.[80]

George Fairfield's migratory path from Robbins and Lawrence in Windsor, Vermont, to the American Machine Works in Springfield, to Colt's Armory as an inside contractor is typical of the region's successful technologist-entrepreneurs. After nine years at Colt's, Fairfield helped develop the Weed sewing machine as an independent venture at Sharps Rifle Company and, in 1876, founded the Hartford Machine Screw Company to manufacture a revolutionary new machine that rapidly and automatically turned out screws of perfect uniformity. Fairfield's final enterprise involved mass producing high-wheel bicycles. Over the course of forty years, Fairfield's odyssey parallels the migration of innovation and capital from north to south along the Connecticut River, and from firearms and sewing machines to machine tools and bicycles.[81] As he gazed from the observatory of the south end of the Hartford mansion he built with Colt earnings in 1864, its panoramic view stretching from Mount Holyoke and Springfield to Middletown, he must have reflected more than once on the remarkable transformation that had taken place along the Connecticut River during his lifetime.[82]

The Hartford industrial concern most deeply rooted in the rich soil of Colt's "training-school in applied mechanics" was the Billings and Spencer Company, founded by two of the region's most peripatetic, enterprising, and creative inventor-industrialists. Charles Billings, like George Fairfield and Richard Lawrence, witnessed the birth of the American commercial machine tool industry at Robbins and Lawrence in Windsor, Vermont, where he served an apprenticeship as a machinist working under Frederick Howe on the machinery supplied for the Royal Small Arms Factory in England. Having learned machine building and mechanical die-casting in Windsor, Billings moved to Hartford in 1856, about a year after the opening of Colt's Armory where he worked at the new state-of-the-art drop-forging facility under the direction of Colt's mechanical superintendent Elisha K. Root. After the Civil War, Billings, in a series of liaisons with Colt alumni, manufactured the Weed sewing machine, from 1865 to 1868, and then, in 1869 with Christopher Spencer, designed and developed a die-casting, drop-forging plant in Hartford for the Roper Sporting Arms Company, reorganized in 1872 as the Billings and Spencer Company.[83]

The company, described as pioneers at drop-forging, attracted "immense business" almost immediately and within a year of incorporation gained international recognition by winning a Medal of Progress at the 1873 World's Fair in Vienna.[84] The art of duplicate die-forging, or drop-forging, a process by which metal is pounded into the intricate shapes of die-cast impressions by the force of

power hammers, achieved its highest standard of accuracy and cost effectiveness at Billings and Spencer.[85] By 1880 Billings and Spencer were praised for "manufacturing 1,000 different articles... all descriptions of steel and iron drop forgings... [including] lathe wrenches, machine handles, thumb screws... screw drivers, drill chucks... crank shafts... vise jaws... parts of guns, pistols, sewing machines, etc." Not very glamorous, but remarkable to a society that a generation earlier still imported almost everything that was not custom-made by hand.[86]

Billings's partner, Christopher Spencer, was a prolifically inventive and creative spirit. Like Sam Colt, Spencer was a homegrown original with roots in the metal and woodworking trades and an astonishing capacity for self-teaching.[87] Having learned a little gunsmithing and cabinetmaking at the knee of his grandfather Josiah Hollister, Spencer apprenticed to a local machinist, Samuel Loomis, also descended from a long line of rural woodworking artisans. Spencer joined Cheney Brothers as a journeyman machinist, and in 1853 he assisted Frank Cheney in perfecting an automatic silk spooling machine that helped revolutionize their business. Spencer then devoted a year to tramping around the industry, studying steam engineering and tool-making in Rochester, New York, and working in the machine shop at the Ames Manufacturing Company in Chicopee before returning to the Hartford area where, in 1855, he helped install machinery in Colt's new armory.

As a skilled marksman, machinist, and gunsmith, Christopher Spencer soon recognized that Sam Colt's revolving cylinder mechanism was flawed and potentially dangerous when applied to long arms. Spencer began searching for a better repeating rifle mechanism. In 1856 he returned to Cheney Brothers as superintendent of the machine shop. Having perfected and patented a design for a repeating rifle in 1860, Spencer appealed to Cheney Brothers for the financial backing to commence manufacturing.[88]

After the Civil War, Spencer joined fellow Colt alumnus George Fairfield in developing experimental machinery for the Weed Sewing Machine Company, located on the premises of Sharps Rifle Company in Hartford. Although his Spencer rifle was absorbed by the Winchester Repeating Arms Company after the war, Christopher Spencer continued to tinker with firearms. In 1866 he helped found, with Sylvester Roper, the Roper Sporting Arms Co. in Amherst, Massachusetts, which relocated to Hartford in 1869. In 1882 he invented a repeating, single-barrel, 12-gauge shotgun capable of firing ten shots in five seconds.[89] Between working for Weed Sewing Machine Company and founding Billings and Spencer, Christopher Spencer produced his greatest invention, which became the basis for another of Hartford's industrial triumphs, the Hartford Machine Screw Company, founded by George Fairfield in 1876. The automatic turret lathe, or screw machine, which Spencer developed in 1873, perfected the mass production of screws.[90] This seemingly simple implement represented one of the most difficult challenges of the machine age. Instead of being hand-forged and filed, screws could now be mass produced by a turret lathe—a multioperational milling machine—with an automatic, rotating machine head that could measure, turn, thread, stamp, and point anywhere from three hundred to four thousand screws per machine per day.[91]

Throughout the formative years of his campaign to perfect repeating firearms, Sam Colt showed a remarkable ability to endure conflict and ignore the suffering his actions caused others. Once established, he proved equally adept at repackaging his past as a heroic struggle. Perhaps the quality that enabled him to move single-mindedly toward a vague abstraction allowed him to view in the abstract such basic moral functions as keeping promises and honoring contracts. To describe Colt as a man of flexible morality is not quite to charge him, as his detractors did, "with almost every offence against morality and the laws of society which can be committed."[92] There were those who loved the man once described by an admirer as "a high-minded, honorable man, above suspi-

cion," to whom "reputation... is dearer than money."[93]

In the final analysis, whether Colt's greatest achievement lies in the system of manufacturing and machinery he installed in his Hartford armory or in the relentless and brilliant promotional war he waged on multiple fronts to create markets and move the product, both accomplishments surpass in significance his role as an "inventor" of firearms. Indeed, the most famous Colt revolver—the fabled Peacemaker, or Colt 45—was not invented until ten years after Colt's death. Unfortunately, because Colt's many panegyrists have accepted uncritically his claims as an inventor and, perhaps, because creating a system of manufacturing and an atmosphere conducive to innovation seems less glamorous than inventing "the gun that won the West," his most significant achievements have been relatively neglected. And since the Barnumizing tendency of the carnival-barking, medicine show man has less cachet, Colt's unprecedented achievement in shaping the reputation of his product has also been downplayed. It matters, because it is still unclear how he was able to transform an inefficient and haphazard process of production, in which batches of twenty or fewer were worked up in assembly line fashion, into a system involving hundreds of specialized machining operations carried out *simultaneously*, thus making Colt's Armory the most sophisticated manufactory of its age, described by historians as "perhaps the first in America" in which "all the elements of modern mass production" converged.[94] Was it Sam Colt who created this system, or was it the machinists and technologists who gathered around his flame, hoping to share in the dream woven by his gift at salesmanship? Either way, Colt created the environment that shifted the revolution in machine-based manufacturing into overdrive.

Colt relied heavily on the Connecticut Valley's interrelated network of machine tool manufacturers for the machinery acquired for the new armory between 1856 and 1862. From the Ames Manufacturing Company he acquired engine lathes and "alterations of barrel boring machinery";[95] from the Springfield Armory, "one compound screw drop machine... less amount due Mr. Root—$100" and "4 compound drop machines [at $1,100 each] less amount due Root—$400."[96] From the start-up Pratt and Whitney, Colt purchased a whopping $4,462 (almost $.25 million today) in "rifling machines" and "16 pistol barrel drilling lathes";[97] chucks and unspecified machinery and tools from Eli Horton and Son of Windsor Locks;[98] and "compound drops & boring machines" from the Pacific Iron Works in Bridgeport.[99] But Colt's major source of machinery was Hartford's George S. Lincoln Co., which supplied "planing machines," "screw lathes," "barrel boring machines," "spindle drills," "milling machines," "index machines," "swedging machines," and more during the years it served as midwife to the future Pratt and Whitney Company.[100]

Colt promoted the belief that "not only guns and pistols, but the machinery that makes them, was invented and constructed on the premises... as a branch of their regular business."[101] How much machinery Colt and Root actually made is unclear. It was probably in 1857, and not before, that Colt decided to manufacture machinery commercially.[102] However, aside from one major contract with the Russian government to supply machinery for the National Armory at Tula, Colt's commercial foray into machine tool manufacturing does not appear to have flourished to any great degree.[103] Furthermore, references to Elisha Root's traveling to Springfield, Lowell, New York, and Philadelphia, and to "amount due Mr. Root" by the Springfield Armory, for what was probably his role in customizing machinery purchased, suggest that, far from being technologically self-sufficient, Colt relied on the regional network to meet the company's long- and short-term needs for new and ever more serviceable machinery.[104]

In 1854 Colt, his attorney Edward Dickerson, and their political allies lobbied Congress (unsuccessfully) for an extension of Colt's patent by arguing that "his entire profits, thus far, have been principally expended in extending, improving, and perfecting his machinery, with a view [to] increased operations." To achieve "accuracy and perfection in [the arms] manufacture," they argued, "can only be done by perfect and expensive machinery. To procure that perfection has been Mr. Colt's constant effort." Colt's allies claimed that "as rapidly as profits arose, he reinvested them to increase and perfect his machinery." The central question was to determine if Colt had yet been "rewarded" by his invention. How much profit and how much capital improvement had occurred became a matter of considerable speculation among Colt's opponents, who were probably close to the mark in assessing Colt's investment in "the machinery, tools, fixtures, and other property... in Hartford... at $450,000" (today, about $25 million).[105] Colt was playing for keeps this time, a game of technological brinkmanship in which the person with the most machines (and hence lowest unit costs), wins, with or without patents.

The spectacle of so much specialized machinery under one roof soon transformed Colt's Armory into one of New England's premier destinations for foreign travelers, budding machinists, and those simply eager to gawk and

30. *Lincoln Milling Machine, 1855–70. Design perfected by Francis A. Pratt for George S. Lincoln and Co., Hartford; illustration from the* Report on the Manufacture of Fire-Arms... Tenth Census, *1883. Colt's Armory used hundreds of milling machines tooled up to perform specialized metal cutting operations in* *repetition and in large volume. This was one of the first commercially successful machine tools of its generation, used in the manufacture of firearms, sewing machines, and other complex iron forms. Its design was perfected in conjunction with the 1855 contract to provide machines for Colt's Armory.*

ponder the wealth that had created this gun-making leviathan. As one visitor reported in 1859, from Colt's "residence... overlooking the whole of the works... he can sit and gaze on the place he has created... a spectacle... more sublime than any that a crowned king can ever behold."[106]

During the ten years beginning with Colt's move to the Porter Manufacturing Company building, he developed three increasingly ambitious manufacturing facilities: Commerce Street (1849–55), the London armory (1852–56), and finally, occupied in the summer of 1855, the fabled Colt's Armory (1855–1994), with its Russian-style "onion dome."

Sam Colt personally described the theory and practice of machine-based manufacturing to the Institution of Civil Engineers in London in 1851, noting plainly:

The machinery requisite for constructing the repeating firearms... [is] composed of the simplest elements.... [The] lock frame is forged by swages, and its shape completed by one blow.... [The] barrel, forged solidly from a bar of cast steel, is bored and completed to calibre, and is then submitted to the various operations of planing [and] grooving the lower projection.... All separate parts travel independently through the manufactory, arriving at last, in an almost complete condition, in the hands of the finishing workmen.... Machines become almost automatons, performing certain labour under the guidance of [even] women, or children.[107]

Of special interest to the civil engineers were the milling machines (fig. 30), which used irregularly shaped "revolving cutters" to shave off particles as the work was fed slowly against them, a method "not employed [in England] to the extent they might be," the "judicious use" of which Colt's "good work... must be entirely attributed."[108] The armory that Colt built and operated on the American system in London was described as "the admiration of all England, resorted to by visitors as a show shop, for its rare perfection and beauty."[109]

But the greatest of Colt's show shops was the armory on Hartford's South Meadows, the centerpiece of a self-contained industrial compound known as Coltsville. Although several accounts of Colt's Armory were published by firsthand witnesses, none surpasses the poetry and impressionistic allure of Mark Twain, who first visited in 1868.[110]

The Colt's revolver manufactory is a Hartford institution.... It comprises a great range of tall brick buildings, and on every floor is a dense wilderness of strange iron machines that stretches away into remote distances and confusing perspectives—a tangled forest of rods, bars, pulleys, wheels, and all the imaginable and unimaginable forms of mechanism. There are machines to cut all the various parts of a pistol, roughly, from the original steel; machines to bore the barrels out; machines to rifle them; machines that shave them down neatly to a proper size as deftly as one would shave a candle in a lathe; machines that do everything but... trace the ornamental work upon the barrels. One can stumble over a bar of iron as he goes in at one end... and find it transformed into a burnished, symmetrical, deadly "navy" as he passes out at the other.... In all that world of complex machinery [no]... two machines [are] alike, or designed to perform the same office. It must have required more brains to invent all these things than would serve to stock fifty Senates like ours.[111]

31. Armory of Colt's Patent Fire-Arms Manufacturing Co., 1856. E.D. and E.C. Kellogg, Hartford. Sam Colt commissioned this view of his industrial compound within a year of its completion and occupancy. Left to right: Colt's Armory, the office, the tobacco warehouse, Germania Hall, and Charter Oak Hall. This is one of the few views that predates the construction of a railroad spur and depot (1857). This building burned in 1864 and was rebuilt almost exactly as shown here in 1866.

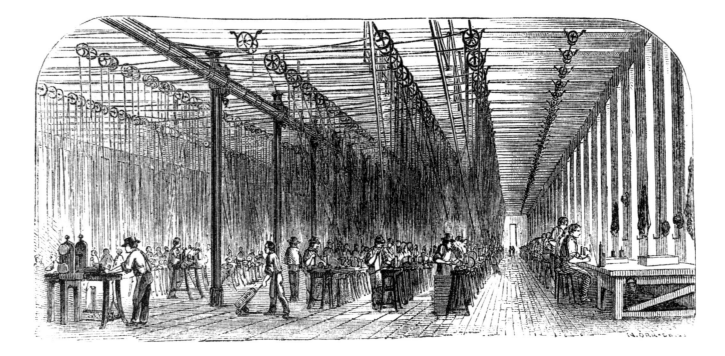

Visitors liked to quote statistics, because, aside from all else, Colt's Armory was most impressive in its sheer scale (figs. 31–33). It may not have been quite the largest building in New England in 1855 (at least one of the factory buildings in Lowell had more volume under one roof), but it was far and away the largest in Hartford and the Connecticut Valley, a fact made all the more "sublime," by its seeming to rise suddenly out of nothing. As Colt doubled the armory's size and achieved "further perfection of his machinery," the effect, just before his death in 1862, was "almost beyond the power of the imagination to conceive."[112] At the height of the Civil War, at peak output, capacity, and work force (estimated at 1,400), Colt's Armory was described by visitors as "8 rooms, 500 feet long, 60 feet wide, filled with men, machinery, and materials" turning out every day 300 U.S. pattern military muskets and "300 army revolving pistols." A Woodruff and Beach-made double cylinder, high-pressure steam engine with a flywheel measuring 18 feet in diameter and a pump "which supplies water to the boilers and the old factory," generating 300 gallons

per minute were contributing elements in the grand spectacle of sheer immensity.[113] "If extended in one line," enumerated a reporter, the armory floors with their "cast iron columns, sixty in number" and shafting, powered by 5 steam engines with an "aggregate of 900 horse power,... would be a mile long by fifty-five feet wide," equal to 6.5 acres.[114]

Colt authorized two detailed magazine stories on the inner workings of his armory. The first appeared in *United States Magazine* in 1857. The second appeared in the *Boston Olive Branch and Atlantic Weekly* in 1859:

At the center of the main building... [is] the steam engine... a miracle of mechanical ingenuity... 250 horse[power].... [Its] fly-wheel, weighing 7 tons, is 30 feet in diameter.... [The] store room... [is] crammed with iron and steel in bars.... [In the] forge shop... 200 feet by 40... [we witnessed] a host of forges in full blast, roaring like so many lions, and yet the air... perfectly pure.... [There we saw a] double row of black columns... hammers of the kind termed "drop," [that are] peculiar to this establishment... [being] raised on the endless screw principle and tripped by a trigger with the utmost facility.... The

32. *Colt's East Armory, Second Floor, 1857. Engraved by Nathaniel Orr (b. 1822); illustration from the* United States Magazine, *March 1857. Here, in a five hundred-by-sixty-foot room, was a row of sixty cast-iron columns, supporting line-shafting powered by a 250-horsepower steam engine. Belt-driven milling, drilling, and boring machines stand in rows between work benches beside the windows where filling and fitting took place.*

33. Colt's Forging Shop, 1857. Engraved by Nathaniel Orr; illustration from the United States Magazine, *March 1857. Under the direction of Elisha Root, Colt's Armory became one of the most progressive forging shops in the world, with numerous gun parts shaped in dies by a single blow from a screw-lifted vertical hammer applied to heated metal.*

amount of forging work done on a single rifle or pistol... [involves] 32 separate and distinct operations.... The Foundry...[is a] vast workshop 500 feet long by 60 [feet wide with]... 112 windows... innumerable gas-burners, warmed by steam, and perfectly ventilated.... [This is] the largest workshop in the world!... Running down the center... [are] long rows of wheels all turning and whirring at once.... [Of the] machine tools,... one group... [is used] in chambering cylinders;... another is boring barrels; another milling the lock frames.... Then there are the rifling machines.... These various machines, about 400 in number... [are] puzzling in the extreme.... [The] "lock frame" passes through 33 distinct operations;... [the] cylinder,... thirty-six separate operations.... The barrel goes through forty-five... other parts:—lever 27; rammer 19; hammer 28... stock 5; in all 454 operations.... [In the] assembling department... every... part... [is] inspected with so much vigilance.... [The] tools to inspect a cylinder alone are fifteen in number.... [The arms] finished... [by] polishing, blueing, case hardening, electroplating, burnishing & c.,... [are then] ready for the prover... [where they are so] severely tested... [that it is] impossible that a defective arm can escape detection.... [At last, the arms head] to the storeroom, ready for sale.[115]

Among the key functions in arms manufacturing, Colt and Root made the greatest contribution in the art and science of duplicate die-forging using drop-hammers, a technique pioneered by John Hall in the national armory at Harpers Ferry during the 1820s, adapted by Root to make axes in Collinsville during the 1830s and 1840s, and brought to near perfection by Root, whose state-of-the-art, compound, crank-driven drop hammer was patented in the 1850s.[116] Drop-hammers shaped such gun parts as the lockframes, levers, pistol barrels, hammers, and triggers in steel dies called swages. The forged parts were milled, drilled (fig. 34) and finished by hand.

Of the gun-stocking machines, first invented by Thomas Blanchard and then brought to perfection by Cyrus Buckland, in Springfield, Colt operated sixteen variations (fig. 35). Of this process, Henry Barnard wrote poetically; "No function in gun-stocking is more magical than cutting the hole for letting in the lock.... A tool springs down fiercely, and seems to be tearing the stock in pieces.... A blast from two brass pipes blows away the

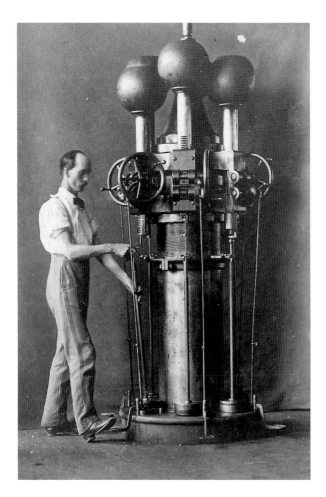

34. Colt's Drilling Machine,
1855–60. Photograph,
ca. 1870. A workman
is shown moving a barrel
into place beneath one
of the four drilling tools
mounted around a center
axis. (Courtesy of the
Connecticut State Library.)

fragments. At the end of forty-five seconds, we behold, hollowed out of the solid wood, a polished cavity of five different depths, and so cunningly fashioned."[117]

In 1852 Elisha Root patented improvements in the turret lathe—predecessor to the screw machines that brought eccentric milling to its highest standard of perfection.[118] One of the fascinating aspects of Root's system of machinery, noted by several contemporaries, is the way the concept of the revolver, or working parts in rotation around a center pole, was adapted to the machinery itself (fig. 36) "How is a revolver made?" asked Henry Barnard. "By revolving... rotation is its law, and the law of its production. From the huge fly-wheel of the engine... to the tiniest circlet of leather for polishing; in all apartments, even in the attics; beneath all ceilings, in all machines, you are reminded of Ezekiel's vision of wheel on wheel, and wheel in the middle of a wheel. Rotation pervades the product and the producing processes.... The revolving cylinder... rifle-ball turns round;... screws, are fixed in position by revolving."[119] Increasingly, as turret lathes, screw machines, barrel boring machines, and drop forges were "made new" in the form of "working parts about a center," a certain spiritual unity began to radiate out from the working heart of Coltsville.[120] As it became increasingly apparent that the machines that made the guns could, perhaps, make anything, the final rotation was in the minds and hearts of a community transformed by a revolution waged not with guns, but with machines.

Fig. 10.

35. *Gun Stocking Machine,*
1855–60.
Illustration from
the Report on
the Manufacture
of Fire-Arms...
Tenth Census, *1883.*
The gun stocking
machine had a
revolutionary effect
on the mechanization
of firearms manu-
facture by speeding up
the pace of production
and achieving almost
total uniformity
in the parts produced.
This machine
was commercially
manufactured by the
Ames Manufacturing
Co. in Chicopee, Mass.,
and represents a type
used at Colt's and
in armories around
the world beginning
in the 1850s.

36. *Jigging Machines, 1857.*
Engraved by Nathaniel
Orr; illustration from the
United States Magazine,
March 1857.

Colt's jigging machine
featured a revolving
wheel to which were
attached many different
cutting tools used
to shape the gun
lock frames.

GUNS, GUN CULTURE, AND THE PEDDLING OF DREAMS

God created men; Colonel Colt made them equal.
FRONTIER SAYING
A musket is an old established thing; it is a thing that has been the rule for ages; but this pistol is newly created.
SAM COLT, 1854
Economy and necessity made Texas vigilant to choose the most efficient arms, and she had chosen Colt's.
REPORT FROM THE SECRETARY OF THE NAVY, 1840
I am proud of the results of my exertions, and can paddle my own canoe.
SAM COLT, 1854

Gun-making is a heck of way to make a living. For as long as there have been people willing to sell arms, there have been critics ready to condemn them. More than a century ago, even in a city made rich through the sale of armaments, critics described the "gun lobby" as "always inventing trouble" and being anxious to "keep the public mind on war, not because they fear war, but because they have cannons to sell."[1] Epitomizing the qualities that have made social pariahs of arms merchants for generations, the character of armorer Andrew Undershaft in George Bernard Shaw's play *Major Barbara* explains his philosophy of good government:

You will make war when it suits us, and keep peace when it doesn't. You will find out that trade requires certain measures when we have decided on those measures. When I want anything to keep my dividends up, you will discover that my want is a national need. When other people want something to keep my dividends down, you will call out the police and military. And in return you shall have the support and applause of my newspapers, and the delight of imagining that you are a great statesman.[2]

Repeatedly accused of bribery, of fomenting war scares, and of conspiring with each other to raise prices, arms manufacturers accept as an occupational hazard that few will ever be grateful for their contributions to national defense. More often they will be vilified.[3] Why would anyone bother? Progress and morality, the twin peaks of Victorian culture. Sam Colt personified the evangelical strain of thought often apparent among gunmakers certain they are making the world safer and more secure by arming the weak and just against the strong and predatory. As Sam Colt put it, "The good people of this wirld *[sic]* are very far from being satisfied with each other & my arms are the best peacemakers."[4]

The middle third of the nineteenth century brough the most rapid period of technological innovation in the history of armaments and warfare. Much like the computer revolution of today, the pace of change was quick, and "innovation... appeared almost endless" as each season brought forth new, lighter, faster, less expensive, and more powerful weapons. Whether in the scientific press, reports from the War Department (a branch of government we now call Defense), or the Hartford newspapers, all concurred that "in no branch of scientific industry have there been greater strides in improvements.... Few branches of manufacture have received so much attention as... weapons of destruction." As witnesses described how, what "five years ago... had been scarcely heard of... already we have grown to consider... obsolete," it became increasingly apparent that "weapons of destruction" were at the center of a revolution not only in the character of technology but in the pace of technological change.[5] For a generation wrestling with the tension between scientific rationalism and religion, the idea that "progress is observable in *everything*" helped reconcile seemingly antithetical tendencies, propelling the notion of perfectibility ever forward. The concept of evolution, a watchword in science and biology even before Charles Darwin's revolutionary *On the Origin of Species* (1859), was applied to all sorts of things, no less to firearms, where the image of the "match-lock, the wheel-lock and the flint-lock," giving "way to the percussion lock," giving way "to the copper cartridge" validated the progressive impulse and in its steadily increasing pace of change hinted that the biblically charged idea of endtime known as the millennium was imminent.[6] "The end," such as it was, passed without notice as science and secular rationalism chipped away at the euphoria of religious expectation, demolishing hope of an impending moment when society, politics, and technology would complete the progressive march to the promised land of a perfect world. But during the middle third of the century, as the two great ships of religion and science passed in the night, *anything* seemed possible.

For those who imagine science operating in a vacuum, history furnishes few better examples of the enormous multiplier effect that occurs when science and technology are in open dialogue with the forces of religion and culture. As religion and science danced to the rhythm of "progress," the nation worked itself into a millennial frenzy, fueling tremendous economic growth, while mowing down incompatible peoples and ideologies in its path. It was not a very pretty picture, and it is almost impossible now to imagine our society, or any society less single-minded than theirs, tackling problems as big and divisive as, to name one of many, slavery. Americans used to be intense. Indeed, if they could see us, they'd likely make it their duty to conquer us. But they'd fail. We now have better weapons.

The United States in the age of Colt was preachy, judgmental, inflexibly moralizing, aggressive, and young.

The missionary zeal evident among the machine builders and city-makers of the age was fueled not only by a belief in the miracle of the nation's successful War for Independence against Britain but also by the evidence, interpreted as God's favor, of the unparalleled growth and expansion that followed. Guns, technology, and the campaign of western expansion were overlapping layers of the same progressive tendency. Each fed and enabled the other, amplifying the perception that Americans were a chosen people.

Issues of race, guns, and violence had special meaning in the 1840s and 1850s, one of the most violent and adventuresome eras in the nation's history. Although nostalgia for the Civil War era would be misplaced, in an odd and sadly heroic way, it is, or was, our Golden Age. It was our youthful adolescence, and there is hardly a feature of national character that did not come into focus at that time. Americans have never grappled with the meaning of nationhood more passionately. Violent, idealistic, and proud, never before or since has American society embraced guns with such a vengeance, nor sanctioned violence on so grand a scale. Die for your country? They signed up in unprecedented numbers. The age of Colt was the age of the gun, an age when a youthful nation first dreamed of glory and empire.

Manifest Destiny—the phrase defines an era. The author was John Lewis O'Sullivan. The year was 1845. The place was the *United States Magazine and Democratic Review*, the house organ of the empire builders and western expansionists. If Emmanuel

37. Westward the Course of Empire Takes Its Way, *1862. Emmanuel Leutze (American, 1816–68); painted for the rotunda of the U.S. capitol, this is the defining mural of the age of Manifest Destiny. (Photograph courtesy of the Architect of the Capitol.)*

Leutze's *Westward the Course of Empire Takes Its Way* (1862; fig. 37) was its image, and Horace Greeley's "go west, young man, and grow up with the country" its slogan, O'Sullivan's *United States Magazine and Democratic Review* was the intellectual vortex of the age of empire. As a journalist, diplomat, and expansionist, O'Sullivan epitomized the political sentiments that brought forth the Mexican War, made Sam Colt's achievements possible, and set the stage for the war of values that was our Civil War.[7]

What came first, the zeal for guns and glory or the campaign of Manifest Destiny and western expansion? Both were symptomatic of the age. The *Hartford Daily Times* opened the decade of the 1850s by announcing how our "young Republic has advanced steadily in the bright pathway of glory and renown" toward a "future" in which the "progress of the race" is "assured.... Despotism will die.... Liberty and order will become the common blessings of mankind.... Our own beloved country... will advance... to National glory.... Onward! then, brethren, in the campaign of 1850!"[8]

The cult of progress and the mantra of Manifest Destiny guided national life in what historians have dubbed the Young America Movement. Combining the qualities of rampant Anglo-Saxonism and expansionism, one of the central ambitions of the intensely nationalistic Young America Movement was the vision for a slave empire stretching west to the Pacific and south to the Caribbean, including Central America, Mexico, and California.[9] Why not? Progress, in the mind's eye of 1850, consisted of pressing onward to create "new empires of trade and wealth... in regions... shut against the world for ages by the iron door of heathen prejudice and custom."[10] Rarely has the rhetoric of national life been more florid and melodramatic than during this period nor, in retrospect, more embarrassing for those uneasy about the way the United States achieved its wealth and power.

To varying degrees, many got into it. Secretary of War Jefferson Davis believed that the acquisition of Cuba was "essential to our prosperity and security."[11] Pumped up with pride after a lopsided victory in a war of aggression against Mexico, even more sober voices like Daniel Webster's assumed a bombastic and moralizing tone, sure that the events of the nation's past and present proved the superiority of republican institutions. For the first time, the United States adopted an aggressive foreign policy, based on the perception of moral superiority and a willingness to judge the virtue of foreign governments.[12] Sounding the part of a nation unaccustomed to respect, the *Hartford Daily Times* boasted in 1851 of how "every encroachment and every insult of any foreign power would be speedily met and punished," a policy described as the "practical Monroe Doctrine."[13]

Gun lust and western expansion were symptoms of a young nation stretching to define its identity as it climbed onto the stage of world affairs. Issues of race, cultural and political sovereignty, and national destiny churned the waters made muddy by the burden of expectations of a generation eager to experience the martial glory of their Revolutionary ancestors. Science, by providing a veneer of respectability to the racist notion of genetic hierarchy, helped validate the disposition to read America's success as a triumph of the Anglo-Saxon race as well as its republican institutions.[14] By assuming its superiority, the partisans of Manifest Destiny lay the foundation for a campaign of expansion that, peppered with a sense of religious mission, transformed alien peoples—blacks, Native Americans, and Mexicans, primarily, but also other Europeans, notably the Irish—into "barbarians" and "savages" whose only options were annihilation or conquest.

The culture, particularly its newspapers, was brimming with testy proclamations of racial destiny. Indeed, the Civil War was the last phase of an intense and divisive national debate that perplexed Americans during the twenty years leading up to it. Perplexity and confusion are what make this unsettling episode in the nation's history so intriguing. The United States was pumped up with a sense of destiny; slavery was the stain on an otherwise "perfect picture" of a nation rightfully proud of its political institutions and anxious to "spread the principles of Americanism" throughout the West and beyond.[15] So long as slavery blighted the nation's reputation, a full-throttle assault on the West—with all that entailed—was impossible.

In a work deeply emblematic of the progressive impulse, Neil Arnott's *Survey of Human Progress* explained how "the human race, unlike the lower animals... has gradually but greatly advanced from... savage to various degrees of civilization," a "fact of progressive civilization... little suspected... until... advances in general science," disproved the "Greeks' and Romans'... opinion, that mankind had degenerated from... a golden age."[16] Or if you prefer C. Chauncey Burr's take on the matter, the "Caucasian race were never barbarian or savage.... All races who ever were savages, are so still.... Is it better to exterminate the Indian, or to civilize him?"[17] These were the straight-faced questions of the day. For those committed to, as the expansionists saw it, spreading good government, commercial prosperity, and Christianity throughout the American continent, there was abundant, widely published, public testimony that foretold "the period when the extinction of the Indian races must be consummated."[18]

Consider the image of Sioux Indians "taking an infant from its mother's arms... drawing a bolt from the wagon, driving it through the body and pinning it to the fence," leaving it "to writhe and die in agony," while she watched in "speechless misery," until the "gloating" Indians "[chopped] off her arms and legs leaving her to bleed to death"; or of the "womb of a pregnant mother ripped open, the palpitating infant torn forth, cut into bits and thrown into the face of the mother."[19] Before long, this combination of anecdote and ethnology had so totally dehumanized the opposition that the destructive march of progress could not only proceed but also claim a moral victory over the forces of barbarism and evil.

While the progressive impulse was as evident among northern industrialists as southern conquistadors, the way West was guided by the South's martial spirit. Curiously, the only Northerners to occupy the White House during the campaign of Manifest Destiny (1845–60) were Democrats with decidedly pro-Southern sympathies, who appointed Southerners to their cabinets. It was New Hampshire's Franklin Pierce who appointed the future president of the Confederacy, Jefferson Davis, as secretary of war. From John Tyler to James Buchanan, the South dominated electoral politics and was primarily responsible for restoring the government's role as an aggressive agent of expansion when, in 1846, President James Polk provoked the attack that led to the Mexican War.

The Mexican War, although as controversial in the nineteenth century as Vietnam was in the twentieth, restored the nation's martial spirit, tipped the scales away from the peace policy that followed the War of 1812, and provided new impetus for arming the nation. This was nothing new in the South, where violence was much more readily accepted as a part of life. With its duels, chivalry, and martial tournaments, its unbroken history of Indian warfare, and the constant fear of slave uprisings, the South had a long tradition of military preparedness and was well suited to lead the campaign of western expansion.[20]

As the nation armed itself in the years leading up to the Civil War, the reputation of improved gunnery as a deterrent to war took hold. The concept of weapons as "peacemakers," revitalized in the rhetoric of the Strategic Defense Initiative and arms buildup of the 1980s, played an important role in antebellum America in the debates about arming a nation with a deep aversion to standing armies and a limited capacity for self-defense.[21] As a concept, it offered hope for reconciling the conflicting ambitions of the expansionist South and the antimilitarist "peace movement" advocates in the North, still vigilant in recalling the costly and seemingly pointless War of 1812 and still convinced that a "chosen people" of virtuous citizens did not need expansive armaments.[22]

An 1841 description of Cochran's repeating gun, Colt's first significant competitor, as "an American peacemaker" is one of the earliest American references to the peacemaking attributes of improved gunnery.[23] The USS *Princeton*, the nation's first propeller-driven warship, was mounted with a twelve-inch iron cannon known as the Peacemaker, made infamous when it exploded in testing before a military commission in 1844, killing the secretaries of state and navy.[24] But it was Sam Colt and his revolvers that brought the peacemaker mythology to the fore, introducing a line of argument that remains controversial and no less effective a century and a half later.

As early as 1852, the *Hartford Daily Times* described Colt's invention as "not without its moral importance." Citing the argument about the invention of gunpowder diminishing "the frequency, duration, and destructiveness of wars," the *Times* concluded that "men of science can do no greater service to humanity than by adding to the efficiency of warlike implements, so that the people and nations may find stronger inducements than naked moral suasion to lead them towards peace."[25] Jabbing at the resistance of pacifists, whose language and institutions it nonetheless adopted, the peacemaker lobby soon grafted the rhetoric of progress and millennialism onto their cause, describing gunnery as a "humane improvement" and arguing that "when a man can invent a process by which a whole army could be killed... the Millennium will arrive, and the lion and the lamb will lie down together."[26] "If a machine were invented and could be readily used, by which a few men could instantly... destroy a thousand lives, wars among civilized nations would cease forever.... The inventor of such a machine would prove a greater benefactor of his race, than he who should endow a thousand hospitals."[27] And finally, the most famous line from the Brooklyn evangelist and antislavery crusader Henry Ward Beecher, whose description of the Hartford-made Sharps breech-loading rifle as containing "more moral power... so far as the slave-holders were concerned than in a hundred Bibles," earned them the nickname Beecher's Bibles in the bloody Kansas war of 1856.[28]

The 1850s was the decade when issues of public protection and defense triumphed over the nation's deep-seated resistance to anything like a standing army. The Revolution and the Constitution valorized the notion of a citizen-army. Then as now, amid the strident debate about gun-control and the Second Amendment protection of the "right of the people to keep and bear arms," the image of a citizen-army achieved the status of national icon (fig. 38). Steeped in the literature of political philoso-

phy and history, and fundamentally distrustful of central government, the founding fathers believed that the greatness of the Roman republic was based upon its citizen army and that its decline coincided with the creation of a standing army of professionals. From Machiavelli they learned that a government of armed citizens was the most effective defense against foreign enemies and the most likely to be united, virtuous, and patriotic.[29] Civilian militias were covered with glory at the Battle of Bunker Hill in the Revolution and in the Battle of New Orleans in the War of 1812, when Andrew Jackson's backwoods riflemen, armed with Kentucky rifles, annihilated a much larger army of British professional soldiers.[30]

Throughout the country, private, civic, and state militias proliferated during the 1850s. In Hartford, the most famous was the Putnam Phalanx (fig. 39), formed in 1858 and named after the hero of Bunker Hill, Connecticut general Israel Putnam.[31] More gentlemen's club than military unit, the Putnam Phalanx was part of a larger cam-

paign by Hartford's upstart Democrats to wrap themselves in the mantle of patriotism. Nevertheless, its formation reflected the growing martial fervor of the period. Sam Colt—a man who loved the glory of a good parade no less than the smell of gunpowder—was a member of the Phalanx. In addition, he successfully lobbied for an appointment as lieutenant colonel in the state militia, rehabilitating and then leading the First Company Governor's Horse Guard in 1853 and five years later sponsoring a private militia company, the Colt Armory Guard.

At the same time, cities across the country campaigned to transform the archaic constable-watch system into a modern uniformed police force that, increasingly and with great controversy, became armed with guns.[32] Initially the police were resistant to the use of firearms, it being contrary to a macho culture that relied on brawn and fists. However, beginning in New York (1845), then in Chicago (1851), New Orleans and Cincinnati (1852),

38. The Minuteman, 1875.
Daniel Chester French
(American, 1850–1931),
sculptor,
Ames Manufacturing Co.,
Chicopee, Mass.
The Minute Man is the
most poignant symbol of

the civilian militia
companies that play
an important role
in national defense.
(Courtesy of the Minute
Man National Historical
Park, Concord, Mass.)

39. Staff and Officers of the
Putnam Phalanx, 1861.
Lithograph; this detail
of the print features D.P.
Francis and I.W. Stuart.
(Courtesy of the Museum
of Connecticut History.)

and in numerous cities across the country, especially following the urban riots and an epidemic of crime that erupted during the economic panic of 1857, uniformed police were for the first time allowed, albeit not authorized, to carry concealed firearms.[33] One of the last products introduced by Colt's firearms before its founder's death in 1862 was dubbed the New Model Police Revolver (fig. 40), a lightweight and powerful .36-caliber, six-shooter with a short, three-and-a-half-inch barrel, easily concealed and relatively inexpensive (today about $660), designed and marketed to appeal to the first generation of police professionals forced to acquire arms at their own expense.[34] From the New York draft riots of 1863, when policemen used revolvers against armed mobs for the first time, to the increasingly horrifying labor riots of the later nineteenth century, the gentle age of the night watchman had clearly passed.[35] Gatling guns were perched on Wall Street during the Great Strike of 1877.[36]

By midcentury, firearms ownership was rapidly losing whatever moral stigma had previously been attached to it as the gun came to be seen not only as a means of ensuring security but as a status symbol and source of entertainment. In New York City, in 1844, it was noted that "the gunshops and hardware stores where firearms are to be found, have had a most extraordinary increase in business," and it was noted the following year that at a high society party in New York "four-fifths of the men were armed with pistols for protection against thieves" and, no doubt, for flaunting their ownership of an instrument typically costing the equivalent of two thousand dollars and more.[37] In 1851 Newport, Rhode Island, was already famous for its "shooting-pistol galleries" and, in what sounds like a scene out of the Wild West, its "pistols... popping at all hours" as an aspect of the "fashionable season."[38] Throughout the 1850s, the reports of Hartford's annual Fourth of July celebrations reveal gun-toting revelers and pistol fire as an annual and uncontrolled nuisance. In 1866 the Hartford press, citing the "hundreds of explosions of pistols... everywhere" reported, matter of factly, a "4th of July celebration mishap," in which "Harriet Beecher Stowe's daughter Eliza" was "accidentally shot."[39]

40. New Model Police
Revolver, 1861,
.36 caliber, 6-shot,
serial #4160
(WA #05.993).

Not surprisingly, shooting sports and recreational marksmanship became increasingly popular. Imported at first by immigrants pouring into the United States from Germany after the failed Revolution of 1848, the sport of marksmanship emerged as an entertainment in firearms manufacturing communities like Springfield, Hartford, and Ilion, New York, where Remington's workers participated in an American marksmen shooting match in nearby Fort Plain in 1853, one of the earliest American shooting contests outside the German American community.[40] The German Schutzenbunde, featuring uniformed marksmen, became a popular component of New York social life almost from the moment large concentrations of Germans arrived, while turkey shoots (fig. 41) emerged as a social custom in smaller towns around New York and New England.[41] Untrained at marksmanship, the American soldier often proved inept at handling the more accurate rifled muskets and breechloaders that came into use during the Civil War. In 1871 the National Rifle Association was founded, not to argue about Second Amendment rights but "to turn the [National] Guard into

sharpshooters," which they accomplished by sponsoring target shooting competitions (fig. 42).[42] Later in the century, celebrated hunters, marksmen, and western heroes such as Frank Butler, Annie Oakley, and Buffalo Bill became national celebrities whose touring displays of fancy gunwork provided crowd-gathering entertainment.[43]

As the demand for guns intensified and the moral argument against them lost force, Sam Colt was well positioned to make the most of the opportunities at hand. Indeed, Colt's greatest invention was not repeating firearms—he had plenty of competition—but the system he built to manufacture them and the apparatus of sales, image management, and marketing that made his gun—and not the equally viable products of his competitors—the most popular, prolific, and storied handgun in American history. Colt was the Lee Iaccoca of his generation, a man whose name and personality became so widely associated with the product that ownership provided access to the celebrity, glamour, and dreams of its namesake. What Colt *invented* was a system of myths, symbols, stagecraft, and distribution that has been mimicked by generations of industrial mass-marketeers and has rarely been improved upon. If we are today drowning in the muck of overhyped, oversold merchandise, thank Sam Colt for lighting the way. Even now, the French word for a handgun is *le Colt*, in the same way that one might describe a photocopy as a "xerox" or a cola as a "coke." This is the stuff Madison Avenue dreams are made of. When one of Colt's congressional allies defended his petition for a patent extension by describing Colt's central ambition as "not so much to amass wealth as to build up for himself fame," he unveiled the central impulse behind Colt's relentless drive.[44] "Colt's idea," it was later said, was "making the world aware that he was in it."[45]

To advance his reputation and the reputation of his invention, Colt drew from an extraordinary grab bag of tactics. First, he drew a bold line between "mere capitalists" and inventors, and carefully crafted his reputation as the latter, even though his legacy at amassing, investing, and deploying capital was more impressive. Colt helped pioneer the theory and practice of American patent law by surrounding himself with attorneys and threatening to sue and outflank his many competitors. Colt used direct advertising as skillfully as any manufacturer of his generation, expanding both the character of the message and the means of delivering it. Colt sought, bought, and gained the recognition of "expert" authorities in the military and scientific communities, and never blinked (or asked permission) before exploiting a testimonial once provided. Colt was a sensationalist who cultivated controversy, virtually coining the practice that led to the cliché "there is no such thing as bad publicity."

Colt adopted the policy, today called creative destruction, of cannibalizing his own product line to stay ahead of the competition by always offering something "new." Aware of the complexity of his product, he also pioneered the use of users' manuals. And, with astonishing similarity to the personal computer market of today, he was relentless in his drive to lower prices and thus scare off or bankrupt potential competitors. He used commission sales agents, lobbyists, consignments, and quantity discounts skillfully and, while tame by comparison with what passes for influence peddling today, was not averse to bribery and blackmail to accomplish his goals. Finally, and to the enduring affection of firearms collectors then and now, Colt used art to bolster the prestige and soften the associations of a product that was, after all, designed to kill people. Colt translated personal glamour and celebrity into sales. When one of his admirers wrote that "Col. Colt is himself a 'Patent Revolver'—is always loaded, and ready for action!" he described the image in the mirror for generations of (primarily) men eager to be like Sam Colt, a man "ready for action."[46] The man and the gun converged toward a single identity. Colonel Colt, as he liked to be called, had become his generation's Marlboro Man.

Colt lived and worked at the beginning of the American industrial era, and one of his great triumphs was inventing a persona that reconciled the tension between the conflicting ideals of the agrarian tradition of the freeman farmer—the symbol of American liberty—with the promise of technology and mass production. How, in a system that required stifling conformity and regimentation, could one preserve the traditions of self-

42. The Schutzenfest at Jones's Wood, 1868. From Harper's Weekly: A Journal of Civilization, *July 18, 1868; detail of a woodblock engraving. (Courtesy of Watkinson Library, Trinity College.)*

reliance and individualism? By cultivating a persona as a "genius" and "inventor" and a "restless boy… [who] preferred working in the factory to going to school," Colt's gift of insight and vision could be understood as a reward for suffering and perseverance.[47] With this persona, Colt helped invent a model of the heroic industrialist, whose success became the embodiment of, rather than a repudiation of, the American dream. For Colt, who adopted the personal motto Vincit Qui Patitur [he who suffers is victorious], the gift of genius was a reward for suffering. For without genius there was no gift. And without the gift—a veritable sign of God's favor—Colt would have been like the corporate predators he avidly distanced himself from. In the closing argument of an 1851 lawsuit for patent infringement against the Massachusetts Arms Co., Colt's attorneys decried the spectacle of "a corporation… exerting itself… to break down one of the most meritorious inventors our country has ever seen" and concluded that this "contest is between a meritorious and important inventor… and… a corporation. Their very name imports a being that can have no merit as an inventor. They can only be the purchasers of the ingenuity of others." In the jury's verdict, "that heroic boy" Sam Colt triumphed over the corporation.[48]

Colt's persona was cut straight from the fabric of the ardently populist social and political discourse of his age. From Ralph Waldo Emerson and Walt Whitman to Abraham Lincoln, the American dream was sustained by an abiding faith in the power of individualism and self-reliance, which was soon translated into a cult of self-expression and self-determination. Prior to the emergence of the industrial, corporate, and urban landscape of the 1850s, Americans had less cause to worry about issues of civic and personal identity. Tied more to family, place, and tradition, the uncertain future was mostly a continuation of a more certain past. But with the rapidly escalating pace of change, the need for self-definition intensified. Walt Whitman's *Song of Myself*, published the same year Colt's Armory began operations on Hartford's South Meadows, elevated personal experience and self-determi-

nation almost to the level of a national secular religion by celebrating the intensely existential quality of the *now* and the *new*. "I celebrate myself," Whitman began.

You shall no longer take things at second or third hand….
You shall listen to all sides, and filter them from your self.
…I am satisfied—I see, dance, laugh, sing:
…What is commonest, cheapest, nearest, easiest, is Me;
Me going in for my chances, spending for vast returns,
…I dote on myself—there is that lot of me, and all so luscious;
Each moment, and whatever happens, thrills me with joy.
…O I am wonderful!
What is known I strip away; I launch all men and women
*   forward with me into* the Unknown.[49]

From the beginning, Sam Colt recognized the euphonic appeal and symbolic potential of his name. "Sam Colt" has all the snap and crackle of the pop icon he became. It was an age of Sams, Andys, and Abes—Sam Houston, Uncle Sam, Andy Jackson, "Honest Abe" Lincoln—not Samuels, Andrews, and Abrahams. The symbolism of the colt was also not lost on Sam who adopted a "rampant" (a synonym for erect, excessive, and dominant) colt as a personal and corporate trademark (fig. 43). It superseded a more benign version, featuring paired colt heads used as ornamentation on the earliest Colt firearms.

By 1856, with a worldwide reputation, having been hosted and toasted by kings, and roasted in verse by the British press as "Colonel Colt, a thunderbolt," and in the midst of building Armsmear, a visionary and idiosyncratic dream house, Sam Colt ruled an industrial empire, conceived as a monument to the rehabilitation of his "good name." Pomp and ostentation, far from undermining his common-man persona, merely confirmed that in America, uncommon courage and virtue, hard work, and free enterprise were bountifully rewarded. On the meadows below was Coltsville, with Colt's Armory crowned by an onion dome, surmounted by a bronze statue of the rampant colt dancing on the globe. With Colt's Guard practicing its drills to the marching beat of Colt's Armory Band, the commercial schooner *Sam Colt* plying the

Connecticut River and Long Island Sound, Sam Colt had successfully transformed his name into an icon.[50]

Anxious to validate his reputation as an inventor, Sam Colt joined prestigious organizations, presented papers before engineering colleagues, competed for awards at city, state, regional, national, and international expositions, and sought appointments that conferred prestige and recognition. The purpose of this activity was, in Colt's own words, to "help make up the reputation of my arms."[51] Evidence of technical research and study is less apparent than diligence at reputation building. Colt's one "scientific paper"—"On the Application of Machinery to the Manufacture of Rotating Chambered-Breech Fire-Arms"—as well received as it appears to have been, is almost comical in its avoidance of the stated subject, dwelling far more on the attributes of the invention than the innovative methods used in manufacturing. Britain's prestigious Institution of Civil Engineers nonetheless elected Colt a member and soon installed a bust of Sam Colt in its Hall of Honor.[52]

Colt participated in dozens of industrial fairs, and while he coveted international awards, he rarely failed to participate in the annual exhibitions of the Hartford County and Connecticut agricultural societies. Colt earned his first gold medal from the Hartford County Agriculture Society "for design of pattern, beauty of work, finish and perfection" in 1849.[53] He was awarded gold medals by the Connecticut State Agricultural Society in 1852 and 1855.[54] In 1853 the Hartford Society praised Colt's "revolving pistols... mounted with ivory, cacao wood, and other handles, ornamented with silver" and awarded him more gold medals in 1856 and 1857.[55] Colt continued to participate in Connecticut industrial and agricultural fairs as late as 1858.[56] Colt was also involved in the Hartford Arts Union, an organization "composed of mechanics, manufacturers, artisans and all others interested in the advancement of the Arts," organized in 1848 and dedicated to the "acquisition, and diffusion among its members of... scientific and useful information" and to acquiring a "suitable room... for the reception and exhibi-

tion of inventions, models, drawings, designs, and all such articles and specimens of ingenuity, skill, or taste."[57]

The first organization to anoint Colt's inventions was the American Institute in New York. Inspired by the French, the American Institute began mounting industrial exhibitions in 1828, and by 1836, the year Colt first participated, had become the most prestigious organization of its type in the United States.[58] Colt was awarded American Institute medals for inventions including guns, waterproof cartridges, and his submarine battery, in 1836, 1837, 1841, 1844, 1850, and 1855. As early as 1837, Colt volunteered for and received an appointment on the committee that organized its annual exhibition.[59]

Colt accumulated numerous awards and appointments during the 1850s. He was awarded a silver medal at the annual fair of the Massachusetts Charitable Mechanics Association in 1853. He was elected a member of the New-York Historical Society in 1838, the National Institute for the Promotion of Science in 1840, and the Connecticut Historical Society in 1854.[60] In 1851 Colt was appointed to a committee of arrangements that oversaw Connecticut's contribution to the Great Exhibition in London, where he won a bronze medal, and in 1855 he was appointed to a state commission for the Universal Exposition of Industry and Art, Paris, where he won three medals.[61] To rally support for Connecticut's participation in the French World's Fair, Colt urged "manufacturers, inventors, and other citizens of Connecticut who are desirous of contributing to the Great Exhibition... [that will] exceed in splendor the Great Exposition of 1851 in London" to step forward and boost the "reputation of Connecticut as an industrial and manufacturing commonwealth."[62] Colt was awarded medals from industrial fairs in London in 1854 and from London's Universal Society for the Encouragement of Arts and Industry in 1856. After his death, he received a final award from London in

43. Rampant Colt, ca. 1860, gilt zinc. As early as 1831 Sam Colt used horse figures as a company trademark, first with an icon of four joined horse heads, second with an outstretched colt on two legs, and finally with the rampant colt that remained a personal and corporate trademark for generations. The most famous of the rampant colts was the zinc statue mounted as the finial on the onion dome on Colt's Armory. This example graced the dome from 1868 until 1988. The first dome-mounted rampant colt was destroyed in the fire of 1864. This may be the copy Colt had custom-made by armory workmen in 1860 as an ornament for the fountain in the garden pond at Armsmear. It was removed from the fountain by 1866 and there is no record of a casting being made or purchased for the new armory building, which was completed and occupied in 1868. The rampant colt is a sculptural tour de force made from a combination of cast and hammer-formed components. In the years that followed, the rampant colt found dozens of uses both corporate and private, emerging as one of the best-known trademarks of its generation. (Courtesy of the Museum of Connecticut History; photography by Robert Benson.)

1862. Colt's medals (fig. 44 and 45) were displayed in a trophy case at Armsmear, as a living memorial to his quest for recognition.[63]

Colt's fame and the steady work available providing contract services to Hartford's arms industry attracted the talent needed to sustain a reputation enlarged, to some extent, by smoke and mirrors. Ultimately the message and the messenger became so indistinguishable that it no longer mattered who made what first or when, or whether trumped-up claims were immediately verifiable. Colt was a visionary who achieved excellence and consistency enough to overshadow all competitors and set the standard for an industry. Having used personal celebrity and style to propel his dreams to the point of viability, Colt proved endlessly resourceful in protecting his hard-won advantage.

From its inception, as defined in the Constitution, the U.S. Patent Office was conceived to "promote the Progress of Science and the useful Arts, by securing for limited Times to Authors and Inventors the exclusive Right to their respective Writings and Discoveries," or as Colt's advocates stated plainly while defending his patent rights, "to benefit the community by giving encouragement to men of genius."[64] American patent law essentially defined invention as a "flash of genius."[65] But the law had not always functioned effectively, and it was not until Colt's era that culture and legislation converged in a crescendo of public support and acclaim for inventors, another expression of the new glorification of individuality. Indeed, at the height of the patent medicine craze of the 1830s, when "scarcely a wholesale or retail druggist... [was] not consciously selling spurious drugs which were a menace to human life," patents and inventors rather reeked of scam and fraud.[66]

When Colt's victory over the Massachusetts Arms Co. was announced, the *Hartford Daily Times* pronounced it

was "about time... the law protected inventors."[67] Indeed, inventors and artists were increasingly lionized as romantic heroes of the industrial age. In 1849 the National Gallery of the Patent Office in Washington was visited by 80,000 people, and in 1850, 2,193 patents were filed and 995 issued, up from a couple hundred just a decade earlier.[68] By 1854 the number of patents issued had doubled to almost 2,000. Concerned that the tide of public opinion was against them, the defendants in the lawsuit Sam Colt brought against the Massachusetts Arms Co. urged that the jury not fall for the "devise of ingenious counsel" in representing his client, Sam Colt, as belonging to an "aggrieved and persecuted class of men." The defendant's attorney noted that "when there was less of science applied to the subject, when invention was undertaken rather by adventurers and fallen into by accident, there was a feeling less favorable to inventions than at the present day... a feeling of hostility... [and] a disposition to disbelieve improvements and to follow the old path," where today, he cautioned, "the public mind has now swung to the other extremity; not a new thought comes in, however absurd, which does not find followers."[69]

Sam Colt used patent law to protect the value of his invention, and few people have ever argued more persuasively or convincingly on behalf of the inventor struggling against the odds. But ultimately, Colt's main weapon against "infringers" was the fame of his product and the pricing advantage he created through volume production. He never fully blocked imitators and, in fact, went to considerable pains to license production in Europe and set up manufacturing facilities in London as an obstacle to European copyists he could not stop by legal means.

The patent secured Colt's reputation as an inventor and, if not quite a seal of approval, was at least a seal of originality at a time when public opinion strongly favored

44. *Colt's Awards and Medals, 1841–62. During Colt's twenty-five-year career as an inventor-industrialist, he received almost two dozen local, national, and international awards and honors.*

Top to bottom: medal, *Industrial Exhibition, Crystal Palace, New York, 1853* (WA #05.1096); silver medal, *Exhibition of the Industry of All Nations, Crystal Palace, New York 1853* (WA #05.1097);

bronze medal, *Universal Exposition, Paris 1855* (WA #05.1101); bronze medal, *Universal Exposition, Paris 1855* (WA #05.1093);

bronze medal, *World's Fair, London 1862* (WA #05.1090).

originality and creative genius. Nonetheless, Colt spent ten years in a frenzy about being robbed of his invention. When, within nine months of beginning arms manufacturing in Hartford, Colt "discovered... that two of my principal workmen are engaged with several other persons in getting up a repeating pistol... [with] hope of avoiding my patents," he flew into a rage, accusing the Ordnance Department in Washington of a "conspiracy" against him.[70] In 1851 the lawsuit he brought against these "workmen" and "other persons," then incorporated as the Massachusetts Arms Co., was widely reported and involved a battery of prominent attorneys. Figuring that Colt would make bushels of money with or without their verdict, the jury awarded him one dollar in damages and the right to serve notice on potential competitors. His attorney Edward Dickerson immediately issued a broadside warning firearms dealers to "desist... from the sale of any Repeating Firearms in which rotation or locking & releasing are produced by combining the breech with the lock... except... as are made by Col. Colt at Hartford.... All rotary arms... are in plain violation of Col. Colt's patent; and I shall proceed against you."[71] It was show business as usual.

Colt's experience giving laughing gas demonstrations on the lyceum circuit, his close study of the patent medicine industry's use of expert testimonials and sustained direct advertising, and the triumph of packaging over content in the presidential campaign that sent Tippecanoe [Gen. William H. Harrison] and Tyler Too to the White House in 1840, made Colt an early convert to the new art of advertising and mass marketing. A decade after Colt's earliest forays into mass marketing, the *Hartford Times* editorialized on the virtues of advertising: "Extensive advertising is morally certain to work a revolution in trade, by driving thousands of the easy-going out of it and concentrating business in the hands of the

few who know how to obtain and keep it," it preached. "He who neglects advertising... bestows the spoils on his... rivals."[72]

Sam Colt was always on jovial terms with reporters, cultivating their favor, contributing to their pet political causes, and ever ready to present a gift revolver for a good story—a not inconsequential gesture, considering that even the cheapest ones were worth today's equivalent of about a thousand dollars. After Colt succeeded in securing volume sales to the British military, a journalist wrote of how he had been "introduced by a brother member of the press to Colonel Sam Colt... [who was] sagacious enough to know that one line in the *[London] Times* would do him more good than a column in any other paper.... [He was] ready to pay any sum.... [I] told Colt that I would make an effort... [if he would] show me his new armory.... I at once sent this letter to the *Times*, asking if it was fair to the British sailors that they should still be limited to the old horse pistol and cutlass.... [The] letter was published... [and] Colt flourished."[73]

Colt recognized that a good secondhand story or a testimonial was better advertising than the direct approach. How much was paid to place such stories and exactly when he adopted the tactic of buying "good copy" is uncertain, but the stories that appeared in the *New York Times* in 1839 about "A gentleman in New Jersey" who had "lately killed thirteen English snipe, in one day, with one of Colt's repeating fowling-pieces" and of the "17 black ducks [that] were shot on Long Island with a single round from one of Colt's repeating duck-guns" were almost certainly bought and paid for.[74]

In 1839 another newspaper reported that "with one of Colt's pistols... Catlin, the artist... recently shot three deer in succession while traveling through... Pennsylvania," a story that marked the beginning of a relation-

45. Top to bottom: bronze medal, *Universal Society for the Encouragement of Arts and Industry, London, 1856* (WA #05.1091); bronze medal, *Great Exhibition of All Nations, London 1851; features portrait of Prince Albert, president of the Royal Commission* (WA #05.1098); souvenir of the Great Exhibition of All Nations, *Crystal Palace, London 1851* (WA #05.1102); Telford Medal, *silver, Institute of Civil Engineers, London 1851; Colt was proudest of this medal* (WA #05.1515); gold medal, *for revolving rifles and pistols, Connecticut Agricultural Society, Hartford 1855* (WA #05.1516); gold medal, *for repeating carbine, American Institute, New York 1841* (WA #05.1511).

ship between the Colt and the celebrity artist, sportsman, and adventurer George Catlin.[75] Almost twenty years later, Colt commissioned Catlin to produce a series of twelve paintings and six mass-market lithographic prints illustrating his adventures on the frontier using Colt's repeating rifles. The prints, published later by Day and Son of London with such captions as "Catlin the celebrated Indian traveller and artist firing his Colt's repeating rifle before a tribe of Carib Indians in South America" (fig. 46), "Catlin the artist & sportsman relieving one of his companions from an unpleasant predicament during his travels in Brazil," and "Catlin the artist & hunter shooting buffalos with Colt's revolving rifle," represent perhaps the earliest pictorial use of celebrity endorsements in product marketing. Catlin, whose trip to Brazil and South America in 1855 was reportedly financed by Sam Colt, emerged as the personification of the adventurous, westward-bound, American male, hunting, riding, shooting, and glorying in the freedom of the open plains.[76]

Once Colt's revolvers gained acceptance, Colt was happy to discover unsolicited news stories that he could excerpt, reprint, and repackage. In 1854, at the height of the Crimean War, he wrote of "searching the papers to see new incidents at the Crimea of the gallant use of my repeaters."[77] In 1860 he wrote his secretary J. Deane Alden, then stationed in Arizona on company business, about searching the newspapers for "notices of Sharp[s] & Burnside Rifles & Carbines, anecdotes of their use upon Grisley Bares [sic], Indians, Mexicans, & c.... When there is or can be maid [sic] a good storrey [sic] of the use of a Colt's Revolving Rifle, Carbine, Shotgun or Pistol... the opportunity should not be lost.... [With any published] notices... send me 100 copies... give the editor a Pistol... [and] Do not forget to have his Colums [sic] report all the axidents [sic] that occur to the Sharps & other humbug arms."[78] That, in a nutshell, was Colt's media policy. Even if they had to make it up, Colt demanded good copy, and he wasted no opportunity to make the competition look bad.

The manufacturers of Colt's firearms have chased military contracts, typically the largest but least stable share of the gun market, with unbending determination for 160 years. Colt doted on military officers, especially those who, like himself, were adventuresome, spirited, and hungry for glory. Before the company even had arms to sell, Colt was in Washington wining and dining military and Ordnance Department officials. Lavish and conspicuous spending, personal charm, and an unfaltering

confidence in the rightness of his every utterance made Sam Colt the kind of person who—love him or hate him—one never forgot. Despite his wife's later characterizations of him as practically a teetotaler and a person "who never used tobacco in any form," it is hard to imagine how a man who bought cigars by the thousand, was once accused of appearing drunk before a committee of Congress, and whose liquor bills and wine cellar inventories were enormous, was not also something of a lush.[79]

Sam Colt was a glutton for publicity, especially, but not exclusively, for the kind that advanced the reputation of his firearms. Believing, as he once said, that "government patronage... is an advertisement, if nothing else," Colt was dogged in his pursuit of government contracts, a fact that quickly strained his relations with the desk soldiers in Washington he grew to despise.[80] Tests and testimonials were the instruments Colt most relied on to secure government contracts. In the first go-around, Colt's revolvers had fared poorly in government tests conducted at the U.S. Military Academy at West Point during the summer of 1837. Unwilling to accept rejection, Colt exhibited the pattern of impulsive misjudgment and ethical sleight of hand that would follow him throughout his career. He planted a story in the West Point newspaper stating how Colt's gun, which fires "18 charges in the incredible short space of 58 seconds,... must be invaluable... for an Indian campaign."[81] The same day, writing as "your old school fellow," Colt petitioned one of the cadets for a favorable endorsement.[82]

Military tests were never as valuable to Colt as combat testimonials from officers in the field of battle. Colt's determination to secure battlefield testimonials led him to the front lines of the Seminole War in Florida, where he personally delivered a consignment of revolving rifles during the winter of 1838.[83] Florida produced Colt's first military contracts and endorsements, though he later complained that the success of his repeaters brought the war to a quick end, thus destroying his market. Having failed to gain approval from the Ordnance Department, Colt issued a report, in the style of an official government document, complete with statistics on the arm's accuracy, power, rapidity of fire, safety, and water resistance. The report featured an endorsement by Lt. Col. William S. Harney, the officer whose Florida dragoons were the first to use Colt's firearms in battle. Addressing his superior officer, Gen. Thomas Jessup, Harney testified that Colt's revolving rifles had earned the "greatest honor to the inventor" and that he preferred Colt's over "any other Rifle."[84] Quick to capitalize on even a vague and indirect endorsement (it is not clear if Colt had permission to cite the letter), Colt circulated an announcement crediting "the Gallant Colonel Harney" who "with less than one hundred men armed with Colt's Patent Repeating Rifles... terminated the Florida war."[85] Years later Harney declared that "I honestly believe that but for these arms, the Indians would now be luxuriating in the everglades of Florida."[86]

It was through Indian warfare, first in Florida and eventually throughout the contested zone of contact between whites and Indians, that the Colt revolver found its market and identity as a tactical necessity for the new style of warfare that emerged in the West. It is not clear that even Sam Colt at first understood the tactical significance of his invention. Throughout the era of the Patent Arms Manufacturing Company in Paterson, New Jersey, Colt's manufacturing and marketing emphasized long arms, not pistols. In Florida, Colt observed firsthand how repeating firearms could revolutionize warfare. This was especially true in what was euphemistically called the "irregular" guerrilla warfare practiced with such terrifying competence by Native Americans and other "savage" peoples, no more respectful of white codes of honor than the whites were of their sovereignty and independence.

As early as 1840, reports from the Florida war zone confirmed that "no man in Florida is safe in his own house" and described the rifle, and especially "Colt's repeating rifle," as the best and "only obstacle to the butchery of his family."[87] By the mid-1850s Colt had successfully appropriated the rhetoric of his frontier testimonials as he described the conditions that made his

46. George Catlin and the Carib Indians. *Lithographic print, by Day & Son., London, 1857, J. McCahey, lithographer, George Catlin, artist.* *This is one of a set of six prints based on the nine paintings that Sam Colt commissioned from the artist-explorer George Catlin. This print, commercially published in London, was described as "Catlin the celebrated Indian traveller and artist, firing his Colt's repeating rifle before a tribe of Carib Indians in South America." Note the multiple puffs of smoke and the amazed expression of the Indians, seeing such rapid fire for the first time. (Courtesy of the Museum of Connecticut History.)*

repeating firearms a necessity. In addressing London's Institution of Civil Engineers in 1851, Colt described the United States as "a country of most extensive frontier, still inhabited by hordes of aborigines." Colt cited "the insulated position of the enterprising pioneer and his dependence, sometimes alone, on his personal ability" for defense, concluding with an account of "the model of attack by the mounted Indians," who would "overwhelm small bodies of American soldiers by rushing down on them in... superior numbers, after having drawn their fire."[88] This concept of warfare had tremendous resonance for a British public then in the midst of war with the "savage tribes of the Cape of Good Hope."[89] As described by the London Institution of Civil Engineers, "The tactics of the Kaffirs were to tease an outpost sentry, at a distance, until they had drawn his fire, and then to spring on him before he had time to reload.... Nothing could be more perfectly adapted to meet these tactics, than the revolvers."[90] As "savage" and "civilized" societies clashed in the age of colonial expansion and empire, Colt's revolver gained an almost sacred identity as a factor in the triumph of Christian civilization over the forces of darkness, whatever and wherever they might be. Henry Barnard described the new calculus of warfare in terms of mathematical risk noting that as "every man armed with one of [Colt's] weapons was armed sixfold... [the] savages... would have seen that... the first man of them who rushed forward with the tomahawk would rush on certain death.... A force so concentrated... cannot but 'demoralize' antagonists."[91]

After a second test in 1840 failed to result in the general adoption of Colt's repeaters by the armed services, Colt petulantly petitioned the secretary of the navy, James Paulding, for reconsideration, citing the "wrong conclusion" that had been reached and the "great difference of opinion" held by "those who have used them most." Colt provided additional testimony, this time from the sergeant in charge of drilling Harney's second dragoons in the use of his guns, who claimed that "in passing through Indian country, I always felt myself safer with one of your rifles... than if I was attended by a body of ten... men armed with the common musket."[92]

Conceived in the North, baptized in the South, Colt's arms reached maturity in the unique climate of the American West. In the years that followed, Colt managed to obtain testimonials from dozens of prominent military officers, enough to cover almost every context and contingency. For the pioneer in Indian Territory and on the Texas frontier there was C. Downing writing in 1840 that "any man of nerve, of coolness, of determined resolution, on the inside of his house... would find Colt's rifle a safer reliance in an Indian midnight attack than ten men... with an ordinary gun."[93] On the eve of the Civil War, Colt printed an eight-page promotional brochure stating that during the past twenty years his arms had been "tested as no other arms have ever been" and citing the "following distinguished officers" as certifying its "superiority." The list of almost one hundred names is a veritable who's who of Mexican War heroes, political luminaries, Ordnance Department officials, and past and future U.S. presidents and executive cabinet officers, including 1848 presidential candidate, secretary of state, and U.S. senator Lewis Cass; secretary of war, U.S. senator, and future president of the Confederacy Jefferson Davis; major general and U.S. president Zachary Taylor; Mexican War generals Gideon J. Pillow and David E. Twiggs; brigadier general and U.S. president Franklin Pierce; U.S. senator from California, western adventurer, and 1856 presidential candidate Col. John C. Frémont; Ordnance Department chief inspector Maj. William A. Thornton; and Mexican War and Japan Expedition hero Commodore Matthew C. Perry.[94]

Of all the endorsements Colt received, none was more significant than that of the legendary Texas Rangers. Founded in 1836 by Connecticut Yankee Steven Austin, the Republic of Texas was a melting pot that played a significant role in the shaping of the American character. Here, in this vast, seemingly unlimited space, plantation lord and Yankee, Native American and Mexican converged to create a civilization different from

47. *Texas Rangers Skirmishing, 1859. From Samuel C. Reid Jr.'s* Scouting Expeditions of McCulloch's Texas Rangers. *This is the earliest published scene illustrating the use of Colt's revolvers in action during the Mexican War, specifically the action around Independence Hill in Monterey, September 1846. (Courtesy of Watkinson Library, Trinity College.)*

that found in either the North or the South, one shaped more by the conditions of environment than by race or inheritance.

The new man forged in the crucible of the West, the liberated sovereign individual of Walt Whitman, was epitomized by the Texas Ranger (fig. 47). Bursting onto the stage of American popular culture, guns a-blazing, the rangers emerged as folk heroes during the Mexican War when their exploits and skill were widely reported in the press. Was this truly a new kind of man? In 1846 the *Hartford Daily Times* ran an article with the title "The Texas Rangers—Who Are They?" The newspaper answered with the story of Col. John Coffee Hays of San Antonio, who held a "high place... in the hearts of the people of the United States" since moving west to Texas in 1839 after bold and heroic service in the Indian wars in Florida. Having already earned a "reputation for coolness, judgement, courage, energy, and a knowledge of frontier life," in 1840 Hays "induced the government of Texas to tender to him the command of its first company of rangers." Under the leadership of this man who "thinks much and speaks little, and that little always to the purpose," the rangers quickly gained renown as a new breed of mounted soldier "willing... to live on parched corn, ride 70 or 80 miles without dismounting" and as the "best light troops in the world."[95] Served up at the dawn of the campaign of Manifest Destiny, romanticized for his skill at horsemanship, marksmanship, and

freelance tactical warfare, the Texas Ranger was the poster child of the new West.[96]

It was one of Colonel Hays's men, Capt. Samuel H. Walker, who first linked the celebrity of the Rangers to the name of Colt. In November 1846 he wrote a testimonial letter better than any Sam Colt ever paid for. Recounting the widely publicized episode known as Hays's Big Fight, a clash between fifteen Texas Rangers and some eighty Comanches in June 1844, Walker registered his high opinion of Colt's "revolving patent arms" (in this case the New Jersey-made five-shot model) and reported that "people throughout Texas are anxious to procure your pistols."[97] According to Walker, the superior firepower of Colt's repeaters intimidated the Indians into negotiating a treaty and canceling attacks on new settlements along the Texas frontier. Walker concluded by asserting that "with improvements I think [your revolver] can be rendered the most perfect weapon in the world for light mounted troops," an emerging branch of military service championed by Walker as the "only efficient troops that can be placed upon our extensive Frontier to keep the various war-like tribes of Indians & marauding Mexicans in subjection."[98]

Colt reaped a publicity bonanza in Samuel Walker, a natural ally who shared Colt's fascination with new technology and new modes of warfare. Together they entered into a collaboration that resulted in a much-improved generation of Colt revolvers and a government contract

large enough to sustain Colt's dream of building an arms manufacturing empire. Walker was already a national celebrity when he visited Hartford in 1847 to consult with Colt on design modifications for his handgun.[99] In addition to recounting Walker's service in the Seminole Indian wars and his key role in Hays's Big Fight, the local press described a confrontation in which he recovered after being "pierced through the body by the spear of an Indian, [which] pinn[ed] him to the ground!" In another incident, he had allegedly been captured by Mexicans and "tortured" before escaping to the mountains, where he "suffered greatly from hunger" before being "re-captured by the Mexicans" and subjected to Santa Anna's suicide lottery. Later, Walker had been awarded a captain's commission in the Mexican War Regiment, the first U.S. troops to test the mode of warfare that eventually earned Colt's revolvers notoriety as the Gun that Won the West.[100]

Tales of breathtaking adventure served as the inspiration for the decorative scenes Colt had engraved on the cylinders of his revolvers (fig. 48).[101] The scenes included a rendering of Hays's Big Fight, a pioneer and Indian fight, a deer-hunting scene, and an 1843 encounter between the Texas and Mexican navies, described by commanding officer Commodore E.W. Moore, who wrote Colt that the confidence "your arms gave the officers and men under my Command... opposed to a vastly superior force is almost incredible."[102] These distinctive engravings reinforced the link in the popular imagination

between Colt's revolvers and the fabled Texas Rangers, as did the occasional visits to Hartford by these bona fide western heroes that Sam Colt hosted throughout the 1850s. It was an association destined to make Colt rich, particularly when the gunmaker's Texas allies began to exert their will in Washington.

By the early 1850s, the former president of the Republic of Texas and, following annexation, its first U.S. senator, Gen. Sam Houston, had become one of Washington's most influential advocates of modern weaponry. Houston and Texas colleague Sen. Thomas Rusk lobbied Secretary of War William Marcy and President James K. Polk "unasked for" on behalf of Colt's revolvers. Rusk reported to Colt that they had urged Marcy and Polk "to arm every soldier upon the frontier with your pistols and prevent a general Indian war."[103] Elated with an unsolicited endorsement from such high-ranking authorities, Colt wrote Houston to say "that your recommendation will contribute vastly more to the successful introduction of my armes [sic]... than the... opinion of a hundred military men who have had no expereance [sic] in their use in the field." Colt described Texas as his "first patrons" and acknowledged that "Texas has done more for me & my armes [sic] [and has] a better knowledge of their use" than the rest of the country combined.[104]

Colt treasured the collection of "distinguished names" who endorsed his arms for military purposes,

carefully preserving them "in the book I keep for that pur-pus *[sic]*."[105] The collection, which also included thanks and acknowledgments for the gifts of presentation pistols, contains letters received from Italian freedom fighter Giuseppe Garabaldi, Sardinian nationalist King Victor Emanuel, Hungarian freedom fighter Louis Kossuth, reli-gious exile Brigham Young, California Gold Rush testi-mony from T. Butler King and William M. Given, and field service testimony from Indian fighter Major O. Cross and Texas Rangers Maj. G.T. Howard and Capt. I.S. Sutton, who praised Colt's revolvers as "the greatest improve-ment in small arms of the age... the only weapon which has enabled the frontiersman to defeat the mounted Indian in his own peculiar mode of warfare... [and] the arm which has rendered the name of Texas Ranger a... terror to... our frontier Indians."[106]

As testimony piled up in Washington, D.C., former Texas war secretary Thomas J. Rusk urged the "govern-ment to introduce" Colt's revolvers, claiming "that one hundred cavalry armed with the repeating pistol would be, at least, as efficient as 300 armed in the ordinary way." Lt. Bedney F. McDonald, who was in Huamantla, Mexico, the day Captain Walker was killed there, testified that he had "seen Col. Hays hold in check more than 500 Mexicans, with 30 men armed with these pistols." With the Mexican War over, testimony shifted to frontier defense where it was noted that those "who have used"

Colt's revolvers in "Mexico, and elsewhere" regard them as "the most efficient, and economical arms that can be employed in all Indian territory, and on our extended line of frontier."[107] Although sales to the American armed services languished throughout the 1850s, by the time of the Civil War, the question was no longer whether to adopt Colt's revolvers, but how fast could they be supplied.

Colt's factories became another of his successful instruments of publicity. It was in London that Colt dis-covered the appeal of a behind-the-scenes look. Tours of his armory became so popular that he issued tickets of admission.[108] Mounted on top of Colt's London factory (fig. 49), adjacent to Parliament, was a fourteen-foot-long sign reading "Col. Colt's Pistol Factory."[109] Brash and outrageous, Colt was eventually ordered to remove the sign, but not before it stirred controversy and heightened his fame.[110]

As soon as his Hartford armory, mansion, and work-er housing were fully operational, Colt bought himself one of the most expansive "news" stories written about an American industrial concern before the Civil War. In May 1857 Colt paid the publishers of *United States Magazine* $1,120 (today equal to $61,439) for copies of a flattering, twenty-nine-page, illustrated feature story that appeared in the magazine's March issue under the title "Repeating Fire-Arms: A Day at the Armory of 'Colt's Patent Fire-Arms Manufacturing Company.'"[111] Repre-

"Repeating Fire-Arms: A Day at the Armory of 'Colt's Patent Fire-Arms Manufacturing Company.'"[111] Representing a genre of industrial profile that became a staple of the prestigious *Scientific American* during and after the Civil War, "A Day at the Armory" set a new standard for length, quantity of illustrations, and sycophancy in industrial reporting. It also provides the most detailed account, and the only illustrations released during Colt's lifetime, of the inner workings of Colt's Armory. Colt traded access, hospitality, gift pistols, and the personal attention of a recognized "great man," for glowing coverage. Two years later, Boston journalist John Ross Dix actually wove Colt's unique form of hospitality into the text of his narrative tour of Colt's Armory, lauding the Colonel as a man "as noted for his hospitality as for his genius" and transcribing Colt's letter accompanying a gift revolver inviting the writer to "accept in order that you may examine its structure and finish at your leisure."[112] A connoisseur was no doubt born that day.

In addition to testimonials and news stories, Colt developed a national network of retail commission sales representatives to whom he offered quantity discounts and provided printed advertising broadsides.[113] The best of the many versions of Colt's printed broadsides combined price and model information with wit and visual imagery. Although Colt's broadsides are today very rare, a bill from 1854 documents the printing of 30,700 "pistol directions" and 20,000 "pamphlets," which Colt distributed through his sales agents, retail outlets, and lobby-ists.[114] At least 9 different illustrated broadsides were produced during Colt's lifetime. The first, published about 1848 to promote the new .44-caliber, six-shot army pistol, features an image of a cocked revolver with illustrations and descriptions of its use and component parts, printed as if written in the inventor's hand, signed in cursive scrawl, Sam Colt, and so personalized: "I… will be glad to attend any cash orders addressed to me in New York city, or at my armory in Hartford, Conn."[115] The world would know that this man's signature was his bond.

Texas and the West may not have generated the majority of Colt's sales, but its aura and associations were soon adopted as key elements of Colt's marketing strategy. An 1857 broadside addressed to the Pioneers of Civilization noted that Colt's "pistols, rifles, carbines and shot gun" were "sold by all respectable dealers throughout the world." Colt probably drafted the text, which described the "Simple Reasons for preferring Colt's Arms to all others." In a curiously defensive, fourteen-point rebuttal of the many criticisms leveled against his invention, Colt noted that "They do not stick fast, refusing either to open or shut without the aid of an axe when heated, as do the guns which open like molasses gates or nut crackers…. They leave no burning paper in the barrel after a discharge, to blow the next cartridge into your face, as do the guns which open from behind…. They are simple in construction, and easily taken care of, as any ranger or cavalry soldier will tell you. Treat them well, and they will treat your enemies badly…. They are always

49. *Colt's London Armory, 1855.*

From Household Words, *Charles Dickens, ed. (Courtesy of the Colt Collection, Connecticut State Library.)*

50. Colt's Advertising
Broadside, ca. 1854.
Possibly printed by
E.B. and E.C. Kellogg,
Hartford.
(Courtesy of
R.L. Wilson;
private collection.)

worth what they cost—in the Far West much more, almost a legal tender! If you buy anything cheaper, your life, and that of your companion, may balance the difference in cost.... If you buy a Colt's Rifle or Pistol, you feel certain that you have one true friend, with six hearts in his body, and who can always be relied on."[116]

The most artistic of Colt's broadsides (fig. 50), published about 1854, substitutes artwork for text and simply describes the product, illustrated with a break-apart view of the components, as "Colt's Patent Repeating Pistols.... Orders... may be addressed to me at Hartford, Conn., or New York City. [signed] Sam Colt." The illustrations feature the three most popular cylinder scenes of Colt's arms in action, an innovative convention of Colt's marketing since the Paterson days. Scenes of Hays's Big Fight, the 1843 engagement between the Texas and Mexican navies, and a stagecoach robbery are flanked by patriotic vignettes of the Battle of Quebec and "man the hunter" triumphantly mounting the head of a lion with Colt rifle in hand. No more alluring use of images and symbols exists in American advertising before the Civil War.

Colt's broadsides typically accomplished several purposes. Some conveyed price information. Some created

allure. All attempted to demystify an instrument of considerable complexity. Parallels with today's computer-based communications revolution are striking. For all the attention given to the arms industry's success in mastering uniformity of parts and volume production, firearms still demanded complex care and maintenance. Guns tended to malfunction and break. It was a challenge to convince people to part with almost a month's wages for an instrument that was not wholly dependable. By the time he revived his arms factory, Colt had clearly internalized the criticism heaped on his Paterson-era revolvers for "getting out of order." Today, when we are able to operate our televisions and radios (and increasingly, automobiles) with little or no maintenance, it is easy to take for granted the hard-won qualities of low maintenance, then almost unprecedented in the history of complex things. By persistently illustrating and writing about the working parts of his firearms, Colt chipped away at the subconscious resistance and eventually succeeded in convincing people that this gun was "simple in construction, and easily taken care of," words spoken as corporate mantra in the dogged pursuit of ever-greater refinements and simplifications. But for those still con-

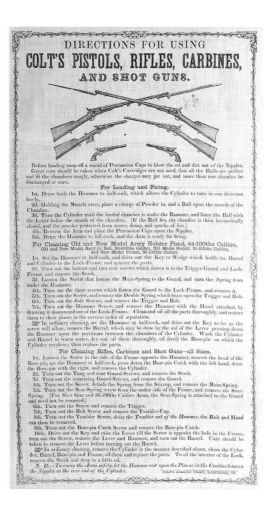

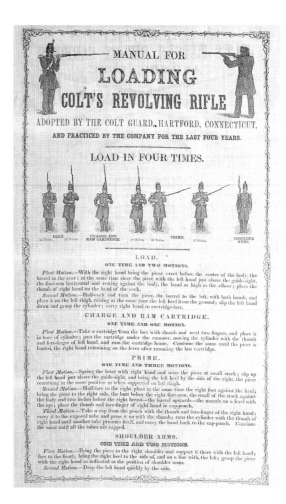

fused, Colt also produced users' manuals, broadsides, and cleaning cloths (fig. 51), reiterating over and over again the principles of gun use, construction, and maintenance. Where modern industry struggles with engineers and specialists interested in dazzling their peers (and the inevitable gadget freaks) with bewildering options, Colt virtually pioneered the field of ergonomics, or user-centered design aimed at product usability.[117]

Colt also mastered the techniques of display and presentation as few arms manufacturers have, before or since. In London he urged dealers and sales agents to arrange his firearms in eye-catching window displays.[118] Colt's display at London's Crystal Palace (fig. 52), was described as a "military trophy" that attracted crowds intrigued by the colorful character of its host and the spectacle of its "symmetrically arranged" collection of pistols.[119] Colt served favored guests cigars and brandy. One journalist noted how Colt "liberally indulged" the "sense of touch," letting guests cock and aim pistols that, he assured them, were not loaded. This was a more carefree attitude than was then common in the sale and display of such costly items.[120] At the New York Crystal Palace in 1853 (fig. 53), "Col. Colt's Revolvers were arranged in the form of a shield and presented a beautiful appearance." Some years after Colt's death, this tradition of showmanship was perhaps in mind when the company arranged an "exhibit... [with] a little bedstead made of pistols," with two "soundly sleeping... beauties, all mother-of-pearl silver," beneath a "satin [coverlet]" for the 1876 Centennial Exhibition in Philadelphia.[121]

Even before establishing manufacturing capacity, Colt lavished considerable attention on the appearance and presentation of his guns, hiring engravers and carvers to decorate barrels, grips, and frames, and purchasing presentation cases made of hand-crafted exotic woods.[122] While in Austria in 1849, Colt commissioned "an artist" to make "pistol cases of a peculiar description of inlaid work called by the Austrians 'Buhl,'" which he adopted as part of the packaging for the finest and most artistic of his engraved and gold-inlaid presentation revolvers (fig. 54).[123] At its height, Sam Colt's Armory tapped an international pool of skilled artisans—mostly German Americans—who provided custom decoration and a jewel-like quality of finish that belied the fact that Colt's revolvers were machine-made.

Far from destroying qualities of art and artisanship, machine production increased the share of total cost that could be practically devoted to embellishment. Between the expectations of the "gentlemen" for whom expensive hand-crafted ornamentation and workmanship was an inherited taste and critics afraid that qualitative compromises accompanied volume production, Colt's use of ornamentation and packaging helped secure the reputation of his arms among purists and skeptics. To Frederick Stegmuller of Newark, New Jersey, he turned for mahogany cases.[124] Colt was one of the first to develop a commercial application for Waterman L. Ormsby's patented "grammagraph," a device for roll-die engraving on steel, used to create the "cylinder scenes" on Colt's revolvers. Ormsby, whose work is one of the most distinctive fea-

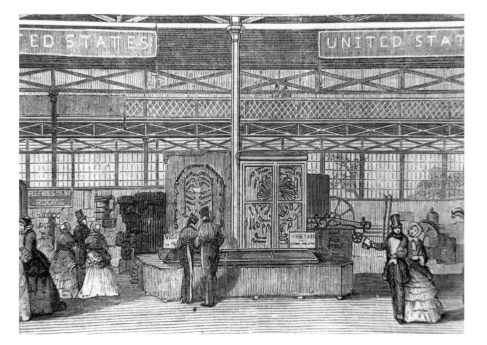

52. Colt's Display at the London Crystal Palace, 1851. From London Illustrated News, December 6, 1851. (Courtesy of Matthew Isenburg.)

53. Colt's Display at the New York Crystal Palace, 1853. From Gleason's Pictorial Drawing Room Companion, *November 1853. (Courtesy of the Colt Collection, Connecticut State Library.)*

54. The Art of Presentation. Top to bottom: buhl box, Vienna, Austria, 1849; described by Colt as the "most beautiful inlaid work ever seen in America," cases like this were supplied with some of Colt's most deluxe gift revolvers, including a pair presented to Britain's Prince Albert at the time of the Crystal Palace Exhibition (WA #05.1254); old model navy, or belt, pistol, .36 caliber, 6-shot, serial #138370 (WA #05.991); old model pocket pistol, .31 caliber, 6-shot, serial #232396 (WA #05.1007); new model pocket pistol, 1863, .31 caliber, 5-shot, serial #6373, fluted cylinder, first solid frame construction pistol (WA #05.999).

tures of Colt's firearms, was a National Academy of Design-trained artist and inventor and the most prolific bank note engraver in America during the 1840s and 1850s.[125] For superb custom decoration, Colt relied on Gustave Young, an immigrant engraver who boasted of having "been engaged in Col. Colt's Establishment... [for] twelve years in the highest walks of art" producing "splendid presents" for "Crowned Heads... [and] illustrious personages," arms engraved and inlaid in gold, steel, and silver (fig. 55).[126] Christian Deyhle, Carl Helfrecht, Augustus Grunwald, Herman Bodenstein, and John Most, all immigrant German craftsmen, were the most prominent of the skilled artisans who engraved steel, and carved ivory and wood pistol grips, and thus helped shape Colt's reputation for beauty and artistic finish.[127]

Colt also realized that the easiest way to make a product user friendly was to make it affordable. The failure to achieve competitive pricing has led to the failure of many a great invention. Colt, having been stung by accu-

sations that his Paterson arms were a bad value, devised a two-pronged strategy for preserving a dominant share of a rising market. One almost wonders, given the outcome, if all Colt's noise about patent protection was partly aimed at taking the competition's eye off the ball of cost containment. Where business often falls into the trap of basing its price on costs, Colt was determined to do the opposite, to keep his costs consistently at or below a level that would maximize sales volume and below the cost his competitors could match and still survive. He knew from constant haggling with government officials what price would optimize volume. For a product that entered the market at $50 (now about $3,875) and bottomed out in 1859, wholesaling for as little as $19 (now about $1,250), Colt ended up with a better and more reliable gun at about one-third the original price.[128]

In 1854, while clamoring for a patent extension, Colt's advocates and attorneys inadvertently revealed his quest of price-based market dominance. The chairman of

55. *The Art of Presentation, Colt's Gold Inlaid Presentation Pistols.* From left to right: old model army pistol, *third model dragoon, 1856, .44 caliber, 6-shot, serial #15821;* old model (1851) navy pistol, *1854, .36 caliber, 6-shot, serial #38843;* old model (1849) pocket pistol, *1853, .31 caliber, 5-shot, serial #71746.*

These pistols are part of a small group of highly decorated presentation firearms, most of which were given to foreign heads of state, beginning in 1854. These were held by the factory as examples of the highest *level of workmanship and embellishment then available in America. The ornamentation is attributed to Hartford engraver Gustave Young (1827–95). (Courtesy of the Museum of Connecticut History.)*

the Committee on Patents acknowledged that Colt's goal was to "perfect his armory by the increase and subdivision of machinery, [so] that he will be able to furnish... a perfect arm at a price which will defy... spurious imitations" and further that "manufacturers of spurious imitations will not find the profit sufficient to encourage the business." Colt's attorney and friend Edward Dickerson noted that "every day... his manufacture is improving, and the price diminishing," to the point where "no one today can make a genuine Colt's pistol and compete with him in the market."[129] Both attempted, unsuccessfully, to mask Colt's pursuit of market dominance and profits behind the rhetoric of patent protection. Colt's also became famous for frequent model changes and the most extensive catalog of styles, patterns, sizes, and surface finishes in the arms industry, an accomplishment that stymied the attempts of competitors to gain a toehold in the market.[130] Only by missing a couple of key shifts in firearms technology—notably the adaptation of metallic cartridges—did Colt squander the advantage created by his superior manufacturing capacity, thus providing Smith and Wesson of Springfield and Winchester in New Haven with a share of the market.

Advertising and competitive pricing were the least controversial of Colt's marketing techniques. The presentation gun was the most. Imagine Ford Motor Company giving cars to the purchasing agents of corporate and municipal fleets coast to coast and you get some idea of how Colt used the presentation revolver. For a product that ranged upward from the equivalent of $1,500, it was no small token to receive a pistol "compliments of Col. Colt" or "compliments of the inventor." And where the conflict of interest was most flagrant, as with an infamous gift to former Virginia governor and then U.S. secretary of war John B. Floyd from "the Workmen of Col. Colt's Armory," Colt simply masked the source.[131]

The presentation gun, specially decorated, handsomely cased, and engraved with an honorific inscription, was a convention Colt adapted from a tradition of presentation swords that began during the War of Independence and flourished during the War of 1812 and Mexican War.[132] The difference was that sword manufacturers like Nathan P. Ames were not usually the presenters. Although many of Colt's deluxe presentation-style revolvers were purchased for gift purposes by others, most of the couple of hundred presentation guns known were gifts made by Colt to curry favor. Colt surely did not invent this form of inducement. But no one developed it further or achieved more sensational results. Colt's presentation arms (figs. 56 and 57) are today prized by collectors of American arms and represent the greatest accomplishment of artistic virtuosity by Colt's workmen, rivaling in painstaking ornamentation, skillfulness, and beauty any form of decorative art produced in America before the Civil War.

How do we understand a man whose hostility toward government officials was so great that he once wrote that "to be a clerk or an office holder under the pay and patronage of Government, is to stagnate ambition.... You had better blow out your brains at once & manure an honest man's ground with your carcass than to hang your ambition on so low a peg"?[133] This same man practically wrote the book on courting influence with those same government officials. The triumph of pragmatism over ideals has rarely proved more decisive, and yet Sam Colt can hardly be accused of setting the rules of the game. It was because of the duality of his character—his "flexible patriotism" and situational ethics—that Colt was able to flip-flop so deftly between the worlds of art and industry, handicraft and machine-based uniformity, personal expression and militaristic workplace discipline. Colt's style of lobbying switched back and forth between a tradition of relationship-based friendship networks and purchased influence, and a rigid obsession with merit-based testing and fair play. In all, Colt was as much a victim as a victimizer, and it was probably as a victim that he first discovered how the game might be played. Never one to abide limits, Colt stretched the rules to the point where he became the target of a congressional investigation for bribery.

56. Presentation Colt Revolver. 1861 model navy pistol, *with accoutrements and case,* 1862, .36 caliber, 6-shot, serial #5726 with bulletmold, cone wrench, percussion cap can, three packages of cartridges, powder flask, and user's manual (WA #32.5).

This is the last presentation revolver issued by Colt's Patent Fire-Arms Manufacturing Co. during Sam Colt's lifetime; it is profusely engraved with foliate scrolls and reads "Wm. Faxon/with Compliments of Col Colt" on the backstrap.

57. Presentation Colt Revolver. 1855 model sidehammer pocket pistol, *with accoutrements and case,* 1861, .31 caliber, 5-shot, serial #25,317. This gun was presented by Colt's Armory Guard to Capt. J.L. Morgan, presumably as a shooting tournament prize.

Although the sidehammer pocket pistol was an economy model revolver, this example is handsomely engraved and features deluxe ivory grips. (Courtesy of the Frank Murphy Collection.)

For a democratic paragon, Colt was ironically addicted to courting favor with men of influence and high social status. His most expensive and impressive gifts were bestowed on European monarchs, whose recognition he craved even more than arms sales. His presentation at the Court of Czar Nicholas in 1854 (fig. 58), and the corresponding gift of gold-inlaid presentation pistols, by most standards a relatively insignificant event in a life filled with accomplishment, was singled out for commemoration in a final memorial—Hartford's Colt Memorial Statue—conceived by a doting wife who knew well the passions that made her husband proud. Was it really necessary that a third visit to Russia in 1858 include a "set of his military arms... elegantly ornamented... presented to the Russian Court" including "large cases... containing samples of all the arms made by Col. Colt... elegantly engraved... [and] gold mounted... for the Crown Princes Constantine, Michael and Nicholas"?[134] And what of the dozens of revolvers provided for Commodore Matthew C. Perry's Japan Expedition, to be liberally bestowed on, among others, "the emperor and several of the princes of Japan, the governor of Shanghai, the King of Lew Chew, and the king of Siam," in 1855? Few of these gifts, worth today's equivalent of nearly one hundred thousand dollars, prompted sales.[135]

At home, scores of gift pistols were presented to military officers and Ordnance and War Department officials.[136] Couched in terms of seeking testimonials, these guns helped create the sympathetic and progressive atmosphere that eventually triumphed over the Ordnance Department's resistance to new weaponry. Once the government contracts began to flow, they provided one of Colt's largest sources of revenue.

On certain occasions Colt used outright bribery to lower the Ordnance Department's resistance. In 1839, with his fortunes sagging in Paterson, Colt proposed bribing ordnance officials, to which his uncle and company treasurer Dudley Selden recoiled in shock, insisting that "I will not become a party to a negotiation with a public officer to allow his compensation for aid in securing a contract with the government. The suggestion with respect to Col. Bomford is dishonorable [in] every way."[137]

In 1851 the plaintiffs in Colt's lawsuit for patent infringement complained that "Mr. Colt... has made his threats that he has $50,000 to expend against us." In 1852, in a sneering comment on the way business was conducted with government officials, Colt wrote to his agent in London, "I rejoice to find you have at last closed arrangements with the government & c. & that you have treated the Babes of the Woods & Forests with a small specemin [sic] of sivilization [sic] in the shape of a Dinner," and who knows what else, "at my expence [sic]."[138]

Playing cat and mouse with government officials was Sam Colt's idea of sport. Eventually it caught up with him. Although he was acquitted, the bribery accusation that was most damaging to Colt's reputation led in 1854 to a congressional investigation "to inquire whether money has been offered to members... or other improper means used to induce members to aid in securing the passage... of a bill to extend Colt's patent."[139] Colt applied for an extension of his patent in January 1854. The Patent Office advised approving Colt's application, but it became mired in debate in the House of Representatives. By July charges that "money has been offered to induce persons to cause members by solicitation to vote for this bill" were out of control.[140] On July 10 a special committee was appointed to take testimony involving the Colt bribery charges. By August it was a national news story, with *Scientific American*, although unfavorable to patent extensions in general, calling Washington a "den of corruption" and basically exonerating Colt, whose crime appears to have been little more than entertaining "the honorables" in the same lavish style he had been doing for twenty years, in this case supplying members' wives with fancy "kid gloves."[141]

Colt's most controversial "marketing" technique involved combinations of bribery, blackmail, bullying, and baiting. Realizing that gun sales thrive in an atmosphere of fear, Colt discovered the value of playing on the fears of adversaries. Although earnestly and fiercely

patriotic, Colt's loyalty was more to ideals than institutions, even when the institution was the government of the United States. It is impossible a century and a half after the fact to know what truly motivated certain words and deeds. The record paints a picture of a man deeply conflicted and easily angered by the slow pace of political decision making. Colt discovered that men react more swiftly in fear than in faith. Angered, in 1841, when the government failed to adopt his invention for a system of harbor defense based on submarine explosives, he sulked and suggested that if his own country did not have sense enough perhaps the governments of Europe might.[142] Frustrated after ten years of trying to convince the military to adopt his revolvers for use by the armed services, he boasted that "the Mexicans offered $100 apiece... during the war," and he sent agents to Mexico as soon as the war was over to make arms sales to the defeated enemy.[143] Accused by the British of aiding the Russians during the Crimean War, Colt defended himself ambiguously noting that "it is not true that I have furnished arms... to the Russian government;... since my armory has been established in London, both it and my own skill have always been at the service of the Government.... It is not my fault if all my facilities are not now devoted to the British Government."[144]

Having lobbied actively and publicly as a pro-Union, pro-Southern, antiabolitionist Democrat, Colt responded to John Brown's Raid on Harpers Ferry in 1859 not by cutting back on arms sales to a militant and rebellious South but by sending an agent "to the western and south western states [meaning the Deep South and Texas]... for the purpose of introducing and making sales of Colt's revolving shot guns and rifles."[145] Colt's sales to Alabama, Virginia, Georgia, and Mississippi in 1860 alone were at least $61,000 (today's equivalent of about $3.35 million).[146] In May 1860 the governor of Virginia sent a delegation to the Connecticut Valley to buy arms and machinery from Colt's, the U.S. Springfield Armory, and Ames Manufacturing Company.[147] For Colt, these sales were not only living by the letter of the law—trade with the South was condoned and unrestricted until the winter of 1861—but an act of faith that arming the South

58. *Colt's Royal Presentation at St. Petersburg, 1854. Bas-relief sculpture in bronze, 1904,*

Colt Memorial Statue, Colt Park, Hartford. J. Massey Rhind (1860–1936), sculptor, Gorham Manufacturing Co., Providence, R.I.

Colt's presentation at the Court of Czar Nicholas I in November 1854 was one of his proudest accomplishments.

might serve as a deterrent to war. Of course, then and now, interpretations of conduct vary widely and are relatively subjective.

Two weeks after the attack on Fort Sumter, the *New York Daily Tribune* launched an assault against Colt suggesting that "the traitors have found sympathizers among us, men base enough to sell arms when they knew they would be... in the hands of the deadly enemies of the Union.... Col. Colt's manufactory can turn out probably 1,000 a week, and has been doing so for the past four months for the South.... That... stopped,... he resorts to the only means left him to carry his purpose, and puts up the price of his arms to a figure hereto unheard of. A power so ill used should be taken at once... out of the hands of so bad a citizen.... Every man who makes arms should be watched, and if he will not work for a fair equivalent for the Government, his manufactory should be taken away from him."[148]

The *New York Times* was no less critical of its industrial neighbor to the north, accusing Colt of "treason," noting that "the Constitution declares *treason* to consist... of 'giving aid and comfort to the enemies,' as is done daily, constantly and by contract, with individuals in Connecticut."[149] Describing the accusations as libelous and "absurd," Colt's friends at the *Hartford Daily Times* rallied behind the Hartford arms industry noting that a recent order Colt received from Tennessee "of course was not filled," and that Sharps' Rifle Company and Hazard's Powder Company, in neighboring Enfield, "will not sell a weapon to the enemies."[150] Colt supplied the *Hartford Times* with a letter he allegedly wrote to his sales agents in January instructing them "not to sell to [South Carolina] or any other State which you know to be in open hostility to the Federal Government" and, while not denying that prices had been effectively raised, explained that doing so was necessitated by the rising costs of sales and distribution.[151] Distrustful, his Republican enemies at the *Hartford Daily Courant* wrote of how "Salmon *[sic]* Adams... superintendent of the new armory in Richmond, Virginia, arrived at Springfield... after his suc-

cessful visit to Colt's Armory" where he was "purchasing large quantities of arms from private manufacturers in this vicinity for Virginia and the seceded states."[152]

The facts do not reflect well on Sam Colt. But Colt was a man motivated at least as much by personal political conviction as by profit, convictions shared by many of New England's protrade libertarian industrialists and by almost half of Connecticut's voters, who gave Abraham Lincoln the narrowest margin of victory of any Northern state in the election of 1860. In hindsight it is easy to condemn them for shortsightedness and profiteering while averting their eyes from the suffering of others. In the context of the time, however, these were deeply perplexing issues rife with uncertainty and a bewildering array of bad and good consequences.

What speaks most of Colt's character is his hyperactive brand of opportunism. By temperament, Colt was the sort of man who instinctively marched to the sound of guns, be it a parade around the block or an impending war halfway around the globe. And once on stage, Colt was not averse to making his presence felt. On more than one occasion, he willfully intervened to intensify conflict in some contested battleground. In 1851 Colt allegedly played a role in arming a soldier of fortune, Narcisco Lopez, and his fellow annexationists, in a failed invasion of Cuba.[153] Kansas, still reeling after a year of armed conflict between extremists on both sides of the slavery issue, was a suspicious destination for three of Colt's top aides, who traveled there on unspecified business in 1857.[154] Stirring up controversy, or as Colt might have viewed it, aiding disputants in their right to self-defense, was all part of a day's work.

With the outbreak of the Civil War, Sam Colt's behavior and reputation waffled between extremes. Was he a patriot or a traitor? One thing was sure, Colt had no intention of waiting on the sidelines for direction. The war was not two weeks old when Colt seized the moment and "offered to raise a regiment, and arm it with revolving breach rifles of his own manufacture."[155] As a prelude to forming Colt's Armory Guard, Sam Colt had sponsored a grand regimental encampment on the South Meadows in Coltsville in 1858, in which eight hundred or more members of the "first regiment of Connecticut Militia" amassed for two days of drills, marches, and activities.[156] This time it was for real. But Sam Colt was now at the end of his rope and, true to form, his behavior was bizarre and self-destructive.

Colt began amassing a regiment—not a company, not a marching band, but a full regiment—before being authorized to do so. It was vintage Colt. Surely, only Colt's money and Connecticut's desperation prompted the Republican governor to grant a commission to arm troops to the state's richest, staunchest, and most outspoken Democrat.[157] He soon regretted it. By May 18, Colt's Rifle Regiment was full, with "four or five companies [100 men each] now in rendezvous at the meadows," who "sleep in Colt's depot… and dine in the Armory Hall." Company A of the 3d Regiment, known as the Mechanics Rifles and composed primarily of men "enlisted from… Colt's and Sharps'," envisioned themselves as a prestigious "flank regiment… detailed for skirmishing" and caused a fracas when they refused "to take the U.S. muskets given them at the arsenal."[158] Testy and vigorous, Colt, whose entire career and life was a case study in the value of "picked men" and "skirmishing," soon announced that the "Barracks… on Huyshope Ave… [would receive] no men under 5'7."[159] The First Connecticut Rifles, or Col. Colt's Rifle Regiment as it was known, aspired to be the best armed and most elite corps on the field of battle. Once again, Colt was marching to the beat of his own idiosyncratic drum, and before long the Republican newspapers were poking fun at "Col. Colt's… regiment of Potsdam grenadiers," an obvious reference to the Germanic atmosphere and amenities Colt had created in the village community around Colt's Armory.[160]

By the end of the month, Colt's Rifles was in trouble. First by enforcing its exclusionary standards of height, the regiment rejected "members of the New Haven fire company… as too small."[161] Next, Colt rankled his own men by lobbying to make Colt's Rifles a regular unit of the U.S. Army, violating their patriotic loyalty to Connecticut. Within two weeks of the regiment's encampment on the South Meadows, the governor revoked Colt's commission and dispersed its members. Most entered the regular army as a rifle corps in the 14th Regiment of Infantry; others, as members of the 5th Connecticut Regiment under Col. O.S. Ferry, and the 12th Connecticut Regiment's Company I, known informally as the Colt Guard Company.[162] Although Colt's dream of a full regiment of the U.S. Army armed with Colt's revolving rifles came to naught, members of the two flank companies of the 14th Regiment were apparently armed with Colt's rifles and may have been the first U.S. soldiers so armed in battle when Connecticut troops shipped off for Manassas, Virginia, the following month.[163]

Most of us will never know what it is like to be in the grip of a dream so consuming and overpowering that you would give anything, your life even, to achieve it. Sam Colt was a brinksman who skirted convention, invented and reinvented himself to meet the needs of the occasion, and rarely missed a beat when opportunity knocked. Sam Colt began crossing the lines of good taste and propriety as a boy, and never stopped hungering for the limelight, even when doing so involved almost senseless risk. A man who mingled easily with and celebrated the virtues of commoners, Colt craved and gained the attention of royalty, making him a sort of rich man's working-class hero. Colt wrote, "If I can't be first I wont be second at anything.… However inferior in wealth I may be to the many who surround me I would not exchange for these treasures the satisfaction I have in knowing I have done what has never before been accomplished by man."[164] Colt was an opportunist in the best sense of the word. His vision was irrepressible, and in the end, Colt came as close as anyone of his generation to inventing the modern industrial system and acting out the dramatic issues of his age.

COLTSVILLE
THE BODY OF EMPIRE

A feeling of responsibility and a zeal to be faithful run... through the whole... establishment, from top to bottom.... It is no holiday affair.
HENRY BARNARD, *ARMSMEAR*

He dared to execute what no man but himself dared conceive.... An enterprise of this magnitude... carried through by...
one man is without parallel in this country.
HENRY C. DEMING, *HARTFORD DAILY TIMES,* 1855

I was asked to part with... land... not... for any public purpose... but simply to enable Mr. Colt to carry out his
"scheme of private speculation."... Supposing the embankment to be finished, who can guarantee that a city shall arise within its circle?
SOLOMON PORTER, 1853

In the winter of 1852, with money now piling up, profit margins high, and having been decorated by science and feted by royalty during his widely reported performance at the Crystal Palace Exhibition in London, Colt returned to Hartford feeling exalted, heroic, and invincible. He shared with a few trusted advisers his plans for a new armory and industrial compound that was to be the seat of an expanding empire on Hartford's South Meadows (fig. 59) and quietly began amassing property. The compound was the most ambitious experiment in economic development and social engineering in the two-hundred-year-plus history of Connecticut's capital city. Had it lived up to its founder's ambitions, Coltsville, as it was dubbed by its friends, with its diversified industrial base and internationally renowned technical college, would have made Hartford a very different place than it is today, certainly a larger, if not more prosperous, place.[1] For better or worse—a real coin toss—Colt died young, with the knowledge that his vision for Coltsville would not materialize. Today, surveying its two hundred mostly barren acres, with Colt's Armory building now cut off from the river and mostly vacant, Colt's Manufacturing Company no longer even a mail stop within its founder's compound, and the zinc rampant colt, its symbolic and triumphant icon, plundered from the rooftop as a trophy for folk art and firearms aficionados, one cannot fail to be moved by the poignancy of passing time and a deep sense of the fragility and impermanence of vision and power.[2]

So much more was believed possible at the time of "the transfiguration"—their word, not mine—of 1855.[3]

The site of a once-flourishing industrial empire that carried Hartford's fame "to the ends of the earth," Coltsville was Sam Colt's most audacious enterprise. "It was the dream of his boyhood," remarked Col. Henry Deming, "that if Providence should ever smile upon his industry... he would here, upon this very spot... rear an establishment which would... be an honor to his native town... a light, a landmark in the weary and disheartening pilgrimage of mechanical genius."[4] Ever since the failure of his first armory in Paterson, Sam Colt had been obsessed with maximizing control over as many facets of the business as possible. What better means to achieve that control than to seek a location where all the components of an industrial empire could be developed in close proximity? Colt selected Hartford's South Meadows floodplain, the largest undeveloped section of the city, with the intention of buying the land cheap and reclaiming it for development by building a dike that would protect it from spring flooding.

Situated near the banks of the Connecticut River, the South Meadows provided ready access to river-based transportation, abundant water, and space to grow. Within ten years, Colt's dream became a reality, an enormous brick armory at the center of a compound that included civic and industrial amenities, commercial, educational, social, and residential facilities for the workers, a gas works, reservoir, and waterworks, a farm and produce warehouse, a river port dock and railroad depot, separate facilities for manufacturing gun cartridges and willow furniture, and, crowning the hillside above, the owner's mansion, Armsmear. All of this linked

59. *Map of Coltsville, 1861, including Armsmear (middle right),* Coltsville *(middle left), the Willow Works (lower middle), armory and workers' housing (right), and* unfinished street plan. *By 1861 the vision was already beginning to fade, as suggested by the dotted-line layout of streets that were never built.* From an engraving by E.B. and E.C. Kellogg, for the book Armsmear, *published by Elizabeth Colt in 1866.*

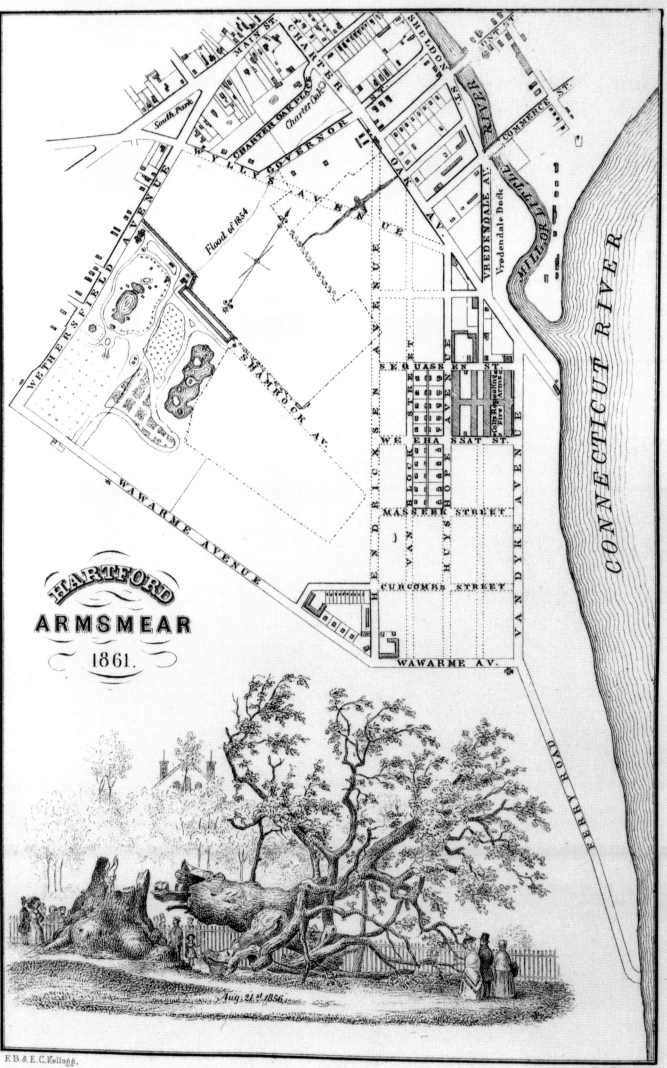

HARTFORD
ARMSMEAR
1861.

Aug. 21st 1856.

F.B. & E.C. Kellogg.

by a crisscross network of streets lined with elms and punctuated with gardens, orchards, and meadow lands suitable for farming.

There was abundant precedent for the kinds of social amenities and civic improvements that Colt planned to introduce at Coltsville. One of the most successful and widely admired of the many factory towns in Connecticut was just across the Connecticut River in South Manchester, where the Cheney brothers "like their prototypes the Brothers Cheeryble in [Charles Dickens's] *Nicholas Nickleby...* [were] notorious for their kind, benevolent, and generous treatment of those in their employ." At the site of their flourishing silk mills, the Cheney brothers sponsored a "library... balls and parties," and "mix[ed] freely with their operatives,... [a] village of happy and industrious people" where "employer and employed each strive to advance the interest... of each other."[5] Cheneyville, which was developed about ten years earlier than Coltsville, was described as "a kind of rural paradise for workmen." Even "more noteworthy than the success of the business," it was described as a "model village," based on "charity and philanthropy, a practical solution of the harmony that may exist between employers and workmen," a testament to the "love of beauty and the artistic sense... in the [Cheney] family."[6] The family could not have described themselves more glowingly, and accounts by newspaper editors of like mind and interest are best swallowed with a dash of skepticism. But the continuous stream of favorable press, more and better than Colt ever experienced, makes one wonder if Cheneyville may actually have resembled "the paradise of New England factory villages," where the mill owners "spared no expense in beautifying the grounds and in providing everything necessary for the happiness of their employees... [including] boating, riding, reading, and dreaming."[7]

Colt's close friend and Democratic crony Col. Augustus Hazard, who founded the Hazard Powder Company in neighboring Enfield in 1843, in a private confessional written almost twenty years later described how

from the moment I first took up my residence in the Town of Enfield, it was my steady purpose to do all in my power to elevate the conditions of things around me.... When first I came to Enfield I took up... a sort of wild & unoccupied water privilege in a part of the town then nearly unoccupied and after years of toil... had the ambition & desire to build up a large manufacturing establishment and lay the foundation of a village that I hoped some day might grow up to a respectable size.... I had pride to bring out, up & develope [sic] some of my own townsmen who... cooperated with me... in promoting the interest of the whole town.[8]

Whether out of genuine affection or because of coercion, "the citizens of Enfield residing in... the vicinity of the Hazard Powder Company" petitioned the town in 1850, to name the place "Hazardville, from respect of our fellow citizen, A.G. Hazard, Esq."[9]

The U.S. Springfield Armory, under military direction, was also known to Colt. Praised in 1852 for "the order, the system, the neatness, and almost military exactness and decorum which pervade every department of the works," Springfield's "definite and strict" work environment populated with "well dressed" workmen, with "walls and floors" kept "neat and clean" and "machinery and tools... symmetrically and admirably arranged" epitomized the qualities of workplace discipline that marked the transition from home-based to factory-based manufacturing.[10] It sounded as neat and orderly as a community of Shakers, whose religious commune in Connecticut was located a few miles from Hazardville and not a day's walk from the Springfield Armory. Industrialists like Colt, the Cheneys, and Augustus Hazard were determined to assuage worries about the moral influence of the industrial workplace. Looking in almost any direction from Hartford, Colt could find powerful evidence that the key to remaking the world of work was discipline and order.

Already one of the most esteemed "provincial cities" in the country in 1852, Hartford did not require Sam Colt to create its civic amenities or to teach it a lesson in civic morality. That year a Hartford resident described with pride and wonder the city's nearly two dozen religious societies at a time when the population hovered around fifteen thousand.[11] When Colt's friend Gen. Sam Houston, a U.S. senator and former president of the Republic of Texas, visited Hartford in 1848, he toured the "Insane Retreat, [the] American Asylum [for the Deaf], [the new public] High School," and the Wadsworth Atheneum, and "spoke the highest praise of them."[12] He might also have mentioned the twenty-five-year-old orphan asylum, the alms house, and the reading room of the Young Men's Institute.[13] The next year Hartford boosters boasted that the "New [railroad] Depot," an architect-designed tour de force built in the "'campanile' [Italian villa] style," of Connecticut brownstone—the "handsomest structure of the kind in the country"—was "progressing toward completion."[14]

Unlike today, when progress is reduced to such misleading and reductionist statistics as the gross domestic product or so many employed, Hartford's Victorian-era civic leaders recognized that civic amenities were the key ingredient of their community's reputation, and they added amenities that would "attract inhabitants of wealth and culture," while caring, within limits, for the helpless and needy. Twenty years into its industrial boom, Hartford was described as "the richest city of its size in the United States," not only because of its "fine agricultural district... railroad center... banking and insurance" industries but also because of its "attractive homes... galleries and museums, handsome parks... cheerful streets... shops, factories, and great commercial edifices."[15] In 1847 Edwin Wesson decided to relocate his rifle factory from rural Northboro, Massachusetts, to Hartford, having endured months of lobbying by advocates of the new Windsor Locks industrial district on the Enfield Canal. His decision was based not only on Hartford's "superior advantages" for metalwork manufacturing but also the new public high school, which he described as "a grand thing for the city," and of course, his prospective employees.[16]

Hartford was not a naturally congenial environment for a hip-shooting soloist who sneered openly at the plodding, consensus-building process by which most of the city's businesses and all of its institutions and amenities had been created. From the Colt rampant, prancing on the globe atop a Russian-inspired Moorish dome, to the constant wrestling with city fathers, Colt's flamboyant and aggressive tactics went down like a lump of coal with Hartford's old guard, among whom his parents and especially his grandfather Caldwell had once been esteemed members. "Showing them" could hardly be described as a reasonable business strategy. But once Colt's arms factory gained momentum, what better thing to do? Was locating an enterprise zone on a floodplain any more preposterous than creating one of the world's largest cities (Los Angeles) where everyone expects an earthquake? Had others invested in Colt's dream and had he consummated his promised gift of the park and technical college, it is hard to imagine how Hartford would not now be richer in many ways for it. Sadly, both Colt and the city ended up losing much because of their mutual animosity.

Looking back, it is difficult to judge the situation. The worst that can be said of the city is that its merchant-traders and financial elite, and the Whig politicians they controlled, were uncompromising with a man who sneered at them, flaunted his worldly associations, and worse yet, scorned insurance and refused to offer them a stake in his company. The worst that can be said of Colt is that he willfully put the city at risk by bullying it into a stake in a risky development proposition in hopes of generating offsetting tax revenue. For a man who bragged of "paddling his own canoe," Sam Colt drove a hard bargain with the city, using his cronies at the Democratic *Hartford Daily Times* to bait and bully his adversaries and threatening repeatedly to stop work and pull up stakes when he did not get his way.

Word that Colt was acquiring property in the South Meadows leaked as early as March 1852.[17] It is now impossible to untangle the ugly maze of actions and accusations that put Colt on a war footing with his landlord and soon-to-be-adversary Col. Solomon Porter. Their feud almost capsized the plans for Colt's Armory, ended with Porter losing his position as president of the State

Bank he helped found, and led to years of bitterness and recrimination. Indeed, the Porter-Colt relationship is a perfect match for the Hartford-Coltsville relationship in which old versus new, Whig versus Democrat, tradition versus innovation, and steady management versus brinkmanship collided.

What a strange game of cat and mouse was played by these two powerful men. Porter, thirty years Colt's senior—a man of and from the Revolutionary era—was described at the time of his death as "one of the last of the old Merchants of Hartford, who flourished previous to the war with Great Britain... [and was] one of the originators of the State Bank."[18] Although, in the end, not nearly as rich as Colt, his estate, valued at nearly $555,000, was one of the city's largest.[19] He was a man who, unlike Colt, made his money the old-fashioned way—he married it and inherited it.[20] His real estate investments included the actual deed for Hartford's city hall, a small empire in commercial farm land, the Porter Manufacturing Company building on Commerce Street, where for five years Colt was the primary tenant, and a mansion with gardens—then the most contemporary and ostentatious in Hartford. It was the house—later valued at $177,000—located near the northwest corner of the South Meadows that would prompt Colt to build a palace next door, dwarfing it in size and spectacle. Envy and emulation make strange bedfellows.[21] Porter's house, with its acres of curving picturesque gardens—Hartford's first—was not only the source of Colt's envy but also an indication that both landlord and tenant, old money and new, had their eye on the same piece of cheese, Hartford's South Meadows.

During the summer of 1853, at the height of the Colt-Porter feud, Hartford's two dominant newspapers poked and jabbed at each other's editorials. With the city council equivocating, Colt threatened to abandon the project and leave Hartford if Porter did not yield.[22] Describing as "extortion" Porter's holding out for $50,000 ($3.2 million today) for a final parcel of undeveloped farmland and his refusal to sell "unless I would... buy his house and upland... and [drop] my lawsuit against the Porter

[Manufacturing] Co.," Colt played on the sympathies of those voters and councilmen who shared the Hartford *Times's* disdain for "the jealous, the envious, and the mischief-making, who ever hang upon a city, like the shadow of evil."[23]

Porter quickly defended himself by insisting that "I purchased... land... in the South Meadows... a year before Mr. Colt conceived the idea... [which I am now] asked to part with... to enable Mr. Colt to carry out his 'scheme of private speculation.'... I named a price... below... that... Mr. Colt has paid for other meadow lands.... [The] offer was rejected." Porter accused *Mr.* Colt of making "a grossly abusive attack... upon me and a member of my family," ridiculed his "visions of an unbuilt city," and sneered that "building cities is not the work of a day." "Mr." Colt? Since May 1850, when Gov. Thomas Seymour commissioned Sam Colt an "Aide-de-Camp to the Captain General of the State of Connecticut, with the rank of Lieutenant Colonel," Colt's friends had called him "Colonel." It was getting mean.[24]

Politics was the backdrop for the conflict that engulfed Colt's South Meadows improvements, and it was with the sense of barbarians at the gate that Solomon Porter waged war against Sam Colt. The year 1853 brought one of the most tumultuous upsets in Connecticut's political history. The state's legendary Standing Order—the multigenerational dynasty of interlocking, intermarried political leaders that ruled Connecticut as a one-party dynasty for more than a century—was dumped more decisively than ever before.[25] Having secured the governorship and small majorities in the general assembly in 1850, Colt's friend and Democratic standard-bearer Gov. Thomas Seymour led the party to its greatest victory when, in 1853, the Democrats retained the governorship by the widest margin yet, gained a two-to-one majority in the House and a four-to-one majority in the state senate, and placed another Colt crony, William Hamersley, in the Hartford mayor's office.[26] It was a full house, and Colt's Democratic friends were now almost punch-drunk with confidence. It was the Standing Order's darkest hour, and yet socially and economically, they still controlled the dominant church, the charitable organizations, the banks and insurance companies, and most of the city's leasable commercial property.

The record may never show the full extent of gifts made, favors done, or coercion exercised by Colt on behalf of his Democratic cronies.[27] But as his empire rose, he devoted more and more time to pet projects, and no pet was dearer to him than the principles and policies of the Democratic Party. Fervently libertarian, egalitarian, and expansionist, the Democrats provided the ideal platform for a rising industrialist. Resentment of entrenched old money, a welcoming hand to immigrants, a live-and-let-live attitude about differing cultures (so long as they could vote), and an insatiable appetite for growth gained the Democrats a huge following in the urban and industrial North, which had much to gain by western expansion and stable relations with the South. German and Irish immigrants flocked to the Democratic Party, which repealed the poll tax, a requirement that had long reeked of class privilege.

The Colt-Porter feud compromised the long-term viability of Coltsville and cost the city a bequest that, had it been made, would have secured for Hartford one of the nation's most prestigious technical colleges. As the *Hartford Daily Times* opined during the winter of 1853 when it appeared that Porter had succeeded in driving Colt away, "Hartford has received some hard blows in its day, but this one, which stopped Col. Colt's improvements... [is] the severest of all."[28]

Alfred E. Burr's *Hartford Daily Times* was Colonel Colt's personal public relations department, riding the issues and hurling epithets at everyone who stood in the way of "progress." As champions of industrialization, the *Times* editorialized about how "manufacturing establishments have done more to increase the price of real estate here, and to give a fresh start to all kinds of business, than all other causes combined."[29] When, in 1851, Colt announced that he was expanding and "intends to employ 150 workmen... almost every one... from abroad," the *Times* applauded, and in 1850 raved that "of all our importations [immigrants from Europe are] by far the most profitable kind." In August 1852, six months into Colt's program of land acquisition, with Colt and Porter already on a war-footing, the *Times* revealed that "Col. Sam Colt purchased valuable building lots in New York last Spring, with the intention of putting up a very large armory there... [and] will not consent to any further embarrassment." The following winter, the *Times* picked the afterglow of the Armory Ball, a citywide celebration hosted by "workmen in the employ of Col. Colt," to applaud Colt's Fire-Arms' continued growth, to honor the groundbreaking for Sharps' Rifle Company, and to announce that the failure to "procure a few acres of cheap meadow land [had] stopped Col. Colt's improvements," thus causing the city to lose "advantages which would have added a few thousands to its population."[30]

In addition to Solomon Porter's South Meadows land, Colt sought a tax abatement for his improvements and a reimbursement, up to $150,000 (today $8.2 million), for building public roads through the compound. Although such requests are routinely extended today, the practice was not widespread in the 1850s, and certainly a request of this magnitude had never been granted in Hartford. Colt's negotiations hinged, in part, on the Democrats' success at the polls, conspicuously coincident with Colt's development.

In August 1853, with the construction season now wasting in idleness, Colt opened fire on multiple fronts, determined to blast through the resistance and proceed. On August 15 Colt spoke for himself through the *Times* and explained that

more than two years ago I commenced to purchase lands in the South Meadows, for the purpose of enlarging my manufacturing operations, and increasing the amount of productive property in the city.... My purpose was soon suspected... [and] persons became purchasers of property so situated as to be indispensable to me... with the intention either of compelling me to abandon my plans, or to pay them exorbitant profits.... Mr. Porter has been applied to, to sell this land.... [He has] always refused... unless I would release him from a large claim... for breach of contract.... Porter protests... because he would be deprived of the overflow of the river... I do not desire to injure Mr. Porter... [and] will purchase his lands for more than it cost him.... I now... offer... to assume the entire responsibility for constructing and keeping in repair all the streets... provided the city will release me from... taxes... except at the rate now assessed on meadow lands.... I hoped to have done the City and the State some service by the expenditure of a large sum [and]... I expected... a benefit... commensurate with the outlay and risk.... If I... differ with... Hartford... I... hope that in some other place... I may be able to execute my plans. Sam Colt.[31]

Blaming the *Hartford Daily Courant* for the "continued din [that] is sounded in the ears of Councilmen and others... [about the] expense [to] which the city *may* be subjected," the *Times* urged the council to trust that Colt's "own great stake in the affair, binds him to a faithful performance."[32]

On August 26 Hartford's Court of Common Council passed resolutions "authorizing the chairman of the Highway Committee to contract with Col. Colt for building... highways in the South Meadows."[33] It refused to grant his request for a tax abatement. That September, with the U.S. secretary of war in town toasting Colt's achievements, Colt addressed Hartford's voters personally, threatening to abandon the whole project and sell his land if not granted the abatement.[34]

Debate intensified in October, kicked off by a *Times* editorial urging the city to act on Colt's contracts and grant the tax abatement. The next day a "resolution authorizing a contract with Col. Colt relative to the abatement of taxes" met loud opposition from opponents who branded the request "huge and monstrous" and the idea that Colt could build the proposed dike and reclaim two hundred acres of floodplain "impractical." Again, the council stalled. Now on the attack and almost in a panic at the thought of losing Colt's Armory, the *Times* predicted that "the day is not far off when we can... empty this Council of its present load" and presented a petition on Colt's behalf bearing 660

names of "intelligence and respectability," including "merchants, manufacturers, mechanics, officers of institutions, our oldest and best citizens."[35]

Four days later Hartford witnessed one of the most spectacular public rallies in its history as "probably 2,000" residents and voters assembled to protest the city's continued resistance to Colt's demands. The *Times* beamed as it described "one of the calmest and most deliberate public meetings" in which there was "a united and determined voice approving of the great enterprise of Col. Colt."[36] A committee presided over by former governor Joseph Trumbull, with former mayor Phillip Ripley and Travelers Insurance Company founder James G. Batterson as vice-presidents and *Times* editor Alfred Burr as secretary, read a letter addressed to the crowd by Sam Colt: "You may... approve of the resolution rejected by the Court of Common Council.... Should such be the result... I will... commence the South Meadow Improvements immediately.... Hartford is the city of my nativity. In her permanent prosperity I have ever felt an abiding interest. At home and abroad... it has been my pride to boast of the sterling character of her citizens, and her proud position as a New England City.... In any enterprise of improvement in which I embark I shall never abuse your confidence."[37]

What a show! The crowd roared in acclamation and Isaac Stuart—the high priest of Hartford's history and civic identity—read a benediction followed by former mayor Ripley who proclaimed, "It is no new thing... to exempt property from taxation where the public are to be benefitted.... Our city cannot thrive or increase in population by trade alone—it must depend on manufacturers."[38] Again the crowd roared. By November, rebuffed and disgusted "after a terrible indignation" in which he "threaten[ed] the city dignitaries," Colt simply "decided to go on" without the abatement.[39]

One of the most remarkable aspects of Colt's South Meadows improvements, which left R.G. Dun's field detectives in awe, was that Colt appeared to be financing the expansion out of cash flow.[40] Already incredulous at

60. The City of Hartford, 1877. Detail from a print, D.H. Bailey and Co,, Boston, 1877, showing Colt's Armory complex very much as it appeared twenty years earlier. (Courtesy of the Connecticut Historical Society.)

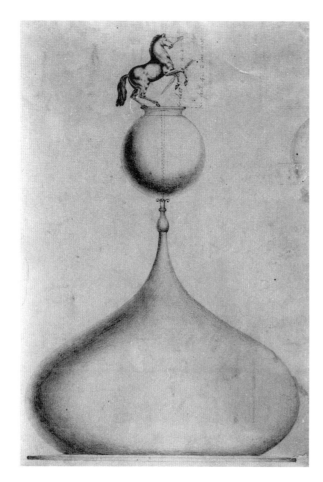

61. *Colt's Armory Dome.*
Pencil sketch, possibly
drawn by Octavius Jordan
(1825–?), ca. 1855.
(Courtesy of the
Connecticut State
Archives, Connecticut
State Library.)

the notion of reclaiming Hartford's floodplain, the city's bankers were probably even more perplexed and annoyed when he did not require their services.

In their determination to credit architects with buildings, architectural historians often fail to fully grasp the fluid character of client-architect relations. Sam Colt was almost certainly aware of the flap, in the 1850s, when Britain's Prince Albert hired not an architect, but building contractors and engineers to build Osborne House, the royal retreat on the Isle of Wight. Given his overwhelming pride in authorship, it is inconceivable that Colt did not personally dictate most of the physical and artistic dimensions of the armory building, widely recognized, then and now, as one of the most distinctive factory buildings erected in nineteenth-century America. Colt's Armory (fig. 60), described at the time as "a factory, in size and massiveness, such as Titan might have dreamed of," was Colt's dream.[41]

During the winter of 1855, Colt hired Hartford architect Octavius Jordan to work up plans.[42] Jordan, a British-trained architect, first established himself in "the Rooms lately occupied by [Gervase] Wheeler" during the winter of 1850 and soon secured a commission to design Allyn House, midcentury Hartford's most celebrated hotel.[43] Jordan later became the first Hartford-based architect to develop a national clientele, boasting in 1864 of the "large number" of "public and private buildings" he had designed and erected "in many parts of the United States."[44] Jordan worked up plans for Colt's concept of the armory's most spectacular feature, its Russian-style onion dome surmounted by a bronze rampant colt dancing on the world (fig. 61). As much a technical tour de force as a work of art, credit for its fabrication unfortunately cannot be established.[45] Having been presented at the Court of Czar Nicholas I in 1854, Colt flaunted his Russian connections by choosing a Russian theme for the armory's design.

Contractors, reading like a who's who of Hartford's building trades—Hiram Bissell, brick masonry, James G.

62. *Colt's Armory from* *warehouse* (right);
the North, ca. 1861. *new construction of the*
Probably R.S. DeLamater, *West Armory can also*
Hartford, photographer; *be seen to the right.*
showing the armory *(Courtesy of the*
behind Germania House *Connecticut State*
(center) and the tobacco *Archives, Connecticut*
State Library.)

Batterson, stonework, Henry J. Huxham and William L. Wright, painting and graining, and Henry Burgess and Son, woodwork—are among the dozens of names that turn up on the various, but incomplete, accounts of the armory construction.[46] On the payroll was Henry A.G. Pomeroy, an "engineer," described as Colt's nephew, who was paid the equivalent today of almost one hundred thousand dollars per year for a range of unspecified services, including the general contracting for both the armory and the Colts' new house, Armsmear.[47] Pomeroy oversaw building construction while William C. Hicks, Hartford's first city engineer, oversaw the construction of the roads, water systems, and the dike.[48] Descriptions of men—up to a hundred or more—living in "shanties" by the river, suggest a scale of operations unsurpassed by even the greatest of public building projects.[49] Colt's Armory was nearly completed and ready for the installation of machinery by the summer of 1855. It was the centerpiece of an industrial compound that eventually included a detached three-story office with an ornamental clock tower, an office boiler house and chimney stack, an office carriage shed, a brick blacksmith's shop, watchman's cottages, a proving house for testing, an armory boiler, annealing shop, joiner's shop, "tobacco warehouse," and a shed for carts (fig. 62).[50]

No sooner had the work begun in the winter of 1854, when Colt faced a trial of near biblical proportions.

Noah's Ark came to mind as the spring snow melt, combined with sixty-six hours of driving rains, raised the Connecticut River to the highest flood level in two centuries of recorded history. The Great Flood of 1854, "the Greatest Storm and Flood of the Century,... tore away... stone and brick... moving a large barn... 100 feet," flooding Pearl Street half a mile from the river banks, west into the city. Commerce Street and Front Street, "now under water," created a "sublime spectacle" as Colt's engineer and "about one hundred" workmen labored around the clock in vain to prevent large portions of the "foundation for the dike" from making "its departure [downriver] for Middletown."[51]

Hartford's nascent manufacturing district was devastated. Fales and Gray's railroad car factory was wiped out. Steamboiler manufacturers Woodruff and Beach sustained heavy damage, and the entire first floor of Porter Manufacturing Company's Commerce Street factory, the center of Colt's existing operations, was underwater and "all the machinery... submerged." Trying to paint a happy face on the disaster, Colt's friends at the *Hartford Times* suggested that, although the flood had "tested to some degree the dike that Col. Colt is building," the embankment, what was left, was in "good condition."[52] Perhaps.

Perched and ready, Colt's enemies struck, writing pseudonymously as "Tax Payer" about the frivolity of

Colt's whole South Meadows campaign. Emotionally drained, Colt summoned up his deepest reserves in assuring anyone that cared that he would not be defeated:

It is not very generous in [the] "Tax Payer" to take an opportunity for attacking me when all my attention is required to protect my property from the destructive flood which is upon us; but I am not yet discouraged, even by the combined forces of the land and water.... When any citizen is willing to incur the risk of contending with the river, such people as the "Tax Payer" do all they can to defeat him. Now, sir, I am not afraid to face the music, or create it; and if the city of Hartford will agree to relieve my property from increased taxation, I will bind myself to exclude the river from the South Meadows; and if the city of Hartford will pay for it, I will agree to dike the Connecticut River from end to end of the city, so that nothing less than Noah's flood can reach the houses which are now inundated.[53]

"I will bind myself to exclude the river," "bind myself to exclude the river," "to exclude the river," echoing, with clenched fist, twenty years of angry perseverance. Colt had defied nature and even now—especially now—was "not afraid to face the music" in the face of nature's revenge. With his machinery under six feet of mud and silt and his workmen paddling boats across the South Meadows acres, Colt dispatched Hartford artist Joseph Ropes to the tower of Center Church to survey the disaster and compile photographs and sketches for a grand panoramic view, a portrait of a community under siege

(fig. 63).[54] In a bizarre act of martial symbolism, the Colonel next rounded up the First Company Governor's Horse Guard, which he and his Democratic cronies had recently usurped, and mounted a full dress drill.[55] With the flood now at high-water mark and Hartford's new Democratic elite parading around State House Square in their handsomely tailored uniforms, Colt boldly determined to battle nature to total victory.

A year later, with Colt's dike now completed and the new armory nearly ready to receive its machinery, another spring flood brought vindication, which Colt and his allies interpreted as a sign of God's favor. It was a moment to savor as Colt's factory worked without interruption, the South Meadows compound "as dry as any meadow where the flood never reaches," while "hundreds of houses in the lower part of the city... [were] flooded" by waters rising almost twenty-two feet above the low-water mark.[56] Colt's friend Lydia Sigourney dashed off an epic verse on the Connecticut River and Col. Colt's embankment, before writing their mutual friend former governor Thomas Seymour to report the triumph of Colt's embankment.[57] The *Times* announced that Colt's dike had "converted that region into valuable property" and "proved to be fully adequate to the use." It was only years later that they confessed that "ever since Colt's dike was built, there has been a leak in the southwest corner," a leak that broke through in 1862 and again in 1869, flooding several acres of Coltsville. Forty

years after work began there was still concern that the dike was "considered unsafe," a cause as important as Colt's premature death for Coltsville's failure to attract the level of population and business envisioned by its founder.[58]

The mid-1850s, however, were years of triumphant celebration, with Colt's friends openly boasting that "no combination of fifty men in our city have done so much to promote the prosperity of Hartford" and of how "one such man in a community is worth a thousand drones who hoard their money, invest it abroad, and vote against everything... calculated to push the city ahead."[59] With employment surging, Sam Colt personally paid more taxes than the three next largest taxpayers—David Watkinson, James Goodwin, and Solomon Porter—combined.[60] Even his enemies began to concede that something monumental had begun.

New amenities were added to Coltsville each of the remaining years of Colt's life. Eventually almost forty duplex and triplex units of worker housing (fig. 64) and a large boardinghouse were built to accommodate an estimated 145 families. In 1857 Colt laid "a track of Rails... to connect the South Meadows with the New Haven and Providence [rail]roads" and built Vredendale Dock, a commercial port facility "for landing coal, iron, [and] timber" (fig. 65), with a harbor on the banks of the Connecticut River "capable of protecting fifty to one hundred vessels through the winter."[61] By 1858, Engine Co.

No. 4, which after the great fire of 1864 was renamed Colt's Fire Engine Company, served the South Meadows community.[62] That year also saw the completion of a modest railroad depot that cost the equivalent of sixty-four thousand dollars to build.[63] In 1859 Colt launched the Union Ferry Company, with a grand new steamboat that ferried between Hartford and East Hartford, where Colt operated a hobby farm on about fifty acres.[64]

The most expansive of Colt's pet enterprises was Colt's Willow-Ware Manufacturing Co. (fig. 66) and its Potsdam Village, the German workers' housing in the south end of Coltsville. Henry Barnard explained that while "imitating the earthen dikes of Holland, Col. Colt was naturally led to borrow also their protective willows, and thus became the first planter on a grand scale of European willows in America." The willows, a French osier esteemed for its fast growth and deep root system, was imported to fortify the walls of the embankment. By 1858 the trees produced a crop of shoots of such commercial value that "a stranger appeared and wished to buy at a high price."[65] Founded in 1859, as much for pride and propaganda as for profit, the willow factory became the centerpiece of Colt's program to promote German immigration and to enhance the Germanic character of Coltsville.

Colt's association with German gunsmiths and mechanics had begun as early as 1847, when he recruited Augustus Fiege and Frederick Kunkle from Eli

63. *View of Hartford to the South, 1855. Detail of oil on panel, Joseph Ropes (1812–55), showing the Connecticut River's floodwaters on the South Meadows, where Sam Colt would build his armory complex (WA #05.54).*

64. *View of Colt's West Armory and Workers' Housing, ca. 1870. Detail from print, Prescott and White, Hartford, photographer; this picture may have been taken from an aerial balloon. Notice Colt's Cartridge Works in the upper-right corner of the view.*

65. *Vredendale Dock, ca. 1868. Probably R.S. DeLamater, Hartford, photographer.*

Whitney Jr.'s Whitneyville Armory. Together with Samuel Knous and his sons, the artist-engravers Carl Helfrecht, Alexander Birkholz, Herman Bodenstein, and Gustave Jung (anglicized to Young), and the artist–ivory-carver Christian Dehyle, Kunkle and Fiege occupied positions of considerable prestige and influence, not only within Hartford's nascent German community but also as two of Colt's first and most valued artisan-contractors. Kunkle, a Prussian-born gunsmith about the same age as Colt, became one of Colt's most trusted aides. In 1860 he was dispatched "to Prussia to find workmen." He was one of the first residents of Potsdam Village and eventually served as supervisor of the willow factory.[66] As symbols of upward mobility, it is noteworthy that both Fiege and Kunkle became rich.[67] Both men epitomized Coltsville's rising Germanic subculture, while serving as intermediaries between Colt and a work force that, by 1860, was about one-third comprised of German immigrants. Indeed, it may have been Kunkle who suggested the manufacture of wicker furniture and baskets from the shoots of the willow trees, with which he was associated from its first mention in the company accounts in December 1858.[68]

Inspired by the image of German basket weavers plying a traditional craft inside the walls of his industrial compound and with an eye to acquiring appropriately skilled labor at less cost, Colt corresponded with his Berlin agent C.F. Wappenhaus about recruiting workmen from the basket-weaving districts of Prussia. Convinced that such workers could only be recruited en masse and then only if the character and form of their village life was preserved, Colt gathered descriptions of a weaver's village near Potsdam, Germany, and decided to replicate its housing and cultural amenities in Coltsville.[69] In 1859 Hartford watched, slack-jawed and amazed, as the first of the "Swiss cottages" was erected in Potsdam Village in a style peculiar to "an American eye," the structure seemingly "turned wrongside out," with exposed exterior staircases and widely overhanging eves (fig. 67).[70] A year later Colt's willow workers converged from a dozen states and nations, speaking at least six languages.[71] For a city dominated by the descendants of English Puritans, many with 225-year-old roots in the community, Hartford found itself unceremoniously plopped into the front seat of the nation's roller-coaster ride with ethnic diversity.

By the time of Colt's death in 1862, Potsdam Village had become a flourishing subculture within Coltsville, complete with livestock, vineyards and gardens, a beer hall, "twelve elegant cottages," and a boardinghouse for "the families of persons employed in the gas, ozier, and pistol manufactories."[72] Located a mile south of Colt's Armory, beyond the insistent pounding of the drop-hammers and the steady whirl and thump of the steam

66. *Col. Colt's Willow-Ware Manufactory. Photograph of Potsdam Village, ca. 1860. Probably R.S. DeLamater, Hartford, photographer.*

engine and belt-shafting, Potsdam Village was a little slice of old-world Europe on the fringe of the new world's most modern factory. It was the microcosmic heart of Coltsville, a symbol of the harmony between old world and new, Europeans and Americans, Catholics, Protestants, and in its second superintendent, Leopold Simon, a Jew.[73] With the willow trees swaying in the breeze, the bleating of goats and chickens, Colt's Alderney cattle grazing in the meadows beyond, and the steady gentle rotation of Colt's Dutch windmill situated nearby on the south corner of the dike overlooking the Connecticut River, Potsdam Village provided a pacific and old-world contrast to the brutally efficient and warlike modernity of Colt's Armory.[74]

Potsdam Village represented a powerful fantasy, empowered by and divorced from the context of its making. It hinted at Colt's ambivalence about industrialization, while welcoming people of diverse cultures to Colt's South Meadows boot camp of the American dream. Potsdam Village embodied Colt's obsession with reconciling the tension between the romantic agrarian ideal and the streamlined, machine-based system of manufacturing that was destroying it. The man who quit the Congregational Church in anger when it modernized its traditional box pews, who made hand-engraved ornament the signature of his machine-made products, and who applied a veneer of history, art, and agriculture onto his machine-age factory village was clearly a man of striking contradictions.

Although a witness noted that the "willow factory was [Colt's] pet," he "caring little whether or not it paid a profit," by inventing specialized machinery and systematizing the processes of production and assembly Colt's Willow-Ware Manufacturing Co. actually made money, achieving the lowest manufacturing costs in the industry and rapidly gaining a market throughout the South and West.[75] The founding of the company was well timed to succeed in the market. Wicker furniture, made of willow and rattan, was first imported into the United States on a large scale from Asia and Europe during the 1840s, with demand taking off during the 1850s as American manufacturers such as Michael Topf, and J. and C. Barrian and Co. in New York and the American Rattan Company and the Wakefield Rattan Company near Boston began mass producing a wide variety of forms and styles, promoted as ideal for summer living and suburban houses.[76] It was Barrian's product line and styles that Colt's Willow-Ware most closely resembled.

By 1864 the willow factory maintained an inventory, valued at the equivalent of more than $1 million, of almost one hundred different products, including chairs, workstands, hampers, office baskets, clothes baskets, cab wagon bodies, cradle bodies, tête-à-têtes, sofas, tables, and chair seats (fig. 68 and 69).[77] The analysts from R.G. Dun credit-rating service declared its "business prospect good" and credit "excellent," and while prone to exaggeration, Colt's father-in-law also predicted that, in 1860, Colt's willow factory would "derive a profit... of $10,000

67. Detail of Swiss Cottages in Potsdam Village, Coltsville, ca. 1851. In addition to one- and two-family residences, Colt built boarding-houses for the armory, willow works, and gas factory workmen.

[the equivalent of $540,000]," suggesting that even while doing good, Colt was able to do well.[78]

Sam Colt's last major building campaign—his family claimed the pressure of it is what killed him—involved doubling the size of the armory to make room for the manufacture of contract U.S. muskets at the outbreak of the Civil War. "Within ten days after the fall of Sumter," Colt had "dispatched an agent to England for procuring machinery."[79] By May 1861 he had proposed the addition and by June, with architectural plans in hand, set out "to double the capacity of his immense works... [and add] twenty more brick dwellings and fifty more frame ones" to the existing stock of workers' housing.[80] When he died seven months later, the work was well underway (fig.70).

Colt dreamed of an expansive industrial and commercial empire that, had he lived, might have made Hartford "the principal center in the United States of manufacturing ordnance"—the "Essen [Germany's industrial hub] of America"—complete with "foundries, forges and workshops to produce everything necessary to equip an army, or navy, or fortress."[81] As Henry Barnard reflected, "the more self-contained, the more... answer-

68. *Willow products made at Colt's Willow-Ware Manufactory, ca. 1860. Probably R.S. DeLamater, Hartford, photographer.*

69. *Settee, ca. 1860. Made by Colt's Willow-Ware Manufactory; a pair of these and a willow table furnished the upper story of the tower at Armsmear. Demand for Colt's willow furniture was brisk in Cuba and the South (WA #05.1603).*

able to the ideal of its founder."[82] Colt increased the self-reliance of his armory by dabbling, without great success, in the commercial manufacture of machinery, wicker furniture, and gun cartridges. In 1857 he built a stylish quasi-domestic Gothic dwelling (fig. 71) for Colt's Cartridge Works, where young single women carried out the dangerous work of loading gunpowder into foil cylinders "about half a mile south of the armory."[83] Colt's farmer's market and "tobacco warehouse" were the center of a profitable tobacco farming operation, built not only to capitalize on the decade's tobacco boom but also to help create an agrarian atmosphere. In 1857 Colt built a brick workshop on the South Meadow's north border facing Hartford, in hopes of attracting "industrious... mechanics... [and] other factories." It soon housed a "large Sash, Blind and Box factory."[84] However, after a decade of floods and controversy, Hartford's industrial development increasingly took to the high ground of the city's west end, where, among others, Sharps' Rifle Company, Pratt and Whitney's machine shop, and the Hartford Cartridge Works were established and flourishing by 1861.[85] For the most part, Colt's Armory remained a solitary monument to its founder's vision.

Aside from the armory, Coltsville's most distinguished legacy was its founder's ambitious program of civic amenities. Despite his bombast and rhetoric, Sam Colt was not without a twinge of ambivalence about the source of his fortune. Raised in a Congregationalist household, Colt was a member of Hartford's First Church until 1853; he was married and buried as an Episcopalian.[86] Colt was a man who knew shame and responsibility, perhaps more than compassion, and it was with a strong moral imperative that he attempted to shape not just the economy but the moral environment of Coltsville.

Like others before him, Colt aspired not only to perfect the system of manufacturing—in this case guns—but also to perfect and create anew a middle landscape between the traditional agrarian and modern industrial states. Imbued with a millennial vision of the impending perfectibility of human institutions, Colt acted out the perfectionist aspirations of his age. As a matter of patriotic pride, Colt and his industrialist peers were determined to avoid replicating "the 'degraded' condition of the European working class," with its low wages, filthy housing, pitiful diet, depraved character, and high crime.[87] Sam Colt is credited with having written,

70. Colt's Armory from the Southwest, ca. 1861. Probably R.S. DeLamater, Hartford, photographer. The nearly completed West Armory appears to the left of the smokestack. The small cottage on the right is a watchman's house, and the forging shop is the building to the left of the cottage. (Courtesy of the Connecticut Historical Society.)

"Money is trash I have always looked down upon.... Your great study should be man—lose no oppertunaty *[sic]* to mingle with the mass[es] & view nature in its most primitive state."[88] Had he lived to reflect on his life, Sam Colt would surely have claimed his effort to create a worker's paradise among his proudest achievements.

Here, in the shadow of Connecticut's righteous Charter Oak, on meadows reclaimed from the assaults of nature, Colt began to consummate his dream for the ideal industrial town, a dream that included parks and fountains, landscaped gardens and farm lands, a museum, church, and school, a place where art, history, and work converged in a spectacle that would elevate "the labor of this Valley upon the platform of intelligence."[89] Today, when history is mostly segregated in the catacombs of academia, it is almost hard to imagine a campaign of civic and industrial development so intentionally charged with historical associations. It was Colt's purpose to keep visitors and residents of Coltsville ever mindful of the historical foundation on which his industrial empire was built. From Charter Oak Hall to Vredendale Dock, Coltsville was to be a living shrine to the legends of Hartford's history. Buildings and streets bearing the names of the great leaders of the Indian, Dutch, and English were to portray the site as an epic drama culminating in its "modern designation as Colt's Armory." Sequassen, Wawarme, Masseek, and Curcombe of the native tribes; Huyshope, Vandyke, Vredendale, Van Block, and Hendricxsen, of the Dutch; and Haynes, Hopkins, and Wyllys of the English signified the place of Colt's empire in the heroic pageant of progress.[90]

The centerpiece of Colt's industrial compound was Charter Oak Hall (fig. 72), a hundred-foot-long, four-story brick building at the north end of Coltsville "devoted to purposes of moral, intellectual, and artistic culture," founded to "marry the forge and work shop to the Reading Room and the Dancing Hall," and dedicated to "the sovereignty of labor."[91] Colt spared no expense in making Charter Oak Hall one of the grandest and most visionary of the village institutes, which historians have described as the "ultimate visual achievement" of America's nineteenth-century factory communities.[92]

Inside Charter Oak Hall were murals "made to express... lessons of history, and of material improvement," one illustrating Hartford as it was in 1640, a second showing the South Meadows as it was when

INTERIOR OF CARTRIDGE WORKS.

71. Interior of Cartridge Works. Engraved by Nathaniel Orr; illustration from the United States Magazine, *March 1857.*

purchased by Colt, and a third panel, still not completed, that would show Coltsville in its perfected state. Plans included a "*reading room,* where newspapers and periodicals of an instructive character are to be collected for quiet perusal in hours not devoted to mechanical labor." It is to be the site of "*lectures...* upon science and art... philosophy and morals, or upon any topic where the purpose shall be to communicate useful knowledge." It also included "a room for *discussion* or *debate...* a room... for the display of such interesting and improving curiosities and pictures as the good judgement of its projector may... select and appropriate." Among the intended elements was Colt's museum of firearms. Finally, there was to be a room "where parties may assemble" at Thanksgiving, Christmas Eve, George Washington's birthday, election day, the Fourth of July, or for fairs and concerts.[93]

Not much of this idealized vision came to pass. Colt alluded to a museum of firearms as early as 1856, and a large collection was gathered by 1861, but it was not developed or installed before his death and never achieved the pubic access he envisioned.[94] Colt's plans for civic amenities faltered when he found the community unable or unwilling to share in the cost of development and maintenance. By 1860, when it was advertised as "ready for dancing parties, offices or lodging... [with] five good stores with cellars... rented low to good tenants," Charter Oak Hall was already as much commercial as philanthropic in its purpose.[95]

Charter Oak Hall was described as "an acorn which he meant should grow [in]to an oak." Sam Colt intended to bequeath one-quarter of his estate so the hall would become the center of a school of mechanics and engineering in Hartford to "surpass the scientific schools... then rising... in Harvard and Yale."[96] It was to be established on the basis of five years' profits from the estate, at which time 30 percent of the fund was to be used to erect "schools, workshops and dwellings," 10 percent for the salaries of "teachers and professors," and 30 percent to acquire "books and treatises on the subject of mechanics & engineering," plus the "necessary tools, implements,

[and] instruments for giving instruction." The remaining 30 percent was to assist "needy and meritorious young men" in gaining "such an education as will enable them to become skillful practical mechanics and engineers," with preference given to "sons of men in my employ, next to inhabitants of Hartford, next to inhabitants of Connecticut, lastly such pupils as [the] trustees see fit to admit if they can pay." Colt's will further specified that "pupils shall dress in appropriate uniforms" and, after "reasonable allowance for healthful sports and recreation," should divide their time "equally between theoretical studies in mechanics & engineering & practical employment in the work shop."[97]

With working capital estimated at $875,000—five years' profits at $175,000 per year—Colt's technical college would have opened in 1867 with today's equivalent of $9.16 million for books, the same each for buildings and scholarships, and the income from one-third that amount to pay faculty salaries.[98] If we figure that additional income from fees and tuition might have tripled the operating budget, a college of 175 students and 15 faculty looks about right and not nearly as small as it sounds, considering that in 1869 the student enrollment at Yale was 518, and Yale's Sheffield Scientific (founded 1861) had only 139 enrolled.[99]

But the dream remained unfulfilled. After six years of relentless haggling with the city over taxes and permits, Colt revoked the bequest.[100] Instead of a great technical college, a South Meadows primary school was established in 1858 and paid rent to Charter Oak Hall until 1871, when its proprietors decided to expand and build elsewhere.[101] Charter Oak Hall also housed "Col. Cooley's grocery store," the armory store, which carried "dry goods, groceries, provisions, drugs & medicines, wooden ware, crockery, wines, liquors & c.," and a retail shop for Colt's willow-ware.[102] Coltsville's benevolent and religious groups met at the hall, including the South Meadows division of the Sons of Temperance, founded about 1861, and the Armory Sunday school, established in 1859, which was the predecessor to the Church of the

72. *Charter Oak Building. Lithograph of Charter Oak Hall, E.B. and E.C. Kellogg, Hartford, 1857. This multipurpose building functioned as a social and learning center for Coltsville.*

Good Shepherd.[103] Mostly, Charter Oak Hall was used for dress balls and as a practice space and headquarters for Colt's Armory Band and Colt's Armory Guard, Colt's two favorite philanthropies. Colt's Armory Band was the most flamboyant and colorful of Colt's philanthropic enterprises. The band, which was "formed among the workmen" and "endowed by Col. Colt... for the purpose of promoting musical art," in 1856, gave Coltsville a Germanic, martial atmosphere.

Until 1946 the band was one of Hartford's most esteemed civic trophies, earning a northeastern-wide reputation by performing at hundreds of community events and touring widely.[104] The 1850s was the golden age of brass bands and martial music. Bands like Colt's introduced a loud and insistent brand of European classical music into the provinces, while providing a vehicle for cultural expression among the nation's rising immigrant population. Colt's Armory Band was especially important in signaling the arrival of German-born immigrants who played a key role in carrying out Colt's revolution in high-tech machine-based manufacturing. With a gift from Colt of brass instruments "made in Boston... of German silver," sheet music provided through Colt's agent in Berlin, a uniform "selected by one of the workmen" and modeled after the uniform of the Prussian cavalry, with "a dark blue frock coat trimmed with green, light blue pantaloons, regulation caps with green pompons, gold epaulets and aiguilettes," and a helmet adorned with a rampant colt hovering over a crossed pair of pistols, the Armory Band personified Coltsville's Germanic flavor (fig. 73). At a German soiree in 1859, the band provided "music which some of the company said could not be excelled in 'Faderland.'"[105]

The band's significance was in part ideological. The American brass band movement burst on the scene in the same age and with the same spirit of Manifest Destiny that made Colt's Armory possible. Traditionalists branded the brass band as "street music" and "all bluster," suggesting that perhaps "the invention of new and deadlier implements of war, which came out about the same time,

73. Colt's Armory Band Member, ca. 1870. Band uniforms, specified by Sam Colt, were modeled after Prussian military uniforms.

had hardened men's hearts," banishing "all the softer companions [the woodwinds] of the savage science." John Sullivan Dwight, a Boston music critic and one of the most astute observers of the brass band movement, labeled brass bands as "military" rather than "civil" music, noting not only its warlike nature but its reliance on military patronage and uniforms. By 1862 there were an estimated 14,832 bandsmen enlisted in the Civil War, men who, in returning to civilian life, helped brass bands proliferate in small towns and cities throughout America.[106] Inherently democratic, the brass band became an instrument of pageantry and public relations and an expression of the martial spirit that prevailed in the leisure hours at Coltsville, but especially inside Colt's Armory, where the grand army of workmen were governed by a military-like code of discipline.[107]

With Thomas G. Adkins's arrival from New York in 1859, Colt's Armory Band moved beyond a standard repertoire of rudimentary marches and quicksteps to become one of Connecticut's premier purveyors of European operatic and orchestral music. Under Adkins the band introduced vocal accompaniment and per-

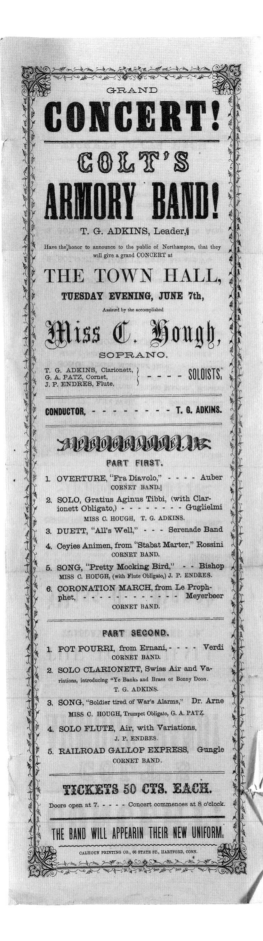

formed scores by Verdi, Meyerbeer, Schubert, and Strauss (fig. 74).[108] In 1859 they thrilled audiences with their performance of "Hen Convention" complete with "cackling, crowing, screeching," the "Railroad Express," in which instruments imitated "the starting, progress, and stopping of a train," and the "Anvil Chorus" from Verdi's 1853 opera *Il Trovatore*.[109]

Colt contributed the equivalent of about three hundred thousand dollars to the operation of the Armory Band in the years before his death.[110] Initially determined to control the band by demanding that "each member binds himself to put his & all other instruments, music & uniform in perfect order" and to "always get permission before... any public exhibition," by 1860 Colt had obviously changed tack and encouraged the band's independence.[111] The band began advertising its availability to "furnish music for parades, balls, parties, concerts, commencements & c."[112]

In addition to the Armory Band, Colt also founded Colt's Armory Guard in 1858, which had a meeting room in Charter Oak Hall, an expensive silk standard "painted with [the] arms of Col. Colt," and the then politically charged words E. Pluribus Unum.[113] Of course Colt's Guard was armed with the new model revolving rifles that it was founded, in part, to promote. In addition to its recreational value, the Armory Guard provided glamour and publicity for Colt's firearms, which it used in target shooting competitions as far away as New York City.[114] The guard's captains, George Chapman and J. Deane Alden, developed special drills to show off the use and handling of Colt's repeating rifles, and it was with great pride that the guard attended the "Third annual target excursion" in Middletown to compete for prizes in 1858.[115] Hartford looked on in bewilderment as Colt's Guard, puffed with martial pride and plumery, marched from Huyshope Avenue over the Front Street bridge through the industrial district along the Mill River to State House Square, where they performed a "dress parade... for the first time in this city." By 1860 Colt's Guard had earned a reputation for the "rapidity, accura-

74. Colt's Armory Band Broadside. Calhoun Printing Co., Hartford, for a performance given in Northampton, Mass., in 1859. Clara Hough, a well-known soprano, accompanied the band. (Private collection.)

cy, and mathematical exactness" of its drills and was described as the "finest in the State," a perfect prelude to the rifle regiment Colt organized at the outbreak of the Civil War.[116] During the war, the contractors, executives, and workmen at Colt's, Sharps', and Pratt and Whitney banded together to form a shooting association and target range that evolved into the Hartford Gun Club (1884).[117]

Together, Colt's Guard and Armory Band provided the veneer of culture, chivalry, and esprit that gave Coltsville its civic identity. Militaristic, pluralistic, fun-loving, and proud, Coltsville was the most foreign and, to some, the most threatening change imposed on Hartford by industrialization. In ten years, Hartford was transformed from a city of old-time Yankees to a city with numerous immigrants, from a city of bedrock political stability to an outpost of the rebellious democracy, from a city of decorous religion to a city in which armed soldiers, speaking broken English with thick accents, marched on the town square armed with quick-fire revolvers. This, more than just amassing a great fortune or making guns, was Sam Colt's legacy.

In keeping with his vision of a harmonious work environment, Colt made sure that leisure and entertainment were available in Coltsville. In addition to music and target shooting, the workers played baseball, informally as early as the 1850s, and by the late 1860s as an organized competitive club.[118] In 1858 there was an Armory Glee Club, and in 1870 the workers formed the Armory Dramatic Association, "composed of gentlemen employed at Colt's" whose "temperance drama 'The Reclaimed'" and comedy "The Loon of a Lover" were performed at Charter Oak Hall.[119] The armory's employees included a dozen or more highly skilled painters, sculptors, and engravers. Among others, William Hawley, a British craftsman who worked in Colt's burnishing and polishing department, painted one of the earliest-known views of Colt's Armory and in 1858 "completed an oil painting for Engine Co. No. 4."[120] Colt, who commissioned photographs of his machinery and workmen as early as 1858, eventually recruited Richard DeLamater as a staff photog-rapher. Among Colt's staff artisans, the German carvers and engravers Carl Halfrecht, Joseph Buckardt, and Christian Deyhle devoted leisure hours to making works of art that attracted critical acclaim. When the *Hartford Courant* reported, in 1864, on the "artistic talent... among the Germans, who... excel in finely executed work," it was not only their professional work but also the art they produced recreationally and semiprofessionally that attracted notice.[121]

Next to Colt's Armory, in the center of Coltsville, was Germania House, a German social club and mutual aid society that may have dispensed medical treatment to members of the German community.[122] Colt was Hartford's largest employer of German immigrant labor, and his workers clearly dominated the city's German community, which numbered about two hundred families by 1860 and included machinists, engineers, engravers, gunsmiths, pistolmakers, willow workers, ivory-carvers, blacksmiths, filers, cabinetmakers, upholsterers, foundryman, a watchmaker; Christian Popp, the German minister; Julius Busch, the artist and art instructor; and a Prussian "music professor."[123] Fidel Bubser, a gunsmith from Wurtenburg who joined Colt's in the late 1850s, eventually parlayed his savings into a restaurant and bowling alley on Mulberry Street, with a bar furnished with nineteen pictures, two stuffed goats, a Swiss clock, stuffed birds, stuffed animal heads, a beer pump and four faucets, and an inventory of three thousand cigars.[124] No subculture was more prominent or played a greater role in Colt's success, which he acknowledged by encouraging their cultural expression and subsidizing their cultural amenities.

As early as 1853, Hartford's German community sponsored an arts festival featuring a "brilliant concert" by the Liederkranz Society and two years later sponsored the "most extensive concert ever given in this city" with "two hundred of the best German singers from New York and Boston" followed by "a grand pic-nic."[125] In 1856 old Hartford was a little uneasy when "Mr. Leopold Bomberger... a German from Texas" visited Hartford to "address the German voters... in their own language," noting that it was the "first time... a stranger was... hired to come to Hartford to address a class of voters in a foreign language."[126] Colt's Armory Band provided the entertainment. The next year the community launched a German-language newspaper, the *Hartforder Zeitung* "for the New England States," and by 1859 had formed a "benevolent society of Germans known as the 'Bond of Friendship.'"[127] During the 1860s Hartford's Germans founded a German-language school, a rifle club, a gymnastic society known as Turnvereins, and beginning in 1869, sponsored the Connecticut Schutzenbunde, a festival that by 1871 attracted shooting societies and choral groups from New York, Boston, and Springfield, Massachusetts.[128]

The most memorable events in Coltsville brought Yankees, Irish, and German workers together with the rest of the Hartford community. Whether it was Sam Colt's lifelong delight in explosives or simply a product of the times, Hartford's Fourth of July celebrations during the 1850s have never been surpassed. In 1854, with the South Meadows and armory under construction, Hartford celebrated the Fourth with "40,000 glasses of hop beer," a "Balloon Ascension," and "a fine display of fire works" witnessed by twenty thousand, with seating for two thousand. Inaugurating a tradition of fireworks on the South Meadows that continued well into the twentieth century, it all took place on the grounds of Coltsville at Colt's expense, estimated at about one thousand dollars (today about fifty-five thousand dollars).[129]

The 1856 Fourth of July celebration was Hartford's greatest ever. A parade began with Hartford's military officers, followed by Colt's Armory Band and "13 Young Ladies personating 13 colonies," followed by other brass bands and fire companies, "Revolutionary Pensioners," the "orator of the day," and delegations from the Young Men's Institute, the Hartford Art Union, the Connecticut Historical Society, Trinity College, St. Patrick's Benevolent Society, and the German Liederkranz Society. Next came the manufacturers: Woodruff and Beach steam boiler

manufacturers, A.W. Roberts Fire Engine Manufacturing Co., Colt's, Sharps' Rifle Co., Webster and Son's pottery, and Batterson's Steam Marble Works, each outvying the other for the most spectacular display. Colt's Armory built a forty-foot-long "miniature house on wheels,... drawn by twelve horses" and "filled with stacks of arms," with "twenty-five mechanics" wearing "masks of the most grotesque and laughable character." Preceding Colt's Pistol Car were 150 marching workmen and on its top, a "colossal Pistol." The air crackled with gunfire as workmen from Sharps' Rifle Co., with a "grand Car" next in line, exchanged "volley after volley" with Colt's workmen. Also represented were Cheney's silk factory, milliners, bookbinders, printers, tailors, Rogers Brothers electroplaters, merchants, furniture dealers, and the "most brilliant car" of all, James Batterson's Steam Marble Works, displaying "the artistic chisel" of G. Argenti, Hartford's Italian sculptor-in-residence and the man responsible for first gaining Batterson's a national clientele.[130] What a carnival! With gunfire, booze, explosives, masked mechanics, music, and money floating down Main Street, Hartford celebrated not only the glorious Fourth but also its birth as an industrial powerhouse.

No less grand were the Mechanics Balls of the 1850s, a self-defining ritual of cultural expression that marked the emergence and acceptance of Hartford's new and prosperous industrial working class. Coinciding with the unprecedented triumphs of Connecticut's Democratic Party, these events were politically flavored celebrations of democracy and free enterprise, opportunities to rejoice in what was for many a first taste of political freedom and upward mobility. The Mechanics Balls were also a stage for exhibiting manners and decorum in a city where many among the old guard found "the brawny arms and dusky faces of the honest-hearted mechanics... offensive."[131] The Mechanics Ball of 1851 and the Armory Ball of 1853 were evenings of pride and splendor, with "dancing, quadrilles, waltzes and reels," conducted with "complete order and decorum.... Young men, full of health and in the vigor of early manhood, intelligent and independent," and "ladies" of "modest and exceedingly neat appearance,... as fine... as an equal number from any class" showcased the rising "industrious, producing classes, without whom," it was said, "we should have neither prosperity nor national greatness."[132]

Among the most poignant images of Coltsville were the moments when the social distance between the Colts and the workmen collapsed.[133] Never far removed physically, the Colts did what most modern corporate chieftains would find unimaginable, they invited the armory workmen into their home, within reach of the silver, glass, tufted furniture, and pictures. Sam and Elizabeth never looked more humane and earnest than during the parties they hosted for armory workmen and their children. In 1858, a year marred by the first major layoffs in Colt's history, Sam and Elizabeth "invited all who reside on the meadows up to his house... [for] refreshments" after fireworks and the day's company picnic. Later that year Elizabeth Colt presented the Armory Guard with its new banner, after which they "partook of the hospitality" at "Col. Colt's residence," where he toasted "the health and prosperity of mechanics" for being "as necessary to successful enterprise as the will and genius of the inventor."[134]

It was Colt's aristocratic bride, Elizabeth Hart Jarvis, who narrowed the gap between employee and employer and who was the creative force behind the Festival of the Armory Sunday School, an 1860 event long remembered by all in attendance. The "Colonel gave the children of his employees a sleigh ride." A sleigh, thirty feet long and decorated with evergreens, was specially built for the occasion and towed the school's 160 children and teachers behind fifteen mules. Afterward, inside the "elegant mansion," Mrs. Colt and her friends served the "children of working parents, of various nations, and differing religious denominations,... oysters, ice cream, ornamented cakes, oranges, coffee," and lemonade from a "magnificent china-bowl" in a room filled with "greenhouse flowers.... When they departed Elizabeth gave each... a cornucopia filled with a variety of sugar plums." This "happy set of little folks" departed, "singing sacred melodies."[135] Coltsville was emerging from the mind's eye to become a real place in the hearts of the real people who lived there.

Coltsville was, first and foremost, a place of work, and a workplace based on a military model of discipline. Adjustment was inevitably creaking and uneven. Influenced by the advancing and occasionally backsliding accomplishments of the superintendents of the U.S. Springfield Armory, Colt and his generation of American industrialists refined and perfected a workplace code of "military discipline," invoking words like "order," "system," "neatness," "military exactness," and "decorum" to describe their efforts.[136] Praising the "feeling of responsibility" that ran "through the whole... establishment" and its "prison like silence," Henry Barnard noted that the workmen at Colt's Armory were "at their posts by 7 a.m." and forbidden "to enter, if tardy even so little as an instant," explaining that Colt's "system of mutual checks, works like the law of gravity."[137]

How did Colt's workmen adapt to an environment in which "one man sometimes tends six machines," with a "silence" like that "enforced at... Sing Sing"?[138] With the exception of Colt's ill-fated London armory, not a single firsthand account has been found that reveals how Colt's workmen weathered life inside Colt's Armory or even what they thought of him, a reminder of the lopsided character of historical records. When business began to sour at Colt's London armory, one of his contractors, Jabez Alvord, although generally admiring, described Colt as "cross and ugly and inclined to be overbearing." Apparently Colt, in figuring he could "do about as he pleased," caused the departure of several key workmen by "trying to crop them on the price for their work" and afterward "rather [wished] that he had them back." A year earlier, after finding the London factory "not so far advanced as was represented to us" and then after continued costly delays, Alvord contemplated leaving Colt, accusing him of "violating his contract" and promising that "if some morning I should touch my — in adieu to his lordship he must not be surprised."[139] The usual grumblings of basically contented workers? Who knows?

It was said that the "steam whistle on Colt's Armory... [could be] heard... [for] twenty miles."[140] As the workmen ambled across the snow on crisp winter mornings, the sun just rising as they crossed the armory's threshold, what did they think of this new world? The steam engine whistled and the belt shafting began to roll. Days piled on top of days, like so many revolver parts, fleetingly touched but only made whole out there, down there, somewhere else. The keystone of the American dream has always been an abiding confidence in the prospect of "upward mobility." Was America really unique in the world in having "no fixed class relations"? As historians have noted, the "prospect of being permanently dependent on factory work... ran counter to the American mythology of the manly, independent producer."[141] Was Colt's "industrial paradise" an end in itself or the means by which machine tenders and immigrants might achieve prosperity and independence, out there, down there, somewhere else?

How good was the money? When Colt moved from eleven- to ten-hour days in the early 1850s, he was way ahead of the curve internationally. In good times an efficient workman in the more highly skilled filing department could make up to $3.75 a day, and in 1866 it was estimated that as many as 154 of Colt's employees were making $5.00 a day, today's equivalent of about $50,000 a year, at a time when taxes were much lower.[142] Women such as Mary Daly, engaged in the dangerous work of wrapping cartridges, were paid only $.75 per day, barely enough to eke out a boardinghouse existence.[143] However, because of Colt's piecework system of compensation, wages varied widely. Colt's boast of paying $5.00 a day is less believable. And while his top department heads and contractors made the equivalent of about $100,000 a year, a single surviving payroll document from 1860 shows machine tenders ranging between $1.25 and $1.75 per day, with more-skilled laborers in the machine shop and handicraft operations such as pol-

ishing and assembly making between $1.50 and $2.25 per day.[144] Comparison estimates throughout the Connecticut Valley suggest that the filers, drillers, and machine tenders probably did not quite average $2.00 per day even in boom times.[145] But in 1858 that equaled about $625 per year, or today's equivalent of about $35,000 a year, significantly more than the $500 per year historians have estimated it cost for a married laborer to support a small family, and far more than the $25.18 per month paid farm laborers without board in Connecticut as late as 1870.[146]

A voluminous body of literature, culminating in Karl Marx's epochal denunciation of capitalism, argues that far more than money was at stake here. Indeed, it was the fear that the United States might replicate the conditions in Britain's textile mills that caused communities like Hartford to resist industrialization. Evidence is overwhelming that by 1850 America had its own share of "dark Satanic Mills," where workers were reduced to conditions of hopelessness and despair. The Connecticut Valley, by concentrating on the highest value-added sector of the industrial economy, largely escaped the conditions that even northern liberals like Horace Greeley acknowledged: "[If] I am less troubled concerning the Slavery prevalent in Charleston or New-Orleans, it is because I see so much Slavery in New-York.... How can I devote myself to a crusade against distant servitude, when I discern its essence pervading my own immediate community."[147]

In William J. Grayson's *Hireling and the Slave*, the most famous of the South's proslavery, anti-industrial screeds, it was argued that "slavery... establishes... kinder relations between capital and labor,... provides for sickness, infancy and old age... an engagement... for life," while the "isolated, miserable creature who has no home, no work, no food... [is] seen among hirelings only." Answering the charge that the courts of law were open to the industrial victim who "suffers wrong and cruelty in England," Grayson offered a sarcastic reply, "Yes, and so are the London hotels: justice and a good dinner at a public house are equally within his reach."[148] When the British social critic John Ruskin spoke of "men broken into small fragments and crumbs of life" by the "division of labor," he spoke for the ages.[149] Tending machines is boring and offers few opportunities for advancement.

Compared with the industrial standards elsewhere, Colt's Armory provided tolerably good working conditions. In the summer, ice-cooled drinking water, pumped in from Colt's reservoir, took the edge off the blistering heat of the forging room and, perhaps, diminished the oppressive stench of the oil that was used on everything: oil on all the wheels, oil on all the bright iron, oil on all the tools, oil even inside the gunstocks—11,377 gallons in one year alone.[150] Colt boasted of the armory's smoke-free forging shop, steam-heat, ventilation, and lighting by 112 windows and "gas-burners to illuminate... for night-work."[151] Night work? Required overtime was unpredictable and unpleasant, and at the height of the Civil War even the *Hartford Daily Times* looked askance when the armory was "driven all day... Sunday."[152]

The safety record at Colt's Armory is astonishing when compared to industry norms, even among his arms-making competitors. Nevertheless, working conditions during the 1850s would have given OSHA inspectors nightmares. The labor press of the period was filled with images of deafening trip hammers, overheated grinding wheels exploding in workmen's faces, and respiratory systems choked with persistent fumes and smoke. Milling machines lacked proper guards and fencing. Workers were expected to apply oil to gears and cogs of moving machinery, and even elevators lacked basic safety catches—all this, combined with long hours at monotonous tasks.[153] Regardless of the psychic effect, discipline not only promoted efficiency, it also saved lives.

The most dangerous of the Connecticut Valley's new industries was gunpowder and cap manufacturing, a prominent side-by-side companion to the region's arms manufacturers. Colt understood the dangers, locating his cartridge works a mile from the armory and staffing it primarily with single, immigrant women. In 1864 Hartford's C.D. Leet and Co.'s cartridge factory on Market Street exploded, and its female workers were seen screaming, "bleeding and disfigured," from the "third story... enveloped in flames." The same year, "girls... perished in the flames" after an explosion at the American Flask and Cap factory in Waterbury, the same place "in which an explosion occurred last December."[154] Gunpowder wagons traveled through the center of Hartford with frightening impunity, and the accounts of explosions at the Enfield, Connecticut, gunpowder works of Colt's close friend and business associate Augustus Hazard were so frequent and deadly it reads like a freak show.[155]

Among the region's high-tech enterprises, deadly accidents were a routine fact of life. Lives were lost when a boiler exploded at Eli Whitney's armory in 1861. In 1864 a fire "originated in the... polishing shop" of the U.S. Springfield Armory, where the floors and woodwork were "thoroughly saturated with oil." At Hartford's Grove Car Works, a workman lost the sight in one eye when he was "injured... by the flying of a knot from a board under a buzz saw." And at Sharps' Rifle Company a workman was "struck by a small piece of metal" that "entered his heart," causing "instant death."[156]

Although Colt's Armory was not immune to industrial accidents, there is no reason to doubt the company's claim that the 1870 death of John Kallaher, killed in Colt's shooting gallery while "testing the Gatling gun," was "the first serious accident and the only one resulting in death that has ever occurred at Colt's. The company takes every precaution against accidents, and the rules are extremely strict."[157] In 1861 N.B. Ford lost a hand in a drop-forging accident, and in 1864 a workman named O'Hearn was "struck by a piece of steel in the bowels," but survived.[158] Certainly there were other accidents at Colt's. But the accounts of Augustus T. Leonard's slipping off the window sill and falling several stories to the ground, Patrick Delany's losing two fingers in a buzz saw because he was "not... used to the thing," and Thomas Morrison's wandering drunk into "Mrs. Colt's barn," where he was "gored [to death] by an Alderney bull," while tragic, hardly suggest gross negligence on the part of the armory.[159] The most tragic and widely reported death at Colt's involved genuine heroism on the part of Edwin K. Fox who, on his third or fourth trip to salvage machinery and tools inside the burning armory in the great fire of 1864, was lost when the burning roof overhead collapsed.[160]

The most controversial accusation directed at Sam Colt involved his efforts to control the voting behavior of his workmen. Despite denials, Colt was guilty as charged. By 1859 the politics of slavery had so divided the Democratic coalition that Colt's friends may have been correct in estimating that "two-thirds of all Colt's workmen... [are] opposed to him and the Democratic Party."[161] The reverse had almost certainly been the case in 1853. In spite of the fact that the impending Civil War made Colt richer than peace, his obsession with preventing it and his paranoid rantings about "disunionists," abolitionists, and "black Republicans" earned him, and the *Times*'s publisher Alfred E. Burr, a position on the isolated lunatic fringe of northern politics.[162]

Their position of noninterference on slavery, although indefensible in retrospect, had widespread northern support until the mid-1850s. As the *Hartford Daily Times* explained it in 1849, when slavery "is made to override all other questions... there are thousands... who will not be driven into that corner."[163] Colt never was driven into that corner, but he was very nearly driven mad by the repudiation of political values that he cherished as much as his proudest accomplishments involving guns and machines. Regarding *his* workmen, *his* armory, and *his* town, Colt eventually crossed the line between benevolent patriarch and autocratic maniac.

Colt's willingness to, as his discharged workmen claimed, "coerce and control the votes of free men"—an "oppressive and tyrannical exertion of the money-power"—was divisive enough to permanently tarnish the sylvan glow that emanated from Coltsville at its founding.[164] Colt defended himself personally, insisting that "in no case have I ever hired an operative or discharged one for his political, or religious opinions. I hire

them for ten hours, upon the task assigned to them, and for that I pay them punctually every month."[165] Discharged Republican workmen claimed that in 1859 Colt "required his employees to assist him to elect William W. Eaton." He had his secretary stand "by the ballot-box," after deliberately timing layoffs to coincide with the election, circulating something called "the 'Seymour book,'... for enrolling members of Democratic clubs," and "intimating... to Republicans that if they signed, they could have work."[166] Even the *Times* acknowledged that Colt's Armory had "stopped his engine... an hour before the usual time" and called the workmen together to listen to a "letter from the Colonel... *suggesting* that they... 'vote for city officers favorable to improvements in his section of the city,'" which the *Times* described as "not... very unreasonable when we consider that he is taxed about $7,000 a year."[167]

The survival of a damningly revealing notebook carefully listing the political affiliations of Colt's employees would seem to belie his protests.[168] Given the climate of the times, it is hardly surprising that Republican workmen, discharged for any reason, would call it politically motivated. Colt appears to have preempted the accusation by including Democrats among those laid off. Initially, at least, Colt was more concerned about local issues than national issues. Given the city's history of skepticism and envy, one can hardly blame him for urging his employees to paddle the boat of industrialization in one direction.[169]

With the 1860 election on the horizon, Colt became increasingly fanatical, to the point where one can reasonably surmise that, had he lived even a year into the Civil War, his politics could have jeopardized Colt's Armory and Hartford's industrial revolution.[170] The South essentially walked away from the election of 1860, giving Abraham Lincoln the lowest margin of support—about 3 percent—ever awarded a successful presidential candidate by any section of the country. In fact, the South's Democratic candidate, John C. Breckinridge of Kentucky, did far better in Connecticut than did Lincoln in the

South, coming in second in a four-way race, behind Lincoln but ahead of the northern Democratic candidate Stephen Douglas, with more than 20 percent of the vote. In Hartford, where the *Times* openly lobbied on behalf of the one candidate least likely to interfere with Southern "rights," Breckinridge scored 22 percent of the vote and Lincoln just over half, about as lopsided a balance as anywhere north of the Mason-Dixon line.[171] Clearly, the politics of slavery and industrialization had driven a wedge through this once-quiet port on the Connecticut River. Terrified that "disunion" might destroy the United States' prestige and influence on budding democracies, while opening the Southern coast to unrestricted trade with Europe, Colt and his Democratic cronies were not quite prepared to believe in the inevitability of war.

It should come as no surprise that New England's pioneering industrialists were loath to attack the values and institutions of the South. Factory towns like Coltsville resembled nothing so much as a Southern plantation with its orderly hierarchy, military discipline, and godlike monarch living in "the big house." This was the unspoken code that made Coltsville so controversial, and it is one of many reasons why the factory village was long ago made obsolete. While Colt's workers were not legally enslaved, it was always more a matter of degree than kind. By attempting to control the daily lives of his employees and by offering himself as a model of virtue rewarded and the "correct" way of life, Colt appears as the model patriarch, with Coltsville an "economic community of interest" in which workers and employers were "bound together by ties of duty and obedience."[172] Like the Southern plantations, Coltsville was organized around the production of a staple crop—revolvers—and exhibited a strict hierarchical pyramid, with a reigning monarch, a "benign but powerful father," who expected obedience in exchange for his "affectionate and protective attitude… toward his dependents."[173]

As historians have noted, Southerners were not alone in seeing the plantation as "the ideal form of American society" and the "final expression of… expansion and conquest," in which groups of dependents looked "to the master for work, protection, and moral guidance."[174] How a social structure so autocratic and essentially un-American could have emerged as a viable alternative to the messy but endlessly resourceful and inspiring chaos of democracy, speaks not so much of Colt's uniqueness or character defects as of the difficulties Colt and his peers faced in adapting industrialization to an existing economic and social order with conflicting values and priorities. By wrestling with the challenge of industrialization, Colt and his peers changed the course of American history and made the United States a richer nation, if not more durable and free.

SAM COLT AND THE CHARTER OAK
THE HEART OF EMPIRE

In the recent rage for rehabilitating history… it is seriously doubted if the charter was ever in the tree at all. Prof. Broadhead discredits the whole story…. "Broadhead be blowed!" exclaimed a Hartford youngster to whom I mention[ed] the historian's doubt.

HARTFORD DAILY TIMES, AUGUST 29, 1879

Samuel Colt never forgave nor forgot how Hartford's social elite ostracized the family when his father's business failed in the depression of 1819. After fifteen years piling up personal and professional failures of his own, Colt returned to his native city bitter, angry, suspicious, and vengeful. As Colt's first biographer described it, "Whoever has made a name in the world… feels it most exultantly when near those once his superiors…. There… he is most envied, and hence slandered and abused, just as Joseph was… among his brethren."[1]

In 1849, with success finally at hand, Colt's father urged his son to be political and invite a few of Hartford's prominent elites to buy shares in the company. Remembering the years of snubs and rebukes from reluctant investors, he shot back: "Now that I am able to paddle my own canoe they want to climb into it. There is no reason why I should carry them, and I will not."[2] When Colt's Patent Fire-Arms Manufacturing Company was incorporated in 1855, only a handful of his most-trusted aides were invited to be shareholders. In a city of banks and insurance companies, Colt financed expansion out of earnings and never carried insurance. Colt's hostility toward inherited privilege, bureaucratic authority, and capitalists (Colt imagined himself an inventor and manufacturer) led to a series of symbolically charged campaigns. The most graphic and poignant of these campaigns centered around the Charter Oak.

The Charter Oak is Connecticut's Paul Revere and Pilgrim's Landing, its Alamo and its Liberty Bell all rolled into one. In spite of the fact that Connecticut's license plates still read Constitution State, there is probably not one resident in fifty today who can tell you why that is so. A century ago the state's unifying myths, especially that surrounding the oak, were almost oppressively familiar. The state capitol (1874–79) is drenched in Charter Oak

iconography, and the oak had already become a parody of itself in 1868 when Mark Twain visited Hartford and satirized the legend:

The Charter Oak… used to stand in Hartford…. Its memory is dearly cherished…. Anything that is made of its wood is deeply venerated…. I went all about town with a citizen whose ancestors came over on the Mayflower*…. He showed me all the historic relics of Hartford… a beautiful carved chair in the Senate…. "Made from the Charter Oak," said he. I gazed upon it with inexpressible solicitude. He showed me another carved chair in the House. "Charter Oak," said he. I worshipped. We went down to Wadsworth's Atheneum, and I wanted to look at the pictures, but he conveyed me silently to a corner and pointed to a log rudely shaped, somewhat like a chair, and whispered, "Charter Oak." I exhibited the accustomed reverence…. He showed me a walking-stick, a needle-case, a dog-collar, a three-legged stool, a boot-jack, a dinner-table, a ten-pin alley, a tooth-pick, a—I interrupted him and said— "Never mind—we'll bunch the whole lumber yard and call it—" "Charter Oak," he said. "Well," I said, "now let us go and see some Charter Oak for a change." I meant that for a joke…. He took me around and showed me Charter Oak enough to build a plank road from here to Salt Lake City.*[3]

Twain's irreverence notwithstanding, many people in mid–nineteenth-century Connecticut took the Charter Oak and its mythology seriously. Sam Colt was one of the oak's greatest champions. He wrapped himself and his enterprise in the bark of the oak, exploiting its associations and proximity to his South Meadows factory village. At stake were definitions of patriotism and progress, as well as the question of whose vision of the future would prevail in Hartford, whose industrial future never ceased to be controversial. Factions squared off for a high-stakes game of capture the flag as Colt and his supporters challenged the old merchant guard for supremacy, with the oak as prize and mascot.

The constitution that makes Connecticut the Constitution State is not the U.S. Constitution of 1787, but the less well-known Fundamental Orders of 1639. In 1638 the brilliant and charismatic pastor of Hartford's First Church, Thomas Hooker, preached a sermon asserting that the "foundation of authority is laid... in the free consent of the people." This sermon became the basis for the Fundamental Orders, described by nineteenth-century historian John Fiske as "the first written constitution known to history that created a government," thus marking "the beginnings of American democracy."[4] Over the years the Fundamental Orders acquired an aura of primacy and significance as "a document far in advance of anything the world had ever seen, in its recognition of the origin of all civil authority as derived... from the agreement... of the governed."[5] Having produced what has been called "the first written constitution in the history of nations," Connecticut became mythologized as the birthplace of U.S. liberty and independence. Connecticut—at least in the minds of its citizens—identified itself as the first true constitutional democracy, based on fine distinctions and willful indifference to context that today only a political scientist could fully understand.

Connecticut's obsession with the priority of its constitution was also rooted in the evolving firestorm over states' rights and the conviction that its peculiar culture and traditions were being marginalized in the drive to forge a national identity. Situated "between New York and Boston," Connecticut has labored long in the shadows of its bigger, richer, more-storied neighbors. That Bostonians made "writing New England's history" a cottage industry exacerbated Connecticut's need to tell its own story and control its destiny in rapidly changing times.[6]

The Charter Oak embodies the spiritual and human drama at the core of Connecticut's constitutional history,

enabling the state to maintain the illusion, if not the reality, of continuous self-government. The story begins in 1662, when Connecticut freemen petitioned the newly restored monarchy of Charles II for a Royal Charter. Ably negotiated by Connecticut governor John Winthrop Jr., a charter was granted that protected Connecticut's geographical and political autonomy, its right of self-government, and its status as a relatively free and independent state.[7]

This charter, "more liberal than the other American colonies," was threatened when Gov. Edmund Andros arrived in Hartford on October 30, 1687, bearing a Royal Commission as governor of the Dominion of New England. He brought with him an armed retinue and instructions to dissolve Connecticut's independent government.[8] Historians dispute whether Andros actually demanded the charter itself. But at the decisive moment, a gust of wind allegedly extinguished the lights in the meeting room, and when light was restored the charter had vanished. Legend has it that Capt. Joseph Wadsworth, dashing through the shadows, stole off and hid the "sacred" charter in the hollow of the great oak.

The facts aside, in the civic mythology, Connecticut had kept the faith under trial. The charter was never surrendered, and neither the written word nor the oak was betrayed, thus marking a symbolic prelude to the Revolution by signifying the courage to resist oppression in defense of liberty. From there on after, the oak and the "divine cavity" that held Connecticut's "sacred charter," belonged to the ages.

The legend of the oak was passed down as verbal tradition for almost a century, until the Revolution renewed the power of its symbolism. Victorian eulogists claimed that the oak was invoked during the War of Independence to inspire men to take up arms in defense of their liberties and to honor the "heroism of the fathers," but there is no record to that effect.[9] A recollection, written at the height of the oak craze, about a pilgrimage in 1791 to see the "celebrated old tree, where the old colony charter was hid" is the earliest documented instance of oak veneration and may be apocryphal.[10] The oak appears in a portrait of Polly Wyllis Pomeroy, painted by Ralph Earl in 1792, but more as a landscape detail than as icon.[11] A member of the Wyllis family who owned the property where the oak stood alluded to its "divine purpose" as early as 1805, while mentioning that in 1797 "the cavity, which was the asylum of our charter... has closed."[12]

The renewal, if not the creation, of the Charter Oak myth can be traced to 1818, a milestone in Connecticut history and the year when Connecticut's centuries-old Standing Order (then represented by the Federalist party elite) experienced political defeat. The new Jeffersonian leadership uprooted the 136-year-old constitution based on the 1662 charter and adopted a new constitution that, among other things, effectively separated church and state and inaugurated a tradition of two-party rule. Despite the associations of liberty, freedom, and the oak, Connecticut was, in fact, slow to embrace democracy.

Civic leader and antiquarian Daniel Wadsworth—the man who epitomized the standing order's aristocratic predispositions—reacted by commissioning the earliest-known portrait of the Charter Oak from a Hartford sign painter and artist named George Francis.[13] This was the first evidence that the oak might emerge as an important symbol in the heightened tension between factions and in the changing notions of history, governance, and progress.

During the 1820s and 1830s, poems, prints, and historical narratives began entering the public domain, heightening interest and awareness of the legend of the oak.[14] But the two events that spawned the oak craze, transforming a local folk legend into an international symbol of freedom, occurred in the 1840s. First, antiquarian, politician, and civic evangelist Isaac Stuart became the "proprietor of Wyllis hill and the Charter

Oak." Then, in 1848, Europe was swept by "messianic republicanism" in a wave of democratic uprisings. With a dash of jingoism from the military triumphs of the Mexican War, the oak was embraced by a celebrity poet—Lydia Sigourney—who had a flare for writing spiritual postmortems, and adopted by a media-savvy inventor-industrialist in search of glory—Sam Colt. The oak craze was launched and would dominate the popular culture of Connecticut well into the twentieth century.

In 1848 Benson J. Lossing, the nation's best-known historian, visited Hartford to sketch the oak. The United States' periodic bouts of missionary zeal intensified following the military triumph in Mexico. Coinciding with a wave of democratic uprisings in Europe, America's predisposition to believe in its divine and exclusive destiny as a chosen people evolved into a swaggering evangelism during the 1850s.[15] Was the United States really the "great and last asylum of the free"? It appeared so in 1849 as democratic uprisings were crushed throughout Europe. At the beginning of the new year, Alfred Burr of the Democratic *Hartford Times* proclaimed that the "torch of revolutionary freedom... has at last gone out... [and] Darkness has dawned on the East."[16] By contrast, in the United States the advance of empire, the rapid spread of railroads and telegraph, and the discovery of gold in California affirmed a sense of impending reward and divine mission.

In 1850 Isaac Stuart, a retired Whig politician, orator, antiquarian, and civic booster, began work on a history of Hartford that he serialized in the *Hartford Courant* the next year. Stuart, the "anointed priest and oracle of the past... the lord, the guardian, and the Clio of the Oak," was one of Hartford's most intriguing, if not eccentric, characters in Colt's era and the man most responsible for propagating the myth of the oak.[17] This man, whom eulogists described as "not what he might have been," was a peculiar but irrepressible champion of state and local history whose "favorite political theories were protection of American industry, and the historic dignity of New England."[18] He lectured around the state and nation providing "sternly accurate historical biographies"—hagiographies by another name—of the deeds of Connecticut's patriots, Jonathan Trumbull, Israel Putnam, and Nathan Hale. He was lionized by the public and was the most sought-after toast-master of his generation in the community.

Not long after breaking ground for his new armory in 1854, Sam Colt befriended the Clio of the Oak. Increasingly obsessed with his vision of Coltsville and anxious to claim that, like Hartford's founders, his efforts to "replenish the earth, and subdue it" were also divinely sanctioned, Colt courted favor with Isaac Stuart. In exchange Stuart bestowed his (albeit bought and paid for) blessings on Colt's enterprise. He also helped by designing a program of place names, landmarks, and symbols, linking Colt's enterprise with the mythology of the oak. And as if the symbolism of transformation were not enough, Colt's choice of Hartford's South Meadows floodplain, an "uninhabited wilderness," as the site for economic renewal seemed almost designed to amplify the theme of rebirth.[19] Stuart's efforts helped Colt counter opposition to his plans by enveloping it in the language of historical continuity and divine sanction.

The spring of 1856 brought Sam Colt a season of accomplishment and vindication. The new armory was up and running smoothly; the first of his workers' housing was occupied in April; sales and employment were at a record high; and he was to be married in June. On Tuesday, May 6, Colt staged a dedication of Charter Oak Hall, the spiritual and symbolic centerpiece of his industrial empire, a structure "reared for advancing the intellectual and aesthetic culture of mechanics," a shrine to progress and faith in the American dream.[20] At one hundred feet long and four stories high, Charter Oak Hall was said to be capable of accommodating one thousand people. The "Armory Band, endowed by Col. Colt... for the purpose of promoting musical art," made its first public appearance at the dedication. As the workmen and their families, not quite numbering Colt's planned biblical thousand, gathered in orderly fashion, Colt, surrounded by his political cronies, publicists, and senior officials,

75. *Charter Oak Relics, 1856–75.*
Top to bottom, left to right:
Colt's empire lamp shade, 1861, carved ivory, white oak, and engraved silver (WA #05.1151);
Armsmear *bound in the Charter Oak, 1867 (WA #05.1574);*
Elizabeth's Charter Oak gavel and box, ca. 1875 (WA #05.1555);
Elizabeth's "leaf from the true oak," 1856.

mounted the stage at the west end of the great hall. William J. Hamersley, a book publisher, state legislator, former mayor, and the man who chaired the committee that approved Colt's South Meadows development, praised how "the name of Samuel Colt is now more widely known throughout the world than that of any other living American inventor." With a great "sense of exaltation" he dedicated Charter Oak Hall to "the Sovereignty of Labor" and praised its "purposes of moral, intellectual and artistic culture."[21]

Isaac Stuart then spoke of the sense of mission and history in Colt's deciding that all the "principal structures,... avenues,... streets, [and] docks" would "receive commemorative names... replete with significance" and expressed his gratitude at being appointed "a sort of lay-pastor for his christenings." Perhaps then as now, Hartford citizens were more bewildered than impressed with street names such as Sequassen (sachem of the Suckiag tribe that occupied Hartford when Europeans arrived), Wawarme (his sister), Masseek, Weehassat, and Curcombe (his successors). More familiar were the streets named for the Dutch fort (Huyshope) and its original commander, Vandyke. In a curious twist of meaning, Charter Oak Avenue was "christened" to symbolize the past "contest for liberty between English and colonists," in much the same way as Charter Oak Hall symbolized Colt's present contest with the standing order for the liberty of Hartford's economic future.[22]

"By virtue of the nomenclature a three field history is signified... which, compounded, forms one of the most remarkable and thrilling pictures... of human experience in the colonization and settlement of this new world... the whole of this great Past... made to link happily with the Present."[23] This new world made to link happily? Colt and Stuart may have been sure. But many others were sure not, and most were probably unsure. And for the workers, well, it was definitely better than the cotton mills and easier than scratching a living from the soil, but for most it would never become a springboard to prosperity.

In a final pirouette of poetic fancy, Isaac Stuart concluded his address by pointing "yonder, [where] in the glory of a patriotic history that is unmatched by aught else of its nature upon earth—stands that Monarch Tree, the Charter Oak!" In its honor he dedicated "this spacious apartment to the World as the Charter Oak Hall."

Tense after three years of unrelenting conflict with city officials, Colt was relieved and proud as Mayor Deming thanked Stuart and acknowledged his right as the "guardian of the Oak" to "distribute its honors." Grateful and optimistic, the mayor spoke of "popular sovereignty" as "the Charter Oak faith." He concluded by claiming that by marrying "the forge and the work shop to the Reading Room and the Dancing Hall," Colt had founded "an institution" destined "to a still higher eminence."[24]

But all was not well in paradise, at least not for the Charter Oak. Three months later, in the darkness of night, a "providential wind" toppled the aged oak, which crashed to ground at a quarter to one in the morning.[25] Hartford responded with a flurry of ceremony, eulogy, and memorialization. Isaac Stuart, perhaps now driven as much by the loss of his claim to fame as by genuine horror, orchestrated a bizarre ritual of public mourning. It was Thursday morning when he roused Colt's Armory Band, which appeared in uniform at noon and "played dirges for two hours over the trunk of the fallen monarch" followed by "Home Sweet Home" and "Hail, Columbia."[26] Never one to overlook the opportunity of an occasion, the "sweet singer of Hartford" sharpened her pen for a eulogy that was on the streets the next day.[27] A daguerreotypist arrived by noon, together with hundreds of onlookers and the inevitable relic hunters, who had become a nuisance to Isaac Stuart even before the oak fell. That evening at sundown the city bells tolled in what must have been a quiet and pensive hour for a community pumped up with pride over an icon that lay prostrate and broken on the hillside.

In the days and weeks that followed the tributes, condolences and requests for souvenirs poured in from all over the country, especially from the South, then waging its own holy war for self-determination. The *Louisville Journal* described the oak as the "sacred tryst-ing place of patriotism" whose "legend has struck root into the national heart."[28] A Bostonian warned of the impending danger to liberty and Union.[29] In Hartford some were more worried that Isaac Stuart would give away or sell off every remaining sliver of the oak.[30] Although reports of "thousands of strangers from abroad" paying homage to the oak were undoubtedly exaggerated, the tree clearly had become an icon of more than local or state significance.[31]

In November what was left of the tree was still embellished with a "national flag, draped in the emblems of mourning" behind the high picket fence Stuart erected to protect the oak from "plundering." According to news-paper accounts he had already dispersed relics "in the form of canes, snuff-boxes, [and] pieces of wood" to "more than 10,000 persons in various parts of America."[32] Mark Twain's joke twelve years later about "Charter Oak enough to build a plank road from here to Salt Lake City" was true enough. Whether Stuart had gone off the deep end or was simply broke, something was clearly wrong by the following winter when Caroline, his wife of twenty-two years, walked off with the children and put "the old Charter Oak place" up for sale.[33]

In the years following, Charter Oak relics would proliferate wildly (fig. 75). Gavels and ballot boxes were popular, as were canes, nutmegs, picture frames, and small boxes.[34] Colt's neighbor and friend, the German woodcarver and pianoforte manufacturer John H. Most, advertised and exhibited "pianos from the wood of the Charter Oak," one being valued in Stuart's estate at four hundred dollars in 1861.[35] Before the end of 1856, the nation's two leading commercial lithographers, Nathaniel Currier and Kellogg Bros., each turned out pictures of the Charter Oak suitable for framing.[36] By October 1856 Charles DeWolf Brownell's masterful portrait of the Char-ter Oak (fig. 76) was ready for exhibition in the Hartford County Agricultural Society fair. This view, which became the model for most subsequent renderings, portrayed the

oak as more grand, heroic, and statuesque than the rather sickly looking tree that collapsed in August.[37]

Joseph Buckardt, a Swiss immigrant and wood-carv-er at Colt's Armory dazzled the public (and no doubt pleased his employer) with an "ingenious picture of the Charter Oak" painstakingly inlaid with wood from the fallen monarch.[38] Whether influenced by Buckardt's cre-ativity or by parallel inspiration, Sam Colt and Isaac Stuart took the oak fetish to the next level by creating the two most astonishing and eccentric works of art to emerge from the Charter Oak craze.

When Elizabeth Colt gave birth to Samuel J. Colt on February 24, 1857, Isaac Stuart and Sam approached John H. Most, their mutual neighbor, about creating the ultimate Charter Oak relic—a cradle to honor Colt's "babies," his offspring and his vision for Hartford. Most and his German shop carver Charles Burger worked feverishly through the winter sawing, shaping, and carv-ing huge branches, burls, and sections of trunk from the oak, which Stuart dutifully supplied.[39] Colt collaborated in the design and provided exotic jewels that he acquired at "the great Asiatic Fair" in the ancient and symbolically charged Russian city of Novgorod.[40] These, together with a burl of the oak, were incorporated as part of the object's bizarre decorative scheme. When it was unveiled in April, the Charter Oak cradle caused a sensation (fig. 77). In a community still dominated by Congre-gationalists of the old puritan stripe, this was the most egregious and self-aggrandizing display of ego that had ever graced the threshold of this once-insular and proper New England town. The cradle, with its eight colt heads, two rampant colt finials, and carved Colt family coat of arms, swung from gnarly oak posts and was shaped conspicuously like a canoe, a reference to Colt's carefully cultivated populist image as the man who "paddled his own canoe."

In what reeked of a final over-the-top act of syco-phancy, Colt's irrepressible friends in the Democratic press so frothed at the cradle that more sober pens at the *Hartford Daily Courant* could no longer remain still:

76. The Charter Oak, *1856.*
Oil on canvas with
a Charter Oak frame,
ca. 1868, Charles DeWolf
Brownell, 1822–1909
(WA #98.10).

Brownell studied painting
with the Hartford artists
Joseph Ropes and Julius
Busch and by 1857 was
an accomplished profes-
sional landscape artist
whose travels to and
pictures of the White
Mountains, Mexico, Cuba,
and the Connecticut River,
earned him acclaim.

Although this picture is
dated 1857, a perhaps-
unfinished version was
first exhibited in 1856,
two months after
the fall of the oak.
Note Colt's Armory dome
on the horizon.

Last evening's papers rioted in a description of a cradle made of the wood of the Charter Oak and presented by the Hon. I.W. Stuart to Mrs. Sam Colt's baby. The [Hartford Evening] Press *is particularly enthusiastic, not to use a more expressive word, over the "rampant colts facing inward, which produce a most spirited effect"—"noble and massive knot of the oak"—"genuine bark"... "Mr. Stuart's own hand writing"—"appropriate verses from Mrs. Sigourney"—"exquisite taste and judgement!" All this is very silly, very undignified, very toadyish! It sounds too much like the nauseous twaddle about Victoria's babies in the English papers, or about the young Emperor, in the French.... In the name of a decent degree of manliness, keep such nursery spooneyisms out of the newspapers.... The tuft-hunting toady piles his agony altogether too high for the democracy of a man whose boast is, "I paddle my own canoe."* [41]

The *Hartford Daily Times* shot back about being "attacked" and then sneered that the "editor of the *Hartford Courant* is picking a quarrel with 'Sam Colt's baby.' We'll bet on the 'baby.'"[42] Of course by this point they had been betting on the real "baby"—industrialization—for years. It would not be the last skirmish involving Sam Colt and the Charter Oak.

Isaac Stuart's letter to the baby Colt, which accompanied the cradle (a hand-written poem by Lydia Sigourney is actually encased on the underside of the cradle's detachable burl) is a masterpiece of idolatry. It captures the spirit and values of the oak craze and of Sam Colt's creed of liberty, justice, truth, godliness, and prosperity:

Master Samuel Jarvis Colt:—

You are a tiny infant now, Sammy, just bursting into life, and cannot read.... But you will grow... and then you will find out that a friend of your father and mother... sent for your use this day a beautiful present... a cradle made from the wood of a very famous tree called the Charter Oak....

A long, long time ago... a very bad man, named Edmund Andros, came with a troop of soldiers... to take away... the Charter of Connecticut.

Now, this Charter made the people... free... It gave them LIBERTY, and LIBERTY, you will live, I hope, to learn, Sammy, is a very precious thing, and ought to be defended at the cost of all the money in the world... [and] sometimes of human life....

Now, Sammy, your cradle is made out of the wood of this... oak, and it should teach you always to remember that

77. Charter Oak Cradle, 1857. Carved white oak, designed by Isaac W. Stuart, carved by John H. Most, Hartford (WA #05.1580).

hero who hid the Charter... and make you follow his example in defending your country whenever it is in danger. Die for your native land, Sammy, rather than let anybody hurt it!

This cradle too should teach you to be wise, and virtuous and honorable, and industrious, just as those good men were who lived when the oak was made so famous.... They adored liberty.... They worshipped justice.... They made truth their idol.... They worked out their own prosperity... "paddled their own canoes."... Your father thought a great deal of this duty and he acted it out.... Sammy, remember it.

But,... best of all... those good men... loved God.... Sammy, love God, love your father and mother,... love all mankind.[43]

In the months immediately following the squabbling over Colt's cradle, Most and Burger completed a second and equally grand Charter Oak masterpiece, an enormous throne, commissioned by Hartford's mercurial common council for use by the mayor. To look at that chair is to easily understand how it might have cost the $375 (about $20,000 today) asked for it. It is a tour de force of workmanship and wood carving, and probably involved as much labor as any chair made in America before the Civil War. Its rustic elements and paired shields representing the state arms and city seal, all tufted and upholstered in leather and of "solid Charter Oak,"

*78. Charter Oak Chair, 1857.
Carved white oak,
designed and carved by
John H. Most and Charles
Burger, Hartford
(WA #05.1580).*

are the melodramatic climax of the oak craze. The back of the chair (fig. 78) picked up the theme of Joseph Buckardt's "ingenious picture," its entire back carved in bas-relief in the shape and form of the oak. Whether the common council was shocked by it or simply balked at the price, the bill remain unpaid through the fall and winter of 1858, even though the chair was completed the previous June.[44] Sneering at the city's "unpatriotic fit of economy," Sam Colt finally stepped in the following March and made another public spectacle by declaring the chair "cheap at any price" and paying one-third premium over the price billed the city: $400 for the chair, and $100 as a gift to Isaac Stuart for his aggravation.[45]

Were the city and state growing weary of the oak? By 1858 some undoubtedly wished they had never heard of it. Indeed, the oak had been a political football from the beginning.[46] It now emerged as a factor in the generations-old rivalry between Hartford and New Haven. Through most of the nineteenth century, Connecticut remained the only state in the Union that could not agree on a permanent location for its state capital. The general assembly continued to meet alternately in Hartford and New Haven until 1879. In 1857, with Stuart's Charter Oak Hill site on the market, Sam Colt and the Democratic *Hartford Daily Times* launched a campaign to acquire the property and link it with "the large ground and Parks soon to be made upon Col. Colt's premises," which together would "make a Park of 100 acres" that might be "thrown open to the public for military and other parades."[47] Looking back a decade later, Henry Barnard wrote of how "it was the patriotic desire of Col. Colt to preserve [Charter Oak Hill] with as many of its historic memorials as possible, for public use and enjoyment... and as a site of a new State House.... [Colt] expressed his willingness to be taxed... and to contribute largely from his own resources."[48] The city, already in the midst of a campaign to create its first public park, may have seen a two-for-one opportunity. The proposal was actually approved by the state senate, and a new state house designed (fig. 79), before failing to pass a vote in the

House of Representatives. Colt and his Democratic cronies lost and a year later were horrified as Charter Oak Hill was cut up into building lots and all evidence of the oak obliterated in preparing grounds for the mansion of a local attorney.[49]

Among firearms collectors, few Colt revolvers are more prized than those Colt began manufacturing in 1857 with grips made of Charter Oak. John H. Most continued to manufacture Charter Oak relics, supplying a Charter Oak trumpet as a prize in the grand fireman's muster, which attracted companies to Hartford from as far away as Providence, Boston, and Brooklyn.[50] In 1858 Colt spent $122 (about $7,500 today) commissioning Charter Oak toddy sticks and crucifixes as gifts for friends.[51] In November, perhaps more willing to trust Connecticut's newly elected Know-Nothing-party governor, Alexander Holley, Isaac Stuart supplied the wood for "a chair of State," designed by the governor's son.[52] This chair was paid for and is still used today by the lieutenant governor in the state senate chamber.

Sam Colt's last tango with the Charter Oak was in the summer of 1860, when he commissioned Christian Deyhle to make a pair of what he described as "lamp shades," designed to frame and display translucent ivory or lithophane pictures. Deyhle was the armory's chief ivory-carver and the man responsible for decorating the grips on Colt's ultradeluxe presentation revolvers.[53] Trained in the German art of ivory and wood carving, Deyhle had arrived in Hartford from Wurtenburg, Germany, in 1856. While employed on the same terms and rate as other contractors, he worked alone. As a Democrat and one of the first residents of Colt's workers' housing, Deyhle personified the qualities that Colt wanted to associate with the new industrial order: loyal, upwardly mobile, skilled, politically inspired but not radical.[54]

For eighteen months, beginning in June 1860, Deyhle put in what hours he could spare, posting expenses for work on the lamp shades that eventually added up to $410.19 (about $24,000 today), and it is not clear if that was for one of the shades or both.[55] Each shade is made

with silver mounts and carved ivory panels mounted amid a frenzy of naturalistic, carved Charter Oak, attached to revolving rods secured to a carved pedestal. These lamp shades are intensely personalized and biographical works of art, three-dimensional representations of Colt's accomplishments and ideals. Virtually without sources, they suggest a cross between a scrapbook and a monument, or a résumé on a pedestal. In anyevent, as an example of American relic furniture, the shades are unprecedented and have never been replicated—a successful, albeit eccentric, demonstration of individuality.

The politics of industrialization are equally apparent in the design and symbolism of the two lamp shades, which take as their themes, respectively, American liberty (fig. 80) and Colt's empire (fig. 81).[56] Was Colt's empire too personal perhaps? Rising from a pedestal enriched with iconographic representations of Hartford's "glorious past"—the 1796 Connecticut state house; the founder, Thomas Hooker; symbols of agriculture, shipping, commerce, engineering, and manufacturing; and the grapes and oak leaves of the Connecticut State seal—the stand provides a tableau of heroic associations. It is pointedly

79. *Proposed State Capitol, 1857. Oil on canvas, Frederick S. Jewett (1819–64). This picture may have been an artistic rendering created merely to suggest the possibilities for such a building on that location. There is no evidence that an architect was ever engaged to carry out such plans. Notice the stump of the Charter Oak fenced off for preservation. (Courtesy of the Connecticut Historical Society.)*

surmounted by an even grander tableau: a shade embellished with icons of Colt's triumph—a rampant colt prancing among the oaks, laurels, and evergreens, the armory with the steamboat *City of Hartford* in the foreground, and the enigmatic figure of a lone boy—perhaps Colt himself—paddling a rowboat across the Connecticut River. The central panel is flanked by carved ivory representations of Colt's arms framed in oak and laurel, the whole crowned by an American shield-breasted eagle. Pistol-shaped silver-plated screws provide support for the intended gas light. The effect is so clearly personal in its associations that, without being inside the head of its maker, its full power is diminished. As much as any work of art on record, this lamp shade is a boldly audacious "song of selfhood" and personal iconology. It sings of the marriage of art and arms, of past, present, and future linked in an ascending line of progress, of the patriotic achievements of American industry, and most of all of the glorious triumph of the self.

The American liberty shade, Sam's gift to Elizabeth for their last Christmas together, provides a tableau of military glory. It is enriched with figural heroes of the Revolution: Generals Green, von Steuben, LaFayette, and Hamilton. Modeled after the architect Robert Mills's Washington Monument in Baltimore, the liberty shade features two equestrian cameo portraits in carved ivory of George Washington and Israel Putnam, beneath the cameo of seated Columbia surrounding a central panel celebrating the valor and glory of the United States' *civilian* army. The Battle of Bunker Hill and the Battle of New Orleans represent two wars, two epochs, but one universal law of freedom and liberty. The drums and cannon, trumpets, globes, shields, mottoes, flags, and cornucopia—exquisitely rendered in Charter Oak—speak a then-familiar language of patriotic associations.

Colt loved it, and it is not difficult to imagine him discoursing at length about its meaning. Here was a highly personalized rendition of the story of American liberty, through which we witness Sam Colt making history in the most literal sense. The specific ordering of icons and

80. American Liberty Lamp Shade, 1861. Carved ivory, white oak, and engraved silver, designed by Sam Colt, carved by Christian Deyhle, Hartford (WA #05.1150).

Just before the age of the stereopticon viewer, lamp shades were fitted with mounts for interchangeable pictures, back-lit by candles or gaslight. This shade was designed with a fixed scene carved in

ivory on the principle of the porcelain lithophane pictures invented in Germany in 1827, which became wildly popular during the 1840s and 1850s.

emblems again reveal Colt, perhaps guided by Isaac Stuart, as a collaborator in the design and conceptualization of an emotionally charged patriotic relic. Especially noteworthy is the emphasis placed on the Battle of Bunker Hill and on Gen. Israel Putnam, Connecticut's foremost Revolutionary War hero, whose leadership at Bunker Hill was a source of rivalry between Massachusetts and Connecticut during Colt's lifetime.[57]

If this wasn't enough oak and glory for one man and one life, Colt's fellow members of Hartford's private, patriotic, gentleman's militia company, the Putnam Phalanx, presented him with a gift that was surely the pièce de résistance of the age of American patriotic reliquary. This "grand historic souvenir," a cane (now lost) made of Charter Oak, was inlaid with wood from George Washington's Mount Vernon, General Putnam's house, and *Old Ironsides*, "our Monarch of naval conquests." Stuart and Most presented the cane to Colt during his last illness "with our warm hope that it may prove to you, both in a moral and in a physical sense, a stay and a support, warming up your spirits with thoughts of a glorious national Past," with hopes that it might "through the associations it inspires, and the support it yields,... expel... every vestige of the dreaded arthritis," which by 1861 had left Colt weak and in pain.[58]

By the end of the nineteenth century, Hartford, the Charter Oak City, in Connecticut, the Charter Oak State, was brimming with monuments, institutions, and symbols—everything from banks and boats to cigars—named after the Charter Oak. Was it a heroic age, an age of heroes manifestly destined to cross the horizon of hope to the promised land of a glorious future? Ironically, they who believed so become stranger to us as we become less sure. Sam Colt believed that industrial civilization was destined for greatness, and he offered to play Moses leading the people to a new promised land. By wrapping himself in the bark of the oak, Colt helped fuel both the oak craze and his own mythology as a symbol of personal rebirth through spiritual pilgrimage to the frontier of the new industrial age.

81. *Detail of Colt's Empire Lamp Shade, 1861. Carved ivory, white oak, and engraved silver, designed by Sam Colt, carved by Christian Deyhle, Hartford (WA#05.1151).*

ARMSMEAR
THE HEAD OF EMPIRE

You can't imagine a more exquisite little architectural affair than Colt has recently got up in front of his house....
It is all iron & glass à la Crystal Palace—with a beautifully shaped dome
& pointed little minarets, somewhat Turkish in idea & seen under a Western sun seemed like a fairy palace.
ISABELLA BEECHER HOOKER, 1857
It comprises everything that architectural taste, fine landscape gardening,
and, above all, money and millions of it, can put on and into such an estate.
HARTFORD DAILY TIMES, AUGUST 29, 1879

Sam Colt was always a lavish spender. But until the summer of 1857 he never occupied a house of his own and was not especially conspicuous in the outward show of wealth. Armsmear (fig. 82), the dream house Henry Barnard described as the "Corinthian capital of the solid structure of his great enterprise," would finally unmask Colt's wealth and reveal, despite its quirks and eccentricities, a consuming interest in innovation, not only in machinery and armaments but also in architecture, gardening, agriculture, and domestic economy. With Armsmear, Sam and Elizabeth showed Hartford and a nation increasingly addicted to the sideshow of their generation's rich and famous, the vanguard of high living. Through Armsmear, it was said, Colt revealed "the possession of tastes which his impetuous temper and busy career had not led his best friends to give him credit for."[1] It was certainly the consuming passion of the last years of his life.

For Sam Colt, a man who had visited most of the nation's great cities and many of the capitals of Europe, the years following his marriage were a period of starkly contrasting domesticity. Sam Colt, the quintessence of the questing male, would now, in a search for rootedness, build a mansion and oversee the affairs of his industrial plantation. With day-to-day operations at the armory increasingly delegated to others, Colt turned his attention to building the social and symbolic amenities of Coltsville and to creating a garden paradise at Armsmear, the mansion that "arms" built on the "mere," or lowlands, of Hartford's South Meadows.

Armsmear was almost literally the capital, or head, of "his great enterprise," one of the most extravagant mansions created by an industrialist anywhere in America before the Civil War. Again we find Sam Colt motivated not so much by pomp and profits as by an almost messianic desire to light the way to an idealized future. Sharing his generation's intense faith in "progress," Colt applied his genius to the challenge of creating a home that validated the law of progress and symbolized his place at the top of its relentless hierarchy. What challenges would Armsmear address? Art and nature, agriculture and industry, past and future, the old world and new would there be married and made one. For Sam Colt, Armsmear was to be King Solomon's temple, and Coltsville his Jerusalem.

Colt had the idea of a grand mansion in mind as early as the summer of 1852, when he began amassing land for his South Meadows improvements and bought eighteen acres on the Wethersfield Road.[2] The groundbreaking for Armsmear was not until three years later. A newspaper account of "Col. Colt's new house" suggests that designs were in hand and construction was begun during the summer of 1855.[3] In spite of the Colts' determination to occupy the house on returning from their bridal tour in December 1856, the painters were still at work the following March; the last of the core furnishings did not arrive until April; and the Colts were unable to move in until June 1857.[4] Armsmear was a hodgepodge still in the midst of major renovations at the time of Sam Colt's death, and it continued to evolve until Elizabeth's death in 1905.[5]

Considering its scale and importance, surprisingly little is known about Armsmear's design and construction. At three stories and with almost two dozen rooms, Armsmear was a rambling, asymmetrical pile of smooth-faced brownstone, heavily bracketed, lined with balustrades, and anchored at the southwest corner by an enormous five-story campanile, or tower, surmounted by an elaborately turned finial. Its central design is the bracketed, Italian villa style, augmented by exotic oriental minarets and domes in steel and glass inspired by the Crystal Palace.

As with the armory, Colt's nephew Henry A.G. Pomeroy served as chief contractor, with Hartford architect Octavius Jordan being paid for services, but too little to supply a full set of plans.[6] Essentially, Sam Colt designed Armsmear and may have taken his inspiration, in part, from Osborne House, Queen Victoria's royal retreat on the Isle of Wight, which was completed in 1846. Osborne House, with its tall campaniles, balconies, loggia, and irregular roofline, was one of the most-esteemed examples of the new Italian villa style of architecture. It was also famously not the design of an architect, representing a collaboration directed by its patron, Prince Albert, and Thomas Cubitt, a London builder.[7]

82. Armsmear, ca. 1858. R.S. DeLamater, Hartford, photographer. The only view of the house before the expansion and relocation of the conservatory in 1861.

Hartford's first encounter with the romantic Italian villa style was its 1849 train station, no less ironic as a style with rural associations applied to a form symbolizing modern technology and speed. American architects began promoting the Italian villa style in the late 1840s. Andrew Jackson Downing, one of its most esteemed advocates, described the style as

the most beautiful mode for domestic purposes... admirably adapted to harmonize with general nature, and produce a pleasing and picturesque effect in fine landscapes.... [Its] strong contrast of light and shadow... give it a peculiarly striking and painter-like effect... characterized... by scattered irregular masses... [that produce a] great variety of outline against the sky... [including] the towering campanile, boldly contrasted with the horizontal line of the roof... arcades... [and] balconies.... [The] style has the very great merit of allowing additions to be made in almost any direction, without injuring the effect of the original structure.[8]

Here we have a style ideally suited to landscape gardening and picturesque effects, which almost invites the patron to indulge in creative adaptation. Design theorists spoke of "grounds... arranged in smooth lawns, graveled walks, and finished gardens," a fountain here, "a few vases or statues" there, all adding "to the pleasing effect."[9]

With Armsmear, Colt reminded the world of three of his proudest achievements: the international celebrity he achieved as America's most flamboyant representative at London's Crystal Palace exhibition in 1851, his audience with Russia Czar Nicholas I in 1854, and his reception by the Ottoman Empire's Sultan Abdul Mejid I. Colt visited Constantinople in 1849, and he was dazzled by the Russian Orthodox influence of Moscow's recently completed Kremlin Palace (1849) on visits to Russia in 1854 and 1856. From the dome on his armory to Armsmear's "tent-like awnings," "pointed arches," and "gold-tipped and dazzling dome," pinnacles, and minarets, Colt's mansion had a "decidedly Oriental" flavor, which was as often remarked upon as its Italian villa form.[10] In 1876, at the height of its allure, Armsmear was compared to a "garden in Damascus... with its delicate, Oriental, capricious dome,... as if the owner had begun by being an English lord, and had ended by being a Turkish magnate, looking out over the Bosphorus.... It is a little Turkish.... [It] is Italian and cosmopolitan, [and yet] the feeling is English."[11]

Next to Iranastan (1848), P.T. Barnum's Moorish mansion in Bridgeport, Connecticut, Armsmear was, simply, the "most exotic house... incorporating some

84. *View of Armsmear, 1865. Harry Fenn (1845–1911), Montclair, N.J. (WA #05.53); Fenn was commissioned to provide illustrations for the book* Armsmear. *This view was the source for the illustration of the title page.*

Eastern features" anywhere in America before the Civil War.[12] Armsmear's most remarkable feature was Colt's conservatory, an "exquisite little architectural affair... all iron & glass à la Crystal Palace," which "somewhat Turkish in idea & seen under a Western sun seemed like a fairy palace" (figs. 83 and 84). Originally located facing Wethersfield Avenue, it was removed, at great expense, to the new "South front" of the house in 1861.[13] The conservatory most resembled Carstensen and Gildmeister's plans for the New York Crystal Palace exhibition building of 1853, where Colt was an exhibitor. At more than forty feet wide, with its symmetrical minarets, central dome "capped by a golden apple," six-foot-high glass panels, and "foliated arches of iron... painted red, yellow, and royal purple," the conservatory was an overwhelming presence.[14] How Colt managed to acquire the materials and plans to emulate this cutting-edge design feature is unknown. But in December 1856 he paid Hartford's George S. Lincoln Company $1,435 from an account labeled "conservatory," and it is likely that Lincoln, who developed a subspecialty in decorative cast iron, built this feature.[15]

From its steel and glass conservatories, to its patented roofing compound, to its lighting and plumbing, Arms-

mear featured state-of-the-art building technology.[16] Armsmear was ablaze in lighting and glass, two areas of building technology that had witnessed the most improvement in the years immediately preceding its construction. Considering that six-by-eight-inch panes of glass were still the norm in the modern houses of the Colts' early adulthood, the three hundred panes of white and clear glass plates and the dozens of "French looking glass plates" were a spectacle.[17] Elegant French chandeliers, which the Colts selected in Paris, and dozens of gas ports gave Armsmear a modern quality of illumination unachievable by all but the rich until the twentieth century.[18] The power for Armsmear's extraordinary lighting was supplied by Sam Colt's state-of-the-art coal-processing gas works, which also supplied the armory and housing in Coltsville as early as 1857. Visitors recalled a "house illuminated by hundreds of gas... lights," and in one year alone, Armsmear's gas bill was equal to about $17,549.[19] In 1862 it was noted that when the Hartford

85. *Library Table from Armsmear, 1857. Rosewood, designed and made by Ringuet, LePrince and L. Marcotte, New York. (Courtesy of Armsmear.)*

gas works was flooded and the city without lights for several days, "Elizabeth's mansion was the only one in the city" with light.[20]

Although the construction and some of the furnishings were provided by Hartford contractors, the interior of Armsmear had a decidedly international feeling. Colt was billed the equivalent of six hundred thousand dollars for furniture, upholstery, carpets, and drapery custommade by the New York office of the Parisian firm of Ringuet-LePrince and Marcotte (figs. 85 and 86), which Colt knew from their displays at both the London and New York Crystal Palace exhibitions. Marcotte's was New York's leading French decorating house, with clients among the nouveau riche of Newport and New York. The firm's New York office, directed by Leon Marcotte, gave the Colts' dining room, drawing room, library, office, and several bedrooms a distinctly French flavor. Chairs in the French rococo revival style, carved, tufted, and "upholstered in green-and-crimson-striped rep, to match the curtains," filled the drawing room, while a French renaissance look prevailed in the library. Gilt cornices, "light gilt brackets," "green reps" upholstery, "yellow-and-white damask curtains lined with flounce, ornamented with white ground border" must have dazzled under the ubiquitous gas light.[21]

The Colts visited the firm's Paris office during their bridal tour, to make changes and choices.[22] Sam Colt had already dictated certain design decisions by providing "chair covers of embroidery & Poch Persian work that I purchased at the great Asiatic Fair at [Novgorod]," and requiring that the furniture, "where there are drawers & cupboards," be secured with "locks and keys... furnished to me from London by Mr. Hobbs," an American lock manufacture whose fame at the London Crystal Palace exhibition in 1851 rivaled Colt's own. Years later, when Elizabeth Colt described Armsmear as "full of associations and memories," she might have added that associations and memories had played an important role from the beginning in its design and furnishings.[23] Indeed, Armsmear, with its affectations of world travel, cabinets of trophies and collectibles, rooms filled with Colt's own willow furniture, and even hardware selected to stimulate memories, was, in part, an autobiographical shrine.

The most eccentric feature of Armsmear was the dozens, perhaps more than one hundred, lithophane porcelain pictures that Colt ordered from Berlin to install in seemingly random patterns in the library, bedroom, and office windows, a practice undocumented and perhaps unpracticed elsewhere in American architecture.[24] Henry Barnard described the "numberless transparencies of porcelain biscuit... set in the sashes... in place of glass" as "embellishments of great rarity," and it was later noted that the "transparencies set in windows... all over the house... take on the form of famous paintings," creating a "veritable art gallery."[25] What remains of the collection suggests reproductions of paintings of subjects

86. *Chairs from Armsmear.*
Designed and manu-
factured by Ringuet,
LePrince and L. Marcotte,
New York. Left to right:
Louis XVI-style armchair,
1860, ebonized maple and
pine with ormolu mounts
and modern upholstery;
drawing room armchair,
1857, rosewood with
modern blue upholstery.
(Courtesy of Armsmear.)

ranging from religious and military to pastoral and classical architecture (fig. 87). When, in 1861, Colt was offered a unique collection of more than thirty paintings featuring works by Rubens, Murillo, Zurbaran, Vandyke, Tiepolo, and other old and modern European masters, he declined, no doubt concluding that he already had representations of great art, that being what mattered most.[26]

Sadly, most of the furnishings were dispersed, and only four of Armsmear's many rooms were photographed during Elizabeth Colt's occupancy. But the $60,000-plus spent at A.T. Stewart's New York department store, the hundreds of thousands spent with Marcotte, and a bill today equal to about $160,000 for interior painting, together with a few contemporary descriptions, provide a picture of opulence unsurpassed in interior New England in its day.[27]

The best description of Armsmear's interior is the estate inventory taken after Sam Colt's death in 1862. Until the age of photography, estate inventories were frequently the only and usually the best source for documenting inside the American home. Colt's, which was organized on a room-by-room basis, provides a glimpse of Armsmear at its consummation, down to the bridle bits in the barns and the individual flowers and flower pots in the greenhouses and conservatory. We learn, from the 516 books in Colt's library, for example, that he and Elizabeth were readers with interests in mythology, architecture, chemistry, military science, American and English literature, European travel, and above all, history. In addition to works by Swift, Goethe, Hawthorne, Emerson, and Shakespeare, there were histories of Mexico, Persia, Holland, Russia, Germany, and England. He owned biographies of Washington, Jackson, and Connecticut's own Revolutionary governor, Jonathan Trumbull, together with assorted histories of the United States, Connecticut, and Hartford. Closer to his personal experience were accounts of Matthew Perry's Japan expedition, Elisha Kane's Arctic expedition, Catlin's Indians, Capt. Randolph Marcy's exploration of the Red River of Louisiana in 1852, and Lt. William L. Herndon's explo-

ration of the Valley of the Amazon, all campaigns in which his revolvers played a supporting role.[28]

In the music room, which Elizabeth Colt later transformed into a reception room, were a piano, settee, two spittoons, and the cabinet of presents, including prized gifts from heads of state in Russia, Japan, Siam, Turkey, and Sardinia. In Colt's drawing room were a cabinet, tables, two sofas, fourteen chairs, two willow flower stands, and the famous Charter Oak chair. Upstairs in the ball room, which Elizabeth later converted into a picture gallery and display area for Colt's cabinet of presents, were fifteen table chairs, eleven ottomans, a chest, and the Charter Oak cradle, enshrined as a memorial since the death of the Colt's firstborn, Samuel Jarvis Colt, in 1858.

Although Elizabeth Colt remodeled the public rooms in the house, the private and secondary rooms remained largely unchanged for forty years. The drawing room (fig. 88) with its "yellow silk" carpet "made especially for it"; the dining room with its "low-backed upholstered armchairs"; the "maze of bedrooms" in blue, green, and rose with "canopies galore"; the billiard room with its "salmon walls" and "floors... of hard wood of various colors, laid in patterns"; their son Caldwell's "smoking room," with its "cherry panelling," "coat of arms... above the fireplace," Venetian-made Nubbin figures, and "semi-Chinese design of gilt paper"; and Elizabeth Colt's "most pretentious" bedroom, "upholstered in red" with its "frescoing of ceiling" and "books piled up" amplify the descriptions and photographs of the public rooms, providing a portrait of "cultivated elegance" in the "antebellum style."[29]

87. Lithophane Pictures from Armsmear, 1857, Royal Porcelain Works, Berlin, Germany. At least seventy-five panes in the window sash at Armsmear were filled with translucent porcelain pictures called lithophanes. Shown here are three of a collection of about thirty preserved in the Colt Collection.

88. Drawing Room at Armsmear, ca. 1868. R.S. DeLamater, Hartford, photographer. Note the portrait bust of Sam Colt on a pedestal (right).

No room was decked out with more splendor than Colt's billiard room, with its two billiard tables valued at four hundred dollars (today about twenty-two thousand dollars), four "willow chairs," a sideboard for dispensing liquor, four spittoons, and twenty framed paintings and prints. Colt's office at Armsmear was furnished with a "walnut secretary," bookcase, sofa, bedstead, mirror, and office table; his office at the armory contained a desk, sideboard, and "law library," a set of mounted "moose horns," fifteen maps, five "views of the City of Hartford," sixteen "red woolen shirts," and a poignant and revealing collection of pictures illustrating the Fourth of July, the first prayer in Congress, Christ weeping over Jerusalem, and Dr. Hawes, almost certainly the Reverend Dr. Joel Hawes, minister of Hartford's First Church, whose international best-selling *Lectures to Young Men on the Formation of Character* (1828) may indeed have constituted a sort of blueprint for Colt's personal morality.[30]

The campanile of Armsmear provided its most spectacular view and was one of its most alluring spaces. Brilliantly frescoed with Mannerist scrolls and flowers in green, blue, red, and brown, with raised panels and alternating red and blue ocular panels, the tower's two stories con-tained two substantial rooms. In the lower story were a bedstead, bureau, cabinet, and chairs. The upper story, with its willow settees, willow table, card table, and shaving stand was very likely Colt's favorite retreat. One can imagine long nights of cigars, brandy, perhaps a game of cards, ending with a restful sleep beneath the stars, waking, washing, and peering out to the east at the armory below, with the river and uplands beyond. To the south was a view of thirty miles; to the north, Mount Holyoke and Springfield; to the west, Hartford's new Central Park and train station with its Italianate arches and towers. The painted panoramic scenes of Hartford during the Great Flood were originally installed above the windows in the tower's upper story.[31] Here at a glance was an exhibition of transformation, a new industrial city of the future unfolding beneath a scene of old Hartford's devastation. Before the flood, after the flood. Before the dike, after the dike. Before industrialization, and now with the great armory, its smokestacks billowing, its wheels rotating, its hammers pounding, its prospects bright. Gazing in majesty—north, south, east, and west—from the pinnacle of his empire, Sam Colt must have known that he had kept his vows and had made anew the home of his ancestors.

For most residents and visitors to Hartford, Armsmear was a mystery experienced only through its gardens and greenhouses, and then mostly by peering into its protected domain from the city streets and factory village beyond. Elizabeth Colt's father, who moved into Armsmear after Sam's death, complained about being left alone during her many absences traveling, observing that "I do not think it right to leave such a house... under the entire control of servants" and that in spite of a "private watchman outside armed with a Revolver... and watch dog,... I feel sometimes a little timid... to be alone... in this immense house."[32]

Armsmear's reputation for extravagance, worldliness, and beauty was most evident from the outside. Even more conspicuous than its paintings, furnishings, and decorative art, the flowers at Armsmear never escaped notice. Sam and Elizabeth loved gardens, and shortly after breaking ground for Armsmear, Colt began making purchases of rare plants and flowers.[33] Purchases continued throughout 1856, and by midyear a British immigrant gardener named James Stubbins was added to Colt's payroll. Stubbins eventually supervised a staff of as many as "thirty men" who "daily roll, cut, and trim Nature to perfection."[34]

With the structure of Armsmear less than one-third completed, Colt hired Copeland and Cleaveland of Boston, one of the nation's first commercial landscape architectural firms. This firm, founded by Robert Morris Copeland and Henry W.S. Cleaveland, although today overshadowed by better-known contemporaries such as Henry Law Olmsted, A.J. Downing, and Calvert Vaux, was, in 1856, one of America's first and most-respected landscape and ornamental gardening design firms.[35] The firm competed for the contract to lay out New York's Central Park, and in the 1860s Cleaveland, following a stint with Olmsted and Vaux in designing Brooklyn's Prospect Park, relocated to Chicago, where he became "one of the masters of American park designing" in the Midwest.[36] Both men were influenced by British landscape design theorists Charles McIntosh and John C. Loudon, whose vision of social reform through art and of rural architecture and landscape as a civilizing force was expounded in numerous books and articles then circulating among American urban reformers. The Colts owned Charles McIntosh's *Book of the Garden* (1855), noteworthy for promoting steel and glass conservatories, expansive greenhouses, ponds, "rustic... bridges, covered seats, moss-houses & c." that "harmonize" with nature, all features incorporated in the intensely compact ornamental gardens at Armsmear.[37]

Copeland and Cleaveland provided designs for a "private garden," a "public garden," and, in a peculiarly

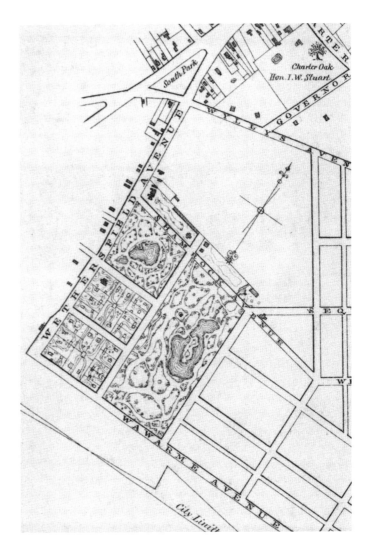

89. *Officers' Compound and Gardens at Armsmear, 1856. From "a Map showing the Lands & Improvements from the South Meadows," in* Armsmear *(1866).*

generous or megalomaniacal impulse, depending on your perspective, "houselots" for the chief corporate officers (fig. 89) Colt hoped might live within the grounds of the executive compound surrounded by gardens.[38] Evidence is overwhelming that, at least in 1856, Colt envisioned Armsmear as the citadel at the center of a compound occupied by a descending hierarchy of corporate officers, relatives, workers, and friends. Apparently neither Elisha Root, nor Colt's stepmother's brother Luther Sargeant, nor Elizabeth's brother John Jarvis bought into it, for they were all living elsewhere at the time of Colt's death.[39]

The private garden was magnificent and over the years served a quasi-public function. When Hartford hosted a meeting of the American Association of Medical Superintendents in 1870, they toured the city's acclaimed public high school, its school for the deaf, Trinity College, Colt's Armory and "the gardens and conservatories of Mrs. Colt."[40] Completed several years before Hartford acquired either a public park or a rural cemetery, the two amenities most influenced by the rural landscape movement of the 1840s and 1850s, the gardens and grounds of Armsmear represent another instance of the Colts as early adopters of the latest technologies and fashions of their time. The masses of earth moved and manipulated to create the gardens and grounds of Armsmear must

have been extraordinary. Henry Barnard alluded to the "bold terracing... of lawn" to the south and the two ponds, one described as almost four hundred feet in circumference and thirty feet deep.[41]

Visitors spoke of the most "serene, and English of views," with their "beautiful marbles, the artificial water—all bounded by... endless wood, through whose green branches" tall-masted ships were seen gently sailing along the "silver thread of the Connecticut... in and out under the encircling belt of hills whose uneven horizon line melts into the sky."[42] In keeping with the tradition of picturesque design, the views from the Colts' gardens were rich in associations and provided numerous contrived and carefully framed scenic panoramas. To one side, the "deer-park, flowers, shrubs, trees," and the "lofty house." To another, "a prospect over the flat below, corn-fields, meadows, grazing ground... [and] monumental elms." To the south, beyond the "ornamental trees... [and] broad orchard,... a glimpse of the armory, and Swiss village," with the steeple on the historic Wethersfield church far off in the distance.[43] The view most admired was the one across the "rustic bridge" (fig. 90) of a man-made lake filled with black bass and goldfish, with swans gliding on its surface, weeping willows hanging in profusion along its banks, peacocks strolling along

90. Rustic Bridge and Gardens at Armsmear, 1865. Harry Fenn (1845–1911), Montclair, N.J. (WA #51.21).

its gravel walks, and in the center, fountains of "water-nymphs" and a bronze colt. The lake was "large enough for a row-boat" (fig. 91) or for skating in the winter. Beyond was "the acme of this cultivated elegance," the "point where Art marries Nature," the "vine-clad summer house," known as Elizabeth's Bower (fig. 92).[44]

The lake embodied the marriage of art and nature with science and technology that was so central to Colt's worldview. The lake, "whose springs and conduits keep it clear and full," was the center of an elaborate system of reservoirs in which "water pumped from the Connecticut River... [was] conveyed to branch pipes into the different buildings" around Coltsville.[45] A pump, capable of moving "three hundred gallons per minute," filled "the two ornamental reservoirs on the grounds" and was then routed to the armory and boilers below.[46]

Whether due to cost or aesthetics, Colt was not content with pumping from the Connecticut River, and he spent the last months of his life drilling a hole into the earth, obsessed with tapping into the natural stream of an artesian well. As Colt's gardens and grounds were under construction, an article appeared in *Scientific American* promoting "Artesian Wells for all Situations," especially "in valleys where a dense stratum overlays springs."[47] Anxious to wed science to the indigenous cir-

cumstances of the Connecticut Valley's geography, Colt aspired to "sink an Artesian well deeper than any in France... in order to get... hot water from... the earth's interior heat, to" among other things, "warm his extensive greenhouses."[48]

Little more than a month before his death "Col. Colt... contracted with Mr. William Logan... to execute" his plan to "sink an Artesian Well on his premises;... to go 2,000 feet unless a stream of water sixty feet high was sooner reached." Beginning on December 5, 1861, Logan rigged up an engine used for pile driving at the armory with a 4-inch-round, 26-foot-long, 500-pound auger that drilled around the clock, reportedly at 5 feet per day.[49] Ten days after Colt's death, Logan had reached 108 feet; a month later, 200. When in March he finally tapped, at 238 feet, "a stream of clear cold water" (hardly a jet, and definitely not hot water), Elizabeth pulled the plug on what was arguably a ludicrous enterprise.[50] It did not, however, prevent her from resuming the search in a different location in 1868, which, after reaching a depth of 1,240 feet was again stopped.[51]

Encouraged by the landscape architects, Sam Colt engaged Hartford's premier monument maker, James Batterson, to supply and install marble figures of Diana and Apollo, the two most ambitious of the eleven sculp-

91. *Swan and Duck Pond at Armsmear, ca. 1870. R.S. DeLamater, Hartford, photographer.*

There were two ponds on the grounds at Armsmear; this scene is from the eastern-most pond.

92. *View from Elizabeth's Bower at Armsmear, ca. 1865. Prescott and White, Hartford, photographer.*

93. Apollo *on the Lake at Armsmear, ca. 1865. Prescott and White, Hartford, photographer.*

tures sprinkled about the grounds at Armsmear.[52] Batterson's *Apollo* (fig. 93), modeled after a classical figure at the Vatican, was described as standing almost thirty-five feet high. Several additional marble figures were installed around the gardens including a "goat suckling a kid," a "mastiff teased by a puppy," "Hebe pouring out nectar for the gods," (fig. 94), and Colt's favorite, a copy of August Karl Eduard Kiss's equestrian *Amazon Defending Herself* (fig. 95), reproduced in various sizes from the original, which was acquired by the Berlin Museum after it won first prize at London's Crystal Palace exhibition in 1851.[53] In addition to the bronze Triton in the conservatory, the outdoor fountain sculpture included A. Kalide's *Boy with a Swan,* Moritz Geiss's *Crouching Venus,* and a copy of the rampant colt from the armory dome (fig. 96).

The most unexpected sculpture on the grounds at Armsmear was the "grove of graves," which the Colts conceived in 1859, following the death of their firstborn son. The marble portrait of the baby, originally commissioned for the "private room" at Armsmear where Sam Colt "gathered the... memorials of those he loved best," portended the death not only of its subject, but of its sculptor, Edward Bartholomew, "who did not... live to complete [it]."[54] James Batterson probably oversaw the completion of the portrait and the creation of a glazed Grecian-style marble niche for its display.[55]

Beyond the gardens was the working farm. It was never clear if the tobacco warehouse, cattle, cornfields, blacksmith's shop, and orchards were the affectation of a gentleman industrialist, or if, having succeed at manufacturing, Colt became determined to create a profitable model farm. Colt's "farm in the meadow" was stocked with fifteen mules, four yoke of oxen, and an extensive inventory of agricultural tools, including a patented mowing machine, ploughs, cultivators, ice hooks, cider barrels, a fanning mill, and ninety-three tons of hay. Across the river in East Hartford, Colt owned a stock farm with a pair of colts named Hetty and Belle, a third named Lizzie (his wife's nickname), seven "imported Alderney" bulls and oxen, geese, hens, pigs, "49 turkeys," and "68 chickens," together valued at the equivalent of about $115,000, not including land; a serious business by any measure.[56]

The Colts stocked their gardens and grounds with exotic pet birds and wildlife. In 1857 and 1858 they pur-

94. Hebe *on the Lake at Armsmear, ca. 1866. R.S. DeLamater, Hartford, photographer. In the background is Elizabeth's bower, the gothic-style trellised* gazebo *reportedly situated at the point with the best views of the grounds and vistas beyond. (Courtesy of the Connecticut Historical Society.)*

95. Amazon Defending Herself, *ca. 1873. Isaac White, Hartford, photographer; bronze figure by August Kiss (1802–65). The exhibition-scale version of this sculpture, dated 1851, was one of the most famous large bronze figures exhibited at the Crystal Palace in London.*

The Colts owned a cabinet-size model (WA #05.1138) and this imposing lawn figure, a symbol of their taste in art and personal memories of travel in London and Berlin.

96. Rampant Colt Fountain *at Armsmear, ca. 1862. Detail of a photograph; probably R.S. DeLamater, Hartford, photographer. The bronze rampant colt originally installed on Colt's Armory dome in 1856 was destroyed in the fire of 1864. In 1860 Colt commissioned armory workmen to make a copy of the original figure out of cast and hammered sheets of zinc. This is the only illustration of the second rampant colt in the location for which it was originally intended. By 1866 it had been*

removed for use on the dome of the armory Elizabeth Colt had rebuilt and occupied by 1868. Standing by the rustic bridge by the lake behind Armsmear are a four- or five-year-old Caldwell Colt and three unidentified individuals who were probably household and garden staff at Armsmear.

149

97. General Jacquiminot Rose. From Kelloggs Series of Fruits, Flowers and Ornamental Trees, *1859, printed by E.B. and E.C. Kellogg, Hartford (WA #25.889).*

chased dozens of turtles, two "Chinese Pheasants," and seventeen geese and ducks, which wandered freely about the grounds.[57] Colt's swans, a gift "from the Duke of Devonshire... worth $100 a pair," were a famous attraction.[58] In 1857 the Colts acquired the deer and fawn that became the basis for Armsmear's deer park, a surprisingly exotic novelty in a then largely deforested New England.[59] "Tame enough to know no fear... the very incarnation of animal refinement," the deer provided amusement for Hartford newspapers that happily reported their success at getting loose and raising havoc.[60]

All this paled next to the spectacle of Colt's greenhouses. Their every aspect was carried out on an enormous scale, and in this effort it appears quite certain that, in addition to providing his family with a year-round supply of exotic fruits, vegetables, and flowers, Colt hoped to open new frontiers in the cultivation and merchandizing of produce for the northeastern market. With greenhouses and grape arbors stretching half a mile along the eastern border of the Armsmear homelot, its scale suggests the commercial aims of Colt's ambition. The operations were divided into two departments, each headed by an English-trained gardener: fruits and vegetables under James Stubbins and later "Mr. Solly," and the flowers under Thomas Maltman.[61]

"Col. Colt's love of flowers" was probably the original impulse behind the greenhouses.[62] From his first acquisition of "Chinese primulas, rhyncospermus...azaleas, forsythia," and "Large Begonia" in 1856, to the prize-winning displays of "Caladium Picteratum... Argentia... Grifinnii... [and] 24 varieties of Verbenas" in the year of his death, the flower department of Colt's greenhouses exhibited an upward spiral of complexity and sophistication.[63] There were cactuses, hydrangeas, fuchsias, chrysanthemums, forty varieties of orchids, and dozens of additional species of plants and flowers that, it was said, "brought back to Col. Colt the scenery of the gorgeous East."[64] It must have been tremendously satisfying to Colt as his friends at Kellogg Brothers printed the first resplendently colored lithographs of his hybrid roses for their "Series of Fruits, Flowers, and Ornamental Trees" (fig. 97). By the time of Colt's death, hundreds of camellias, azaleas, roses, rhododendrons, and geraniums grew in his greenhouse.[65]

Colt also actively promoted his produce by entering it in the annual competition of the Hartford Horticultural Society.[66] First it was pineapples, and eventually cotton, rice, numerous varieties of grapes, bananas, strawberries, cucumbers, peaches, cherries, squash, peppers, nectarines, apricots, oranges, lemons, and figs enough to "confound... ideas of chronology... all the year round."[67] In 1859 Colt showed giant strawberries of a type known as Trollope's Queen Victoria at the Hartford Horticultural Society exhibition. In the year of Colt's death, Solly, "the manager of the fruit department," exhibited five varieties of grapes, four varieties of peaches, nectarines, "blue Italian and perpetual Marseilles Figs," cherries, squash, peppers, large cucumbers, and "a splendid Pineapple."[68]

Observers were impressed by the sheer quantities of produce. In 1862 the gardener harvested 150 dozen peaches from the orchards adjacent to the greenhouses, and the winter of 1861, Elizabeth Colt's father, Rev. William Jarvis, boasted of how "Mr. Stubbins sold $500 worth of cucumbers; this month he can sell $1,000 worth" and estimated that "the hot and cold graperie" alone would produce "a profit of $5,000 when... under way."[69] Keenly aware of the market value of such luxuries, the following winter Jarvis boasted of eating several quarts of strawberries daily, "two or three dollars worth... as they would sell in New York" and of how "Elizabeth distributes them freely... [to] invalids."[70] Although prone to exaggeration, even half the profits Jarvis estimated in 1862 would involve sales valued today in excess of $100,000. Further evidence of the commercial scope of Colt's greenhouses is provided in the inventory of his estate, which contained 11,059 plants, including 4,000 strawberries, 600 tomatoes, 92 peach and fig trees, 350 grape vine slips, and 71 pear and plum trees.[71] Elizabeth Colt continued to exhibit at the annual fair at least until 1870, winning prizes from time to time and helping gain Hartford's fledgling Horticultural Society a modicum of prestige.[72]

While Sam clearly aspired to turn even his gardens and greenhouses to profit, Elizabeth (fig. 98) earned a reputation for gift giving. "To the poorest, saddest rooms; wherever there is a rough path to be smoothed, a sad life to be consoled... thither travel the luxuries of Armsmear."[73] In 1868, when Colt's old friend Augustus Hazard died, Elizabeth "directed that all the white flowers in her green-house be sent to the family."[74] Easter was filled with spectacle at Coltsville's Church of the Good Shepherd as "rare and beautiful flowers... from the extensive greenhouses of Mrs. Colt" were piled high on the altar.[75] For the golden wedding anniversary of a friend, "Mrs. Sam Colt sent a large basket of magnificent roses, pansies... lilies, callas, and hyacinths." For several years Elizabeth personally stage-managed an annual festival for the children of the armory's Sunday school. In 1868 she decorated a room at Charter Oak Hall. At one end was a Christmas tree, ablaze with "jets of gas... between the twigs... festoons of crystal and sugar... from branch to branch... [and] hundreds of brilliant balls, gilt nuts, toys, and dolls." Before it, were long tables "covered with... bouquets of rare flowers," beneath a "large portrait of Col. Colt... surrounded by a wreath of evergreens and flowers... surmounted by a cross of pure white camellias."[76]

98. Elizabeth Colt,

Horticulturalist, ca. 1866.

R.S. DeLamater,

Hartford, photographer.

(Private Collection.)

For most of its private life, Armsmear was "Mrs. Colt's house," and increasingly, it reflected her passions and tastes, and was filled with her things. Taking into account absences from Hartford, Sam Colt actually spent fewer than four years at Armsmear before his death. While the private rooms at Armsmear remained unaltered during Elizabeth Colt's forty-three-year occupancy, its public rooms were rearranged, reshaped, and redecorated into the 1880s. Only in her sixties did Elizabeth Colt slow down the pace of collecting and renovating.

She completed the expansion and reorientation of the house that was underway at the time of Sam's death, spending the equivalent of sixteen thousand dollars to add something called a "mastic nook" in June 1862.[77] In 1866 she converted the second-floor ballroom into a private art gallery. In 1871 she carried out an extensive architect-designed remodeling of her library. She made additional changes in advance of the lavish reception given at the time of her son's twenty-first birthday in 1879.[78] In 1883, following an eighteen-month tour of Europe, she spent today's equivalent of about $350,000

creating a "reception room" on the first floor in the former location of the music room.[79]

Unfortunately, with the exception of the picture gallery, which was first photographed shortly after its completion, the only visual record of these public rooms was made toward the end of Elizabeth's life. Unlike Sam Colt's estate inventory, which describes each room in precise detail, Elizabeth's inventory does not provide room-by-room descriptions. A visitor writing several years before the 1883 reception room renovations described the "endless treasures of illuminated and illustrated books, its jewels, pictures, marbles, and *objets de virtu*."[80] Elizabeth Colt's collecting reached its most intense and eclectic state in the late 1870s and early 1880s.

From the outside, the conservatory was Armsmear's most "oriental" feature. Inside, connecting with the library and reception room, the conservatory featured a bronze Triton fountain, shooting a jet of water high into the dome above a red-and-white marble-tiled floor (fig. 99). A chandelier of Venetian glass, a continuously rotating display of flower pots "brought day by day... when its flower is at its acme," and hummingbirds "flitting through the transom windows" created a "dreamland of poetry."[81] A visitor to Armsmear spoke of how the

99. *Conservatory at Armsmear, ca. 1864. R.S. DeLamater, Hartford, photographer.*

Colts' "winter garden," its conservatory and greenhouses, cheated "Fate of its revenges, life of its hardness," as if to "say to our winter cold and summer heat, Man has conquered."[82] Descriptions of "masses of roses," stands "laden with the choicest flowers,... beautiful wreaths and bouquets of natural flowers," and a "conservatory full of rare exotics" confirm that "Col. Colt's love of flowers waxed stronger in him to the end of his days."[83] For Elizabeth, flowers—at home, as gifts, at church, and in all seasons—became a personal signature and a lifelong reminder of the gentler aspect of her late husband.

Elizabeth's library (figs. 100 and 101), which was remodeled after designs by the architect Edward Tuckerman Potter, became the primary display area for her cabinet collections. Eventually, the room was stuffed with trophies and treasures representing a lifetime of memories and a world of associations. At a time when Hartford collectors such as Irving Lyon, Henry Wood Erving, Gurdon Trumbull, and Stephen Terry were engaged in their pioneering collecting of American antiques, Elizabeth Colt's collections were scrupulously international, to the point of shunning things both old and American-made. The architect's design for the library cabinets enshrined the concept of worldliness by incorporating a sort of iconographical encyclopedia of knowledge and internationalism as a decorative theme. Its fourteen banks of cabinets, reminiscent of the work of Philadelphia cabinetmaker George Henkels, feature carved sculptural portraits and allegorical representations in mahogany of the four continents and of the arts, sciences, and professions, each figure with the representative tools of his trade. Several among the twenty-three carved figures—Beethoven, Benjamin Franklin, William Shakespeare, and Samuel F.B. Morse—are portraits of distinct individuals; at least half a dozen of the rest probably are as well, but have not been identified.[84] Here we have about the library a pantheon of great men, past and present, American and European, real and allegorical—a veritable shrine to Western civilization and a symbol of Elizabeth Colt's implicit dominion over the empire of

100. *Library at Armsmear, 1903. West end, showing the custom-designed mahogany credenza crowned by a figure of Ruth. Photographed by DeLamater Studios, Hartford.*

101. *Library at Armsmear, 1903. East end, showing the mirrored overmantle carved with the Colt-Jarvis coat of arms. Photographed by DeLamater Studios, Hartford.*

knowledge. Needless to say, the furniture was custom-made and custom-designed by an architect whose signature was just this sort of complex, hand-crafted, allegorical ornamentation. He had already applied it in the design for the Church of the Good Shepherd, the Colt memorial church Elizabeth commissioned in 1867, and Potter would eventually carry out a third commission, a masterpiece of allegory and ornament, in the Caldwell Colt Memorial House of 1894.

The library was the most personal and autobiographical of the quasi-public rooms in Armsmear. As Elizabeth Colt's picture gallery was used more and more for entertaining, its cabinet collections were removed to the library. Here were some of the presents from heads of state that Colt prized so greatly. Here was the carved ivory and Charter Oak lamp shade commissioned at enormous expense as a last gift to Elizabeth the Christmas before Sam Colt died. Here were some of the Venetian glass she had purchased on her visit to Italy in 1881, the malachite blocks, ornaments, and dinner bell that the Colts acquired in Russia in 1856 (fig. 102), the reproduction columns from the Roman Forum she bought in Rome on her tour of Europe in 1882, the enormous painted Bohemian cameo glass vases, and the velvet and metallic embroidery upholstered, carved throne (fig. 103), a copy of an original in the doge's palace in Venice and the most outstanding of her grand tour acquisitions.

In this shrine to worldliness were treasures amassed by world travel. At one end of the room was an enormous mirrored and canopied overmantle, carved with the Colt-Jarvis coat of arms; at the other end, Potter's monumen-

tal mahogany credenza, surmounted by a mirrored triptych with a central niche containing a porcelain figure of Ruth (fig. 104), the Moabite ancestor of Jesus, whose marriage to Boaz symbolized the transcendence of tribalism. Decorative art or personal iconography? At the base of the mirrors were two outstanding hand-painted earthenware vases (fig. 105), probably purchased at the Centennial Exhibition in Philadelphia in 1876. To either side of the mirrors were marble portrait busts of Elizabeth, which Sam and Elizabeth commissioned from Hartford's most-esteemed native sculptor, Edward Bartholomew, in 1858. An architecture of personal memories, a tableau of faith and learning, a shrine to worldliness, and an affirmation of her identity as a woman, this, on top of a twenty-foot-square Persian rug, was Elizabeth Colt's library.

Elizabeth Colt's readings were no less reflective of her autonomy and increasingly self-confident femininity than her collections. The dated inscriptions on the flyleaves of many of her more than four hundred books, still

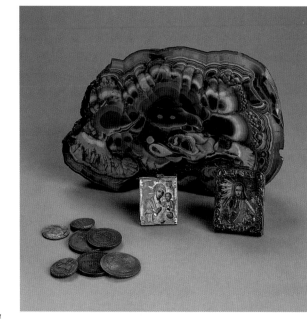

102. Russian Relics, 1856. Polished malachite (WA #05.1562); icon of the Virgin Mary (WA #05.1548); icon of the head and hands of Jesus (WA #05.1547); coins (WA #05.1575).

103. Copy of Grand Chair from the Doge's Palace, ca. 1882. Carved walnut with modern upholstery, Venice, Italy; purchased from Moisa Rietti in Venice during Elizabeth Colt's European tour in 1882 (WA #05.1582).

intact in the library, suggest that she spent much of the first year after Sam's death reading.[85] The library was well stocked in the work of contemporary American, English, and Scottish poets, as well as works by contemporary women writers, such as Charlotte Bronte, Hannah More, and Harriet Beecher Stowe. Religious books were supplemented by revealing expressions of incipient feminine consciousness such as Mrs. Ellet's *Queens of American Society* (1867), Mrs. Jameson's *Characteristics of Women: Moral, Poetical, and Historical* (1848), and a work titled *Eminent Women of the Age* (1869). She owned more than seventy-five of Sir Walter Scott's Waverley novels, and collected works by Dickens, Thackery, Edward Lytton, Thomas Gray, and Washington Irving. There were art books acquired while she was assembling her picture gallery and just prior to her grand tour of Europe, including Charles C. Perkins's *Historical Handbook of Italian Sculpture* (1883), Wilhelm Lubke's *History of Art* (1868), and Wornum's *Epochs of Painting* (1864). Travel literature included *The Gates of the East* (1877) by family friend and brother to the library's architect, Henry C. Potter, *Bancroft's Tourist's Guide to Yosemite* (1871), and *A Complete Pronouncing Gazetteer... of the World* (1858). There were books on gardening and history and dozens of titles personally inscribed by authors and friends, including Hartford's most illustrious writers, Lydia Sigourney, Charles Dudley Warner, and Harriet Beecher Stowe. Finally, there were a number of books in which she helped the authors write, or perhaps paid for, biographical profiles of Sam Colt, notably *Appleton's Cyclopedia of American Biography* (1887), Henry Howe's *Adventures and Achievements of Americans* (1859), and a collection of biographies titled *The New England States* (1897).

The reception room (figs. 106 and 107), which replaced the long east-west music room on the first floor in 1883, had a distinctly more feminine aura than the library or picture gallery. Gold, white, and yellow were its prevailing colors, and it was the one room in Armsmear that was practically devoid of objects or memories of Sam. Aside from a couple of the Marcotte chairs, reuphol-

104. Ruth, *1870.*
Parian porcelain,
designed by William
Brodie (1815–81),
manufactured by
W.T. Copeland and
Sons, Stoke-on-Trent,
England (WA #05.1123).
Brodie was a
prominent sculptor
and member of
the Royal Scottish
Academy. Copeland
perfected the formula
for Parian porcelain,
regarded in its time
as the "next best to
marble" for sculptural
figures. Elizabeth Colt
may have identified
with the biblical
character of Ruth.

105. *Royal Doulton Pottery.*
Salt-glazed stoneware,
Lambeth, England.
Left to right:
vase, 1878
(WA #05.1407);
vase, 1875, by
George Tinworth,
probably purchased
at the Philadelphia
Centennial
(WA #05.1408).

106. *Reception Room at Armsmear, ca. 1901.*

107. *Reception Room at Armsmear, ca. 1901.*

108. Infant Bacchus, *1855. Marble, designed by Anton Hautmann (1821–62), Munich, Germany (WA #05.1124). Hautmann studied at the Munich Academy under Ludwig Michael von Schwarthaler, the principal master of classical sculpture in southern Germany in his day.*

stered to conform to the new color scheme, a German marble sculpture of the infant Bacchus (fig.108), acquired during the bridal tour, and the eerie presence of Edward Bartholomew's postmortem marble portrait of Samuel Jarvis Colt, almost everything in the new reception room postdated 1880 and showcased art and objects acquired during and after Elizabeth's tour of Europe. There was a pair of Sévres-style, so-called Richelieu vases standing almost six feet high on matched pedestals, complete with histories of ownership by French royalty, for which she paid dearly in 1889.[86] Even grander, and half again as expensive, was the five-foot-high, Sévres-style porcelain vase and cover (fig. 109), which when originally mounted on a two-and-a-half-foot-high ebony stand decorated with elephant heads, was a commanding, if somewhat bizarre, presence in the room. With its gilt-bronze mounts and hand-painted scenes of eighteenth-century aristocrats prancing around their gardens, the vase was a tour de force of a sensibility that is now gone, and, perhaps blessedly, forgotten.

Armsmear's reception room epitomizes the kind of blunt ostentation that was the trademark of America's Eurocentric nouveau riche during the 1880s. At Armsmear the only hint of the refined and reformed taste of the British-inspired "aesthetic movement," is the linear grid of the room's ceiling fresco. Visitors were probably more impressed with the room's wall-to-wall nonrepeating Aubusson carpet, which probably cost as much as a small house. The room's monstrous gilt rococo mirrors, wall brackets, and pedestals, the tallest standing six feet high and all being expensive acquisitions during Elizabeth's stay in Venice, are so grotesque as to appear almost comical, even more so by serving as a support for refined, hand-painted French faience or as a backdrop for diminutive, saber-legged chairs. Compared with the unified elegance of the lady's reception room (now a period room at Wadsworth Atheneum) across town in Maj. James Goodwin's Woodlands—Hartford's other "millionaire's mansion"—Armsmear's reception room is more intriguing than beautiful, although Duchoiseuil's bronze figure of an American Indian maiden paddling her canoe (fig. 110) is an aesthetic masterpiece that, for Elizabeth, must also have been rich in symbolism.[87] Surely Elizabeth had not forgotten how Sam Colt had "paddled his own canoe" and the powerful symbolism of defiance, purpose, and self-affirmation that was bound up in this splendid metaphor of American enterprise. In a small cabinet were a few treasured personal mementoes—bridal portraits in miniature of Sam and "Lizzie"; a miniature portrait of her beloved brother Richard Jarvis, a lifelong resident at Armsmear; a portrait miniature of her adult son, Caldwell, also a lifelong resident at Armsmear; a gold locket of George Washington's hair (which she bought in 1871 for the equivalent today of more than ten thousand dollars); small medals and mementoes of Sam Colt's fame; and a miniature painting on copper, slick-finished, striking and lovely, of night-lit cherubim watching tenderly as the baby Jesus is nursed by his mother. Sentimental, even melodramatic, Elizabeth Colt was a woman unashamed of her emotions and affections. God and family were always close at hand and near to heart. What her reception room lacked in aesthetic sophistication, it gained through an intimacy of emotion and association.

Elizabeth Colt may not have been a great interior designer, but her patronage and interaction with painters, writers, illustrators, sculptors, and architects created monuments in American art and milestones in the history of women's leadership in cultural patronage. Was Elizabeth Colt America's "first woman patron of American art"? Her husband, who liked to be "second at nothing," might have cared. I am not sure she did. But Elizabeth Colt did care about excellence and about feelings and faith. These were the qualities that infused the artistic shrine that became her picture gallery (fig. 111). The picture gallery was created by a forty-year-old woman, recently widowed after a six-year whirlwind romance with one of the most famous and charismatic men of his generation. It is a memorial shrine to his fame, to their lives together, and to their shared passion for transcontinental experience. At its unveiling in 1867, Elizabeth

109. Rococo-style furnishings from Armsmear. Left to right: covered vase, *ca. 1870, Sévres-style, soft paste porcelain, probably made by Samson Porcelain Manufactory, Paris. Painting attributed to J. Pascault (1782–1847), mounts attributed to A. Bailly (WA# 05.1260). This vase was originally displayed on a teak stand decorated with carved elephant heads. The ensemble was exhibited at the Paris World's Fair in 1888;* pedestal, *oak and gilt, ca. 1880 (WA# 05.1103). Elizabeth Colt purchased a pair of these and a pair of gilt wall brackets from Venetian art dealer Moisa Rietti in April 1882;* covered vase, *ca. 1860, Sévres-style, soft paste porcelain, also probably by Samson (WA# 05.1259). One of a pair reportedly owned by Mme. de Parabere and exhibited in Paris in 1867 and in 1889; the painting has been attributed to Bertrand. The fashion of eighteenth-century rococo-style Sévres porcelain led to the production of many high quality fakes and revivals. The two vases shown here are the largest among a group of seven comparable pieces collected by Elizabeth Colt, who believed them to be rare antiques of illustrious provenance.*

Colt's picture gallery was one of the most contemporary and lavishly appointed art galleries in New England and the only one created independently by a woman.

Elizabeth and Sam may have discussed the idea of an art gallery in 1861 when he corresponded about acquiring a collection of European pictures through a political crony stationed in Portugal. Sam Colt first talked of building public collections of art and history in conjunction with Charter Oak Hall as early as 1856 but made no progress in that direction before his death. Elizabeth conceived her picture gallery in 1864, in conjunction with the portrait of Colt that she commissioned from the artist Charles Elliot. The year following its unveiling, Elizabeth Colt visited private and public galleries in New York, consulted with Hartford and Newport family friends and artists Frederic Church and Richard Morris Hunt, and acquired her first pictures. By the end of 1866, it was well enough underway to be alluded to in Henry Barnard's memorial biography of Sam Colt. A year later, in time for her sister's wedding, the picture gallery was ready to be unveiled and it served as the focal point for one of the most lavish ceremonies in Hartford's history.

Facing one another across a skylit corridor, Sam's and Elizabeth's portraits were on a scale typically associ-

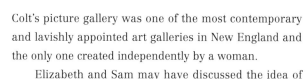

110. Indian Maiden in Canoe, *ca. 1873. Bronze on Eastlake-style ebonized pedestal, by Duchoiseuil, Paris, France (WA #05.1129). Although little is known about this sculptor, his work was featured in the renowned Paris Opera House.*

ated with royalty or heads of state. On the walls between them were symmetrical stacks of contemporary European and American paintings. Bathed in natural light during the day and sparkling with gaslight at night, the picture gallery was a tasteful ensemble of color, texture, and visual narrative that became one of Hartford's most admired and sought-after locations for parties and receptions. With its variously colored hardwood floors, large and intricately patterned oriental rugs and furniture, immense swagged canopy and cornice drapery, and fabric-lined walls and upholstery, all color-coordinated in dark green, Elizabeth Colt's picture gallery provided a spectacle of dazzling theatricality. The effect, while comparable in style and content with picture galleries created by her peers, was more personalized than most and contained enough biographical and autobiographical elements to give it a shrinelike essence with slightly morbid overtones.

The picture gallery's public debut was the marriage ceremony of Elizabeth Colt's sister Hetty Jarvis to Cyprian Nichols Beach just before Christmas 1867. With more than fifty international guests, including "Russian Commissioners... in full uniform" and prominent ladies such as "Mrs. Junius Morgan of London," the "lavishness in dress... diamonds and other precious stones" was a display such as "had never been seen in Hartford."[88] As thirty coaches shuttled guests back and forth between Armsmear and the city's hotels, Elizabeth presided over the affair with "queen-like grace," dressed in a "magnificent robe of black velvet," with a "coronet of diamonds in her hair." That same week, Elizabeth "caused a species of altar to be constructed" in front of Sam Colt's portrait, in order, it was claimed, that he might "seem to assist—to 'give her away.'"[89] As one hundred of an estimated thousand guests crowded into the picture gallery to watch, Hetty Jarvis, a "magnificent looking woman," decked in pearls and diamonds, stepped up to the altar in front of Sam Colt's portrait, "a marriage bell of rare flowers" over her head, and "like a dream of fairy land" was joined in matrimony by Connecticut's Episcopal bishop. After the ceremony, Colt's Armory Band took the stage, and Armsmear, with flowers piled on every mantle and billowing out of every vase and with tables spilling forth a bounty of exotic New York-catered foods, came alive with sound. Guests waltzed into the evening hours as the billiard room adjoining the picture gallery filled with presents "until it seemed as if the whole of Tiffany and Co.'s stock had been transferred to Hartford."[90] As the silver, coral, pearls, embroideries, paintings, dressing and jewel cases, and books piled up, Elizabeth and Hetty's father

111. Picture Gallery at Armsmear, ca. 1901.

speculated that the "gold and silver alone" was worth $30,000 (today, more than $1 million) and that the event "surpassed anything of the kind ever before witnessed in this country." He boasted proudly of the generosity of the new husband who "secured to Hetty an income of $10,000 [today about $350,000] a year for her private expenditure."[91]

The Gilded Age had arrived. The hot new book in the stores was *Queens of American Society,* and women's rights leaders Susan B. Anthony and Elizabeth Cady Stanton were in town for a lecture and public forum in Hartford's premier public hall.[92] The Civil War was over, but the culture and class wars, and the endless jockeying for position among America's elite. had just begun, with American women leading the charge. Life at Armsmear, like manufacturing at Colt's Armory, was carried out on a grand scale. Indeed, until the Newport mansions eclipsed it in the 1880s and 1890s, few homes anywhere in New England rivaled Armsmear's excess.

The dozens, perhaps hundreds, of small gatherings that filled the social calendar at Armsmear mostly passed without notice. By destroying her personal correspondence, Elizabeth Colt protected the inner life of Armsmear and its residents from scrutiny. It was said that the "lady of Armsmear has but to indicate her wishes" and Colt's Armory Band was on hand "to add the charms of good music to… her lawn-parties, or of a moonlight evening to serenade her guests."[93] From the Armory Band to the much respected choir at the Church of the Good Shepherd, from Elizabeth's picture gallery to Sam's patronage of the American artists Edward Bartholomew and George Catlin, the Colts assured that music and art would have a prominent place in the life of their industrial community.

In 1870 Elizabeth hosted "a grand concert" at Armsmear by Dimitri Slaviansky, the tenor and former director of the Grand Opera at St. Petersburg, Russia, who was then on tour in America. Slaviansky and his thirteen-man orchestra, "clad in fancy costumes of their own nationality," entertained one hundred guests in the picture gallery followed by "a magnificent breakfast."[94] Earlier that year, Elizabeth Colt hosted a "masquerade ball" for her eleven-year-old son in which guests appeared in costume and were entertained by Colt's Armory Band.[95]

In 1879 Colt's Armory Band kicked off the second greatest social event in Armsmear's history with a "surprise serenade" for "Mr. Caldwell H. Colt" outside his bedroom window at eight o'clock on the morning of his twenty-first birthday.[96] As Sam Colt's sole heir reached his majority and came into his inheritance, the drum-roll of expectations must have been almost unbearable. An estimated fifteen hundred people were invited to Caldwell's "reception," described as "one of the most brilliant parties which was ever given in Hartford." Armsmear "easily held the large attendance without… discomfort,"

112. Caldwell's Twenty-first Birthday Reception, 1879. From Frank Leslie's Illustrated Magazine, *December 13, 1879. (Courtesy of the Connecticut State Library.)*

with each room "beautifully decorated with flowers" and the number "21" crowning the entrance to the reception room in red roses (fig. 112). A highlight of the reception had Caldwell leading his society friends from Newport, "with powdered hair and in costume," bounding down the stairs from the picture gallery "arm and arm," singing "the Nursery Rhymes from Mother Goose," and dancing "double quadrilles" to the orchestra below.[97]

The serenade may have been a surprise, but the party certainly wasn't. During the entire preceding week, Armsmear was "in the hands of skillful decorators," who depleted the gems of Elizabeth's greenhouses to create a "fairyland." The parlors, drawing room, library were "transformed... into bowers... fragrant with... [the] odor of roses, mignonette, lilies, violets" and "a colt of bronze, resting on a high bank of flowers." With bands performing in the reception room and picture gallery, there was "dancing... [and] waltzes everywhere." In the dining room, silver epergnes were filled with tropical fruit while guests dined from a menu featuring "terrapins, salmon en mayonnaise, boned game and turkeys, pâté de foi gras, salads and partridge."[98] It was a fitting tribute to Sam and Elizabeth's sole heir, whose pampered life epitomized the culture of wealth and privilege associated with the Gilded Age (fig. 113).

Caldwell's story is not a particularly happy one. Smothered with attention, pampered and provided, to little avail, with the best education money could buy, Caldwell, Elizabeth clearly hoped, would lead Colt's Armory into the next generation and restore its fading reputation for inventiveness and enterprise. In 1888, at the age of twenty-nine, Caldwell was made a vice-president of the company, and the *Hartford Daily Times* applauded his "energy and executive ability," suggesting that "his influence in promoting the prosperity of this great company will be great."[99]

Such expectations must have sounded bizarre to anyone who knew the young yachtsman and bon vivant. From his earliest years, family correspondence was

tinged with dismay at the way young Caldwell Colt's "every wish" was gratified by a doting mother and aunts, from the double-barreled shotgun his mother gave him at ten, to the succession of private tutors, to the adventures that took him to Denver and the West at thirteen for six weeks, hunting and camping in Wisconsin at fourteen, to Europe as a teen, and at age twenty, the first of his many trips to Florida.[100] Psychologists could, perhaps, have a field day with the man whose "two... most conspicuous qualities," as described by Mark Twain, were "his devotion to his mother and his splendid masculinity."[101]

Caldwell was no freak. Among the rapidly expanding class of Gilded Age millionaires, Caldwell became a stereotypical icon of foppish indulgence, whose fame was earned by his courage, audacity, and heroism at sports and what today might be called "attitude." After Caldwell received his inheritance in 1879, his family grew increasingly concerned at his leaving "the management of his affairs to his mother," noting, with more regret than surprise, that he "does not take kindly to business."[102] What he did take kindly to was sailing, hunting, and

113. Portrait of Caldwell Hart Colt, 1894. Oil on canvas, Eastman Johnson (1824–1906), New York. To the end, Elizabeth Colt used artistic patronage to validate the aspirations of her generation's artists. Although Eastman Johnson was one of the most esteemed American genre painters of the Civil War era, he ended his career painting portraits, a less glamorous, if not less remunerative pursuit. Johnson portrayed Caldwell as a gentleman yachtsman on the deck of the Dauntless, *a flattering representation of the yachting culture of the Gilded Age. (Courtesy of the Church of the Good Shepherd.)*

gambling. Like his father, Caldwell had a flair for the sensational, once reportedly raising a flag from the mainsail of his ship challenging Britain's yachtsman to "any course, any time, for any sum."[103] Caldwell had not yet come into his full inheritance when the newspapers reported him instigating a bet in Newport with "Mr. DeForrest Grant to shoot at 50 birds for $500." One can only imagine what the workmen at Colt's Armory thought of the young heir gambling the better part of what for them was a year's wages on birds.[104]

Caldwell was a champion of lost causes, admired by his friends as a "sportsman for the sport's sake." His motto, "Give Fish and Bird a Fair Chance," his enthusiasm for entering "contests when he knew his own boat had no chance of winning the victory," and his reputation as a lavish entertainer suggest that the burden of parental expectations was heavy indeed.[105] And yet there was something touchingly heroic, if not a bit crazed, about his embarking in 1887 on an almost suicidal attempt to defeat the *Coronet* in one of the great transatlantic races in American yachting history. Wagering $10,000 (today $350,000), Caldwell cast off in the *Dauntless* from New York in a storm, taking sixteen days to cross the Atlantic and arriving thirty hours behind its rival, whose owner remained comfortably at home in New York waiting for his hired crew to report their victory. Caldwell was on board through an ordeal in which the crew, driven night and day through often terrible weather, ended up subsisting on champagne and wine after their water supply was lost.[106]

Racing the Atlantic in winter from New York to Ireland was the Mount Everest of Victorian yachting culture, and even though he lost, Caldwell Colt became an American yachting legend. The next year he was made vice-commodore of the prestigious New York Yacht Club, and in 1892, at the age of thirty-three, was made commodore of the Larchmont Yacht Club, joining the undisputed elite of U.S. sporting society.

Caldwell's fascination with the world of fast boats and fast living was spawned in 1868, his first summer in Newport, where, at the age of ten, he caught his first glimpse of the *Dauntless*. Although Newport hosted races and attracted yachtsman from New York and Boston as early as 1844, it was in the late 1860s and early 1870s that the race for the Newport Cup and the subsequent America's Cup, became an annual ritual and the centerpiece of Newport's posh summer season. Caldwell and his mother were in Newport in August 1868, when the New York Squadron, led by Commodore James Gordon Bennett and his famed sloop *Dauntless,* arrived in Newport Bay.[107] Under a previous owner, the *Dauntless* had been a champion racer. Under Bennett's command it would defend the America's Cup and, in 1870, achieve a worldwide reputation by racing the *Cambria* in one of the first transatlantic contests.[108] America's yachting culture in the seventies was all flash and glamour.

Caldwell acquired his first yacht, the sloop *Lizzie,* at eighteen and eventually owned five yachts, including the sloop *Wizard,* the sloop *Atala,* the ketch *Oriole,* and the schooner *Dauntless.*[109] In 1881, when Caldwell Colt bought the *Dauntless* (figs. 114, 115 and 116), he was buying a piece of history.[110] He toured Europe and the West Indies and regularly visited Florida, where he was a member of the St. Augustine Yacht Club. In all, he spent at least $250,000 (today $8.75 million) on yachts, equipage, and crews, averaging an estimated ten months a year sailing.[111] He racked up some victories (fig.117), but more often, what was reported in the Hartford newspapers was embarrassing. In 1884 the *Dauntless* sprang a leak off the coast of Spain, and later that year Colt's sloop *Wizard* sank after drifting in the tide and smashing against a bridge abutment in Elizabeth's ancestral home of Saybrook, Connecticut. In 1886 the *Dauntless* collided with a fishing vessel off Faulkner's Island while returning from a race in Marblehead; then apparently sailed on toward New York without ascertaining damages to the other party.[112] These episodes and the compulsive betting and extravagance cannot have been a comfort to Elizabeth, and while Caldwell's drinking, high-living, and womanizing exist only at the level of folklore, there is probably a flicker of truth in his reputation as a lush.[113]

115. Yacht Dauntless.
Photographed by
George A. Stewart, from
Representative American
Yachts *by Henry*
Peabody (Boston:
H.G. Peabody, 1891
[WA #44.201]).
The Dauntless *became*
famous under the
command of its owner
the newspaper publisher
James Gordon Bennett,
who raced it across
the Atlantic twice and
raced for the America's
Cup in 1870.

114. Caldwell Colt and
Officers of the Dauntless,
1887. Photographer
unknown, New York City.
This photograph was the
primary source for
Eastman Johnson's
posthumous portrait
of Caldwell Colt.
Left to right:
Caldwell Colt, an
unidentified helmsman,
and Colt's professional
skipper Samuel "Bully"
Samuels. Samuels was
a former clipper ship
skipper who achieved
fame in 1870 when
James Gordon Bennett
engaged him to skipper
the Dauntless *in a*
transatlantic yachting
race against the
Cambria. *In 1887,*
at the age of sixty-four,
Samuels came out of
retirement to skipper
the Dauntless *in a*
legendary transatlantic
race against the
Coronet, *driving it*
through hurricane-
force winds.

116. Cabin of the Dauntless, ca. 1892. Several of the yachting trophies, lamps, furniture, and tableware illustrated here are preserved in the Colt Collection. With its coal grate, tufted-upholstered interior, and decorated woodwork, Caldwell's cabin in the Dauntless *is a portrait of Gilded Age opulence (WA #44.232b).*

117. Dauntless Silver Trophies. Left to right: loving cup, 1888, Whiting Manufacturing Co., Providence, Rhode Island; this was converted into a lamp used at the dining table in the officers' cabin of the Dauntless; *yachting trophy, 1887, Tiffany and Co., New York; one of Caldwell's most outstanding yachting victories was the Queen's Jubilee race hosted by the Royal Nova Scotia Yacht Squadron in 1887.*

Asserting "dominion over the earth," whether through art, flowers, politics, guns, or machines, was a big part of what motivated Sam Colt—the intense, rootless, visionary idealist—and his wife. Armsmear was both a home and a manifestation of its builder's quest for empire. It was famous and infamous in its time. In addition to commissioning a memorial biography–house book about Armsmear, Elizabeth encouraged essayists and magazine journalists to write about it, and thus to keep the memory of its founder alive. It did not hurt that she was willing to spend what it took to amass dozens of flattering and revealing photographs and illustrations for their use.[114] In 1876, at the height of Centennial fever, popular essayist and travel writer Mary Elizabeth Wilson Sherwood published a lavishly illustrated article on Armsmear as the seventh in a series on the homes of America in the *Art Journal,* the most prestigious of what might be called the "arts and lifestyle" branch of the American publishing industry. And in 1879 Armsmear was featured in a popular book on the homes of America.[115] When Sam Colt died, the house and homelot were appraised for the equivalent of about $10 million, bought and paid for by a mercurial industrialist, his Episcopal wife, and a world more gun-crazed than our own.[116]

Dominion and empire are ephemeral, and by the end of its life as a private residence, Armsmear's glow was subdued as passers-by described its "somewhat stern and gloomy" appearance and the "cold-looking exterior of the west front.... Thirty or forty years ago," the observer noted, "the panorama... must have been entrancing," before the trees, "grown to full maturity,... [shut] out the scene of the inventor's activity and triumph."[117] In the end, Armsmear was isolated not only from the city but also from the factory village, over which its dominion was never fully secure. In 1881 it was discovered that the seedlings of Charter Oak, which Colt had planted with such pride and optimism in 1857, "proved to be... swamp oak."[118]

And what of the residents of Armsmear? What did they think of this "long, grand, impressive, contradicting, beautiful, strange thing"? What of Miranda Robinson Anderson, the black woman and probable descendant of the Hart family's slaves from Saybrook, who nursed the Colts' four children and lived, with her brother, as a member of the Armsmear family?[119] Or Mary Degerinan, the Slovanian housekeeper, who spoke German, Russian, French, and English and so charmed Elizabeth during the Colts' bridal tour in St. Petersburg that she was urged to move four thousand miles from home to Armsmear, where she met and married the Colts' English gardener Thomas Maltman, on one of the few gay occasions during Elizabeth's year of loss and mourning.[120] Or of Elizabeth's parents (fig. 118), brother Richard, and niece Isabelle Colt, all residents of Armsmear at or soon after the time of Colt's death. Or of Elizabeth Kelly, twenty-eight, the Irish seamstress, or Mary Brennan, thirty-six, the Irish cook, or the nurses Mary McGraff and Hannah Conden, or Henry Champion, thirty-two, the black waiter,

bearing the name of an illustrious old Revolutionary War general and Connecticut West Indies trader.[121] These voices and the impressions of the thousands of residents and visitors who annually passed inside and out, beyond the protective hedges of the palace built by Hartford's most famous native son, are mostly silent.

Memories are silent. And in the end memories alone sustained the queen of Armsmear, who wrote in the last months of her life of being "impatient to be home again.... Though there is no one to welcome me... [Armsmear is] so full of associations and memories of my dear ones, that I do not feel lonely there."[122]

118. Elizabeth Colt's Parents at Armsmear, ca. 1868. Probably R.S. DeLamater, Hartford, photographer. Elizabeth Miller Hart Jarvis (1798–1881) and Rev. William Jarvis (1796–1871). Elizabeth's parents and sister became residents of Armsmear shortly after Sam Colt's death in 1862. The parents are seated at the Marcotte furniture in the picture gallery at Armsmear in front of a portrait of Elizabeth and Caldwell. (Private collection.)

FRONTIERS OF CIVIC CULTURE
SAM AND ELIZABETH, ART PATRONS

The salvation of art in this country... depends on the women.
PROFESSOR HAWKINS, *HARTFORD DAILY TIMES*, NOVEMBER 24, 1877

More than guns, machines, or their home, Armsmear, a passion for art, a yearning for self-improvement, and the hope of being remembered for "good works" were at the core of the partnership that bonded Sam and Elizabeth Colt. Their patronage of art, architecture, and sculpture helped them define for themselves and their community the meaning of creativity and invention and the role of privilege and worldliness in the industrial civilization they helped build. Eventually, Elizabeth Colt's role as a patron, collector, philanthropist, and institution builder would overshadow her husband's accomplishments in these fields, because of the enduring identification of Sam Colt with guns and manufacturing and the longer years she devoted to such work. It misreads the man. Curiously, Colt's estate, detailed to the precise number of flowers in the greenhouse and bottles of wine in the cellar, was devoid of guns, and there is little evidence that as an adult Sam Colt *ever* hunted or even carried firearms. His aspirations were bigger than guns.

What did Colt aspire to do, asked his biographer Henry Barnard.

Rear me a palace... within which the great masters of every fine art shall clothe... the familiar with golden exaltations;... dedicate... a school, where art and science shall unfold their mutual relations more harmoniously than ever before... [where] theory and practice, each supplying what the other lacks—shall embrace each other. There, whatever seed of invention lies latent... shall take root... and bear fruit.... The useful arts—one and all—shall become so interpenetrated with science, that they shall... be recognized as equal to... the "fine" and "liberal" [arts].[1]

By midcentury, the impulse to use art to mitigate the unsettling effects of industrialization was apparent throughout the nation as cities created fledgling galleries, patronized painters and sculptors, and made pictures in parlors and dining rooms a convention of genteel living. With its vibrant community of young painters, sculptors, and architects—British, Italian, and German, as well as American—Hartford became a significant center of this burgeoning cultural phenomenon. Among the city's prominent artists during the 1850s were Frederick S. Jewett, hailed as "the best Marine Painter in this country"; George F. Wright, whose legacy is Connecticut's Hall of Governor's portraits; Charles deWolf Brownell, an accomplished landscape painter whose best-known work is a widely reproduced portrait of the Charter Oak; and John L. Fitch, one of the few local artists patronized by Elizabeth Colt.[2] In addition, Hartford boasted two illustrious native sons, the youthful soulmates Edward Bartholomew and Frederick E. Church, both of whom achieved international acclaim as artists during the 1850s.

For Sam Colt, art was raw creative impulse refined through perseverance and knowledge. In an age when few people traveled far from their place of birth, Colt gained a bird's-eye view of America, crisscrossing the United States and Canada with "Dr. Coult's" laughing gas demonstrations on the lyceum circuit. Briefly connected with museums in Cincinnati and Baltimore, Colt understood how such newly founded institutions, bearing the names "athenaeum," "lyceum," and "museum," were instruments of the cosmopolitanism and worldliness that, if applied to the raw talent of American genius, could not fail to multiply national power and prestige. These institutions, based on the principle that access to knowledge is the keystone of a stable and prosperous democracy, reflected Colt's own dedication to the unity of knowledge.

Colt's patronage of art, although a consistent theme throughout his life, was sporadic and impulsive. In addition to gathering around him a network of German carvers and engravers such as Gustave Young, Carl Helfrecht, and Christian Deyhle, the most frequently cited instance of Colt's art patronage was the paintings of Colt firearms in action he commissioned from the renowned artist and frontier adventurer George Catlin.

Sam Colt opened his first Hartford arms factory a few years after the railroads arrived at what was then a declining commercial port on the Connecticut River. River and rail access to markets and materials also attracted James Batterson, another nascent art patron, who moved his family's monument company from the marble regions of western New England to Hartford in 1846. Batterson, who shared Colt's industrial vision for applying machinery and mass marketing to traditional craft, began manufacturing "monuments and gravestones... chimney pieces, mantles, center table... and counter tops" and within a decade had established a national clientele not only for table tops and gravestones but also monumental sculpture, mausoleums, and public buildings. Like Colt, Batterson had a keen instinct for public relations and an ambition to reshape Hartford's reputation through the transformative power of art and machines.[3]

With a shared interest in art and the challenge of imposing large-scale manufacturing on a conservative and resistant community, Colt and Batterson became fast friends. In 1849 Batterson served on the founding executive committee of the Hartford Arts Union, an institution "composed of mechanics, manufacturers, artisans and all others interested in the advancement of the Arts."[4] Colt and Batterson attended Union meetings and competed for prizes in the exhibitions of the Hartford County Agricultural Society.[5] In 1854 Batterson hired G. Argenti, a master stonecutter from Italy, to expand and oversee the customwork and design operations of his monument works. Argenti raised Batterson's capacity to a level

where, for the first time, the firm could bid for and win monument commissions of considerable prestige and expense. During the summer of 1854, Argenti worked on the first public sculpture erected in Hartford, a twenty-foot-high brownstone obelisk with marble bas-relief panels illustrating the story of Thomas Gallaudet's founding of the first American school for the deaf.[6]

Perhaps it was the death of his brother Christopher in 1855 that prompted Sam Colt to commission a monument and purchase a family burial plot at Hartford's Old North Cemetery (fig. 119). G. Agenti was undoubtedly responsible for finishing the monument (signed "Batterson") that, at the time, was as impressive a work of sculpture as Hartford's fledgling monument industry had produced.[7] Decorated with garlands of carved flowers, tendrils of ivy and winged angels, the monument stands about sixteen feet tall and is surmounted by a marble madonna. Here, in marble, is one of the earliest uses of Sam's design for the Colt family coat of a which he adopted about 1852 to varnish a lineage was decidedly less than aristocratic. The commis: though never cited among Colt's accomplishments, e: lished a pattern that characterized Sam and Elizab art patronage for the next half century. Both encour the artists, mechanics, sculptors, and builders patronized to expand their repertoire of skills and ri higher levels of accomplishment. At the armory, skill mation and technological innovation created pr Among the community's emerging artists and art in tries, skill formation generated prestige. The work Colt commissioned about 1855 and the far grander r ument Elizabeth commissioned ten years later ena James Batterson's monument works to reach new l of capability and prestige, rising eventually to the t its industry nationwide.[8]

119. *Colt Family Monument, 1855. Old North Cemetery, Hartford, commissioned from James G. Batterson Monument Works.*

120. *Portrait Busts of Samuel Colt and Elizabeth Colt. Reclining figure of Samuel Jarvis Colt, 1857 (WA #05.1121, 05.1122, and 05.1118) by Edward Sheffield Bartholomew (1822–58). Photographs taken before Elizabeth's death reveal the following locations for these sculptures at Armsmear: Sam Colt in the drawing room on a marble pedestal; Elizabeth Colt in the library on one of the cases; Samuel Jarvis Colt in the reception room atop a large cabinet.*

Colt's patronage of James Batterson continued up to the time of Colt's death. Batterson provided ornamental details and architectural elements used in constructing the armory.[9] He furnished sculpture for the gardens and grounds at Armsmear, and eventually the stones under which Colt and his children were buried.[10] Batterson and Colt could not have been more alike in their disposition toward art, industry, and civic reputation. Until his death in 1901, James Batterson—best known today as the founder of Travelers Insurance Company—remained one of Hartford's most tireless boosters and civic leaders. While still in his twenties, Batterson commenced what became a lifelong involvement in politics, lobbying on behalf of Colt's South Meadows improvements. By the mid-1860s, while still in his thirties, Batterson had become Hartford's most voracious collector and advocate for fine art.

Thanksgiving week 1857 Colt joined a group of seventy civic leaders in hosting a banquet to honor two of Hartford's most illustrious native sons, Edward Bartholomew and Frederic Edwin Church.[11] Bartholomew was lionized by townsmen eager to stake Hartford's claim as a metropolis on the balance of its wealth and civic amenities, and Colt identified especially with the young man who was described as having "risen, by his own energy and the force of his genius." At the banquet Colt commissioned three portraits: of himself, Elizabeth, and their son, Samuel Jarvis Colt (fig. 120). To the amazement of the seated guests, the artist declared his readiness to begin the portraits on the spot and proceeded to manipulate the clay in the shape of Colt's face. This turned out to be one of the last commissions of the artist's life. Six months later the sculptor was dead, and the figure of the Colts' young child, also soon deceased, was found among the unfinished work in his studio.[12]

For Batterson, Bartholomew's death was a heavy loss. Both were the same age, and Batterson later claimed to have supplied Bartholomew with the statuary marble and tools used in his earliest work.[13] Colt launched a campaign to procure the effects of Bartholomew's studio, pledging one-tenth of the five thousand dollars (the equivalent today of about twenty thousand dollars) raised by himself and seventy civic-minded individuals.[14] Batterson was dispatched to Italy to purchase a collection of Bartholomew's work for Hartford's Wadsworth Atheneum.[15] He arrived in Rome bearing a letter of introduction from Sam Colt to Hiram Powers, the most famous of America's expatriate sculptors and a close friend of the deceased artist.[16]

The collection, featuring several marble busts, marble replicas of Bartholomew's masterworks *Sappho* and *Eve Repentant* (fig. 121), and a variety of plaster cast models for full-figures, busts, and bas-reliefs, went on display at the Wadsworth Atheneum in the summer of 1859 and remained one of Hartford's most popular art attractions for decades, an appeal heightened by the fact that Bartholomew was a native son and by the aggressive civic-mindedness of men like Colt and Batterson.[17]

Sam Colt's patronage of art also included commissioning Joseph Ropes to paint a panoramic view of Hartford during the Great Flood of 1854 (figs. 122 and 123).[18] From 1851 until 1856, when he established a studio in Rome, Ropes was one of Hartford's most prominent artists and art instructors. A disciple and imitator of Thomas Cole, Ropes attracted notoriety during eleven years in Italy among the community of American expatriate artists and in his subsequent career in Philadelphia.[19] Colt also commissioned two handsome landscape views of the armory in 1857 (fig. 124) and a lavishly illustrated profile on the armory that appeared in *United States Magazine*. The paintings, bearing the signature initials "A.C.," are possibly the work of Alonzo Chappel, a prolific New York portrait and history painter. Colt later commissioned portraits "of the whole family," from Matthew Wilson, a "celebrated" New York portrait painter.[20] Colt owned a variety of large and small paintings and prints, and bronze and marble statuary installed inside and out at Armsmear and on the dome of his armory.[21] Although

121. Eve Repentant, *1856.*
Marble sculpture
by Edward Sheffield
Bartholomew, 1822–58
(WA #1858.1).
In a letter dated
March 11, 1856,
to his friend, artist
Frederic Church,
Bartholomew wrote,

"My Eve is not yet quite
finished, my studio is
crowded from morning
till night with visitors
to see it. I had no idea
it would ever make
such a sensation."

Colt's art patronage and collecting never reached the level of accomplishment achieved by his wife, it was wide ranging and eclectic. Had he lived longer he would surely have developed this side of his life in surprising ways.

When Sam Colt died in 1862, leaving Elizabeth Hart Colt one of the wealthiest women in America, the task of fulfilling his philanthropic vision fell to his widow. Compared with those of department store baron A.T. Stewart and contemporaries like Cornelius Vanderbilt and George Peabody, the Colt fortune was relatively small. But the estimated $3.5 million Colt amassed after emerging from debt in 1849 was an accomplishment rarely matched in America before the Civil War.[22] And it was more than enough to cut a sizable swath through Hartford's cultural landscape.[23]

As Victorian Hartford's first prolific art collector (and her deceased husband's friend), James Batterson undoubtedly had some influence on Elizabeth Colt. In 1864 Batterson returned from Europe with a collection of paintings, hoping to aggrandize his city.[24] Batterson reportedly assembled the collection with the aim of selling it to Hartford's moribund art gallery, then managed by the Hartford Arts Union in a room in the city's Wadsworth Atheneum. Although the collection was never purchased, it became the first major art exhibition in the Atheneum's history when it went on public display later that month.[25]

Whether compounded by its cost or by Batterson's presumption in acting without authorization on behalf of the trustees, the collection fell victim to public taste, which remained vigorously chauvinistic and pro-American until the late 1860s. With only a single American picture, a western scene by Albert Bierstadt, Batterson's "200 paintings… [of the] Italian, Dutch, Flemish, French, German, Dusseldorf, Belgian and English schools," mostly "old-masters," was ten years and several steps removed in cost and sensibility from Hartford's core audience.[26] About one-third of the collection was work by European "modern masters," Van Schendel, Tschaggeny, Achenbach, and Verboeckhoven being especially popular with American collectors. Biblical subject matter was the most widely represented. There were also European cityscapes and ruins, scenes from mythology and literature, allegorical pictures of the seasons and muses, scenes of antiquity, pastoral views of sheep and shepherds, still lifes of game, a few Dutch peasant genre scenes and interiors, and one battle scene.[27]

A far greater influence on Elizabeth Colt than James Batterson, and the fuse that ignited the art craze nationally, was the exhibition of fine art at the Sanitary Commission's Metropolitan Fair in New York in April 1864. Not until the Centennial Exhibition of 1876 would the nation witness a more spectacular event or one that had so decisive an influence on American art and philanthropy. After the war, Horace Bushnell described the public mood in 1864 as "everything we have for public love," and he no doubt had the Metropolitan Fair in mind when claiming "there was never before a fiscal campaign to match the sublimity and true majesty of the spectacle."[28] Its purpose was to support the U.S. Sanitary Commission, an outgrowth of the women's Central Association of Relief for the Sick and Wounded of the Army, founded in New York in 1861, to provide nurses, hospitals, and ambulance services and to make sanitary inspections of Civil War military camps.[29] In February, advertisements for the Metropolitan Fair in Aid of the Sanitary Commission began appearing in the Hartford newspapers, and in March, its founder, the Reverend H.W. Bellows, made a promotional appearance at Hartford's Allyn Hall.[30] Opening with a prayer and a military parade on April 5, by the time the Metropolitan Fair closed two weeks later, the organizers had raised the equivalent of $43 million.[31]

Sam and Elizabeth Colt had contributed to the cause of soldier's aid as early as 1861, and Elizabeth was already involved with Hartford's Soldier's Aid Society in 1863,

122. Panoramic Views of Hartford. *Oil on panel, painted in 1856 by Joseph Ropes (American, 1812–85) to document the Great Flood of 1854.* Below: *view to the north; the Connecticut State House and United States Hotel, where Colt was then living, appear in the lower right corner of the picture (WA #05.55);*

123. Opposite page: *view to the west; prominent landmarks include the Hartford Asylum for the Deaf* (center left), *the Mill River, and the Hartford railroad station of 1849* (right) *(WA #05.56).*

when she performed as Catharine of Aragon "appealing to Heaven... on bended knees" in a program of theatrical tableaux that raised $1,000 for the cause. In 1864 Elizabeth accepted the presidency of the Hartford Soldier's Aid Society, the first of many philanthropic causes over which she presided during the next forty years. Amassing gifts from Hartford's leading manufacturers—silks from Cheney Brothers, rifles and revolvers from Sharps' and Colt's, and several "boxes of prime cigars" from Essman and Haas—Elizabeth was joined on the Ladies' Committee for New York by her sister Hetty. After expenses, the Hartford Table raised $6,500, selling presentation guns, more than $1,300 worth of Colt's willow furniture, a copy of "[Thomas] Cole's picture of 'Montevideo,'" Cheney Brothers' "American Foulard silks" and neckties, Connecticut "election cake,... nutmeg bracelets,... fancy articles," and one-dollar subscriptions for a piano made of Charter Oak that was presented to Connecticut governor Buckingham.[32]

Conceived as a multimedia extravaganza "worthy of the great city in which it is to be held... democratic, but not vulgar, elegant but not exclusive, fashionable but not frivolous," the Metropolitan Fair featured an Old Curiosity Shop, a department of arms and trophies, an International Hall with displays "of all nations and tribes," a Knickerbocker Kitchen, where mince pies and waffles were served by attendants in colonial costume, and an art gallery with "between six and seven hundred thousand dollars worth of paintings" providing what many considered "the finest and most extensive exhibition of fine arts ever offered in this country." The display, assembled by artists Jonathan Sturges and John F. Kensett, was a noteworthy affirmation of the rising power of American art, fea-

turing New York's most celebrated painters and such acclaimed works as Emmanuel Leutze's *Washington Crossing the Delaware,* Frederic Church's *Heart of the Andes* and *Niagara,* and Albert Bierstadt's *Valley of Yosemite.* With an estimated four hundred pictures and sculptures theatrically amassed floor-to-ceiling, the art exhibition at the Metropolitan Fair helped launch an unprecedented collecting binge.[33]

During breaks from her work at the Hartford Table, Elizabeth Colt may have toured August Belmont's "private collection of Modern European pictures," which he opened to the public during the fair.[34] She may also have visited the Dusseldorf gallery, still popular fifteen years after its founding, or the recently opened branch of the Paris gallery Goupil's, representing European and American artists whose work she later acquired. She was certainly no stranger at the Studio Building on Tenth Street, where Frederic Church, John Kensett, Albert Bierstadt, William Beard, John Beaufain Irving, and William Bradford, all artists eventually represented in her collection, kept studios and entertained clients.[35]

Elizabeth Colt had embarked on the first phase of the campaign to create a private picture gallery before attending the Metropolitan Fair in April. In December 1863 she wrote Frederic Church thanking him for "communicating my wishes to Mr. [Charles Loring] Elliot," who began work on the monumental portrait of Sam Colt later that year (fig. 125). Already immersed in the book project, Elizabeth Colt was rapidly gaining experience negotiating with artists, architects, and writers, and while she voiced some concern to Church about the cost, she confessed that "if he can make it such... as I have in mind, twice the sum would not be enough."[36] Elliot, then

124. View of Colt Works with Steam Boat from the South, Colt Meadows, *1857. Oil on canvas; one of two paintings commissioned by Sam Colt to document the completion of his armory (WA #05.51).*

125. Portrait of Samuel
     Colt, 1865.
     Oil on canvas by
     Charles Loring Elliot,
     American, 1812–68
     (WA #05.8).

The picture frame, with
its Colt coat-of-arms
and rampant colt figure
was custom-made for
the painting.

one of the leading portrait painters in America, had "nearly finished" the portrait of Sam Colt when he arrived in Hartford in February 1865.[37] Sam's portrait reveals him posed with symbols of accomplishment, a membership certificate to the London Institute of Civil Engineers in his hand, a gilt brass vase from the King of Siam, a display case filled with revolvers, and a view out the window of his armory.

Delighted with the portrait, for which she had supplied photographs and carefully chosen the personal mementoes (fig. 126), Elizabeth commissioned a second portrait of herself and her son, Caldwell (fig. 127).[38] The artist completed it, along with a portrait of Frederic Church's father, Joseph, during a two-month stay in Hartford.[39] In May, with his work done, the newspapers boasted that "the portraits executed by Mr. Elliot in this city" were "the finest in existence," proof "that our country is hastening to the front rank in the arts."[40] Elizabeth is seated, prayer book in hand, on a tufted sofa, the work of French designer Leon Marcotte, while her son rests beside her, gently clutching the pull string on a toy cannon, one of Sam's prized relics from the Mexican War. This picture of opulence and maternal affection captures the revolver king's widow at the point where she embarked on her philanthropic career. Aside from a portrait of Matthew Vassar, similar in composition and scale, which Elliot made in 1861, full-length, life-size portraits like these were almost unheard of, no less where the subject was a woman. Elizabeth Kornhauser has suggested John Horsley's *Portrait Group of Queen Victoria with her Children* (ca. 1854) as a possible source for the portrait of Elizabeth and Caldwell. The analogy is apt; Queen Victoria also spent much of the 1860s memorializing, at great expense, a recently deceased husband.[41]

With the twin portraits installed in the second-floor ballroom at Armsmear in the fall of 1865, Elizabeth Colt began amassing the collection that would transform the space into Hartford's most acclaimed private picture gallery. In its first incarnation, made ready for the wedding of her sister in December 1867, Elizabeth Colt's pic-

126. Gilt Silver and Enamel
     Presentation Vase.
     Made in Siam ca. 1855
     and given to Sam Colt
     by the second king of
     Siam, this vase appears
     on the carved cabinet

in Elliot's portrait of Colt
and is also illustrated
in Armsmear. *It is
emblematic of Sam Colt's
success in marketing
his firearms worldwide
(WA #05.1543).*

ture gallery was a memorial shrine in which Sam and Elizabeth faced each other across a forty-foot-long room flanked by the recent work of America's most acclaimed painters. The actual sequence of acquisitions is unknown, and the earliest photographs of the gallery were not made until about 1869 (figs. 128 and 129), when its masterpiece, Frederick Church's *St. Thomas in the Vale, Jamaica,* was finally completed and installed (fig. 130). But it is clear that what began as a shrine to American nature, creation, and creativity, expanded over the next twenty years to include a wide range of subjects, balanced almost equally between Europe and America, a tribute to the transatlantic dialogue in which the fine arts played a conspicuous and formative role.

Elizabeth Colt consulted with Church, Elliot, and the artist-architect Richard Morris Hunt about lighting and upholstery, and it was at this point that the "Resht work" embroidery Sam purchased at the Asiatic Fair in Novgorod, Russia, in 1854 finally found a use covering the gallery's stools and a sofa (fig. 131).[42] Church probably recommended works by his Studio Building peers, although the names Kensett, Beard, Bierstadt, and Bradford, among the most popular American painters of the time, needed no introduction. The centerpiece of the gallery and, next to Church's *Jamaica,* her most expensive picture was *The Angel's Offering* (fig. 132), a portrait of the holy family by the French realist Hugues Merle, acquired more for its subject matter than as evidence of a taste for French art.

Elizabeth immersed herself in study, visiting artists' studios, art dealers, and private picture galleries in New York. In May 1866, with the builder "ready to commence work," she wrote to Church to ask his "advice upon the mode of lighting the gallery at night," relaying Hunt's recommendation of "Frink's Reflectors," which she apparently saw in Samuel Avery's New York gallery, but complaining that "since seeing them, I have felt undecided, whether to adopt those... or light it in the more usual way, like Mr. Johnston for instance."[43] Having already commissioned *Jamaica,* she wrote Church in October

127. Portrait of Elizabeth Colt and Caldwell Colt, 1865. Oil on canvas, Charles Loring Elliot (WA #05.9).

128. *Picture Gallery at Armsmear, 1869. R.S. DeLamater, Hartford, photographer; from a stereo photograph. The small upholstered seats placed around the perimeter of the room were recycled from the room's former use as a ballroom. An elaborate drapery system hung on a brass frame* (top center) *controlled the lighting in the skylit room.*

129. *Picture Gallery at Armsmear, 1869. R.S. DeLamater, Hartford, photographer; from a stereo photograph. Colt's billiard room and table appear through the door to the left of his portrait.*

130. Vale of Saint Thomas,
Jamaica, *1867.*
Oil on canvas,
Frederic E. Church,
American
(1826–1900),
for Elizabeth Colt
(WA #05.21).

131. Ottoman and Lounge,
1856. Carved walnut
with original embroid-
ered rescht-work
upholstery, attributed
to the Paris–New York
decorating firm
of Ringuet-LePrince
and Marcotte
(Wadsworth Atheneum).
A "lounge" (right) and
eleven "ottomans" (left)
appear on Sam Colt's
1862 estate inventory
in the ballroom at
Armsmear. Sam Colt
originally intended
to use the upholstered
lounge in his dressing
room. Having acquired
the embroidered fabric
at the Nizhni-
Novgorod Fair in
Russia in 1854,
Colt asked decorator
Leon Marcotte to
have the lounge,
an armchair, and
twelve small
ottomans, which
Colt referred to as
"chamber chairs,"
upholstered in
the fabric.

concerned about color and upholstery and urged him to visit Goupil's "to see 'The Angel's Offering' by Merle, which I have bought."[44]

The picture gallery was unveiled sometime around Christmas 1866, just as the first copies of *Armsmear* came to hand. Having grown accustomed to broken promises and delays, Elizabeth was perhaps not entirely surprised when, a year later, Church's picture was still unfinished. It may not have been delivered until 1869, when the Hartford newspapers first reported it "among her paintings," gleefully announcing that it was "said to have cost $15,000," equal to twenty years' wages for a machine tender at Colt's Armory.[45] The picture gallery, with its dark-green upholstered walls, multicolored "oriental" seating, three octagonal skylights with immense green canopy, patterned hardwood floors, and "Colt arms engraved on a shield over each door," was indeed a spectacle.[46] But with fewer than twenty pictures to hang and two or more "cases containing the superb presents given [to Colt]... by foreign potentates" (fig. 133), the medals of honor (fig. 134), "curiosities collected from every part of the world," and the melted and fused remains of revolver parts from the armory fire of 1864, Elizabeth's picture gallery was, at first, as much a personal cabinet of curiosities as an art gallery.[47]

It may not have been long before she removed the cabinet treasures and the custom-built cabinet containing the memorial arms collection to the adjoining hallway. Although she continued to buy pictures into the 1880s, existing documentation and the fact that many of the European pictures were painted in the mid- to late 1860s, suggest that the collection was substantially complete by the mid-1870s, when she shifted her attention to decorative arts. When the gallery was photographed for the last time about 1900 (fig. 135), it was stacked floor-to-ceiling with more than forty pictures, almost two-thirds by contemporary European realists, an enormous ceramic vase (fig. 136), and the bronze *Crouching Venus* brought indoors from its original location as an outdoor fountain. Not without quirks and eccentricities, Elizabeth Colt could not be accused of failing to stamp her personality on this most public of private rooms.

132. The Angel's Offering. *Hugues Merle, French, 1823–81 (WA #05.45). This painting and* Church's Vale of Saint Thomas, Jamaica *(130, previous page) were the centerpieces of the long walls of the gallery.*

The pictures in Elizabeth Colt's gallery were primarily the work of painters mostly now reduced to footnotes in the history of art. Taken together, however, they represent a portrait of their age and provide a revealing glimpse into the heart and mind of the woman who assembled them. Most of the American painters whose work she acquired were exhibitors at the Metropolitan Fair, including Church, Bierstadt, Kensett, and Gifford. The next year, prices and acclaim for the work of American painters skyrocketed, with Church, Bierstadt, and the Arctic marine painter William Bradford enjoying unprecedented attention and patronage. Journalists raved about how the "liberal encouragement" that had "greeted our artists of late… impelled [them] to extraordinary efforts" in elevating "the land we love to its proper position."[48] Frederic Church was singled out in particular for the "untiring enterprise with which he seeks his subjects in all kinds of remote and difficult regions," for his skill at rendering atmosphere, and for the nation's pride over his fame in Europe.[49]

For Elizabeth Colt, Church's deep spirituality and Christian faith, evident in his depictions of land "untouched since the time of creation," were alluring inducements. *St. Thomas in the Vale* is, above all else, a testimony to the sustaining power of faith. With its parish church in the distance and its atmosphere pierced by radiant sunlight breaking through the clouds in the moment after a passing storm, the image Church created is saturated with symbols of redemption and hope. Like Elizabeth Colt's, Church's faith was tested by the back-to-back loss of two children in 1865.[50] He traveled to Jamaica, in part, to escape the pain of familiar surroundings. Situated at the center of the long wall, facing Merle's portrait of the holy family, Church's *St. Thomas in the Vale* anchored Elizabeth Colt's picture gallery to the themes of faithfulness and renewal.

Religious overtones also infused paintings of the American wilderness and of the pristine and unblemished American West. This genre first appeared in the 1820s in the work of Thomas Cole and reached a melo-

133. Gifts from the Emperor of Japan that Commodore Matthew Perry brought back for Samuel Colt in 1855 included a roll of Japanese silk brocade (WA #05.1384), and a large lacquer box (WA #05.1487).

134. Turkish treasures, commemorating an 1849–50 trip to Turkey included an illuminated copy of the Koran (WA #05.1551); the Turkish medal of honor, Imperial order of Sultan Abdulmecid (WA #05.1532); a snuffbox, presented by the sultan; Swiss porcelain enamel with a replaced, Turkish-made, diamond-encrusted lid (WA #05.1538); an amulet worn and blessed by the Howling Dervishes (WA #05.1528); a turkish jambiya knife and scabbard with coral and turquoise jewels set in gold (WA #05.1054).

dramatic climax during the 1860s with Albert Bierstadt's western paintings. Bierstadt's *Valley of the Yosemite* (1864) attracted the most attention at the Metropolitan Fair, prompting him to turn out several small views in 1866, including one Elizabeth Colt acquired for her gallery (fig 137). Bierstadt, who first traveled to Yosemite in 1863, raised American landscape painting to an apex of melodrama and theatricality with his increasingly outsized views, culminating in the *Domes of the Yosemite* (1867), which, at 15 x 9 feet, was the largest picture painted in the most active year the American art market had ever seen.

In its first phase, mountain landscapes by American painters dominated Elizabeth Colt's picture gallery. One of the first pictures she commissioned was Sanford R. Gifford's *A Passing Storm in the Adirondacks* (fig. 138).[51] In 1867 she paid $1,450 (today about $50,000) for *Mount Washington from the Conway Valley* (fig. 139), purchas-

ing it directly from Kensett, whom she met in conjunction with the Metropolitan Fair.[52] In 1866 she paid a comparable amount to Hermann Fuechsel of New York for his *Alpine Scenery—Lake Gosau*, a reminder of her swing through the Tyrolean Alps on her bridal tour ten years earlier.[53]

Whether acquired as mementoes of places visited or as a gesture of hope for enduring life's "passing storms," Elizabeth Colt's mountain pictures resonated with emotion. In 1868 she purchased the *Coast of Labrador* (1868), an arctic sunrise by the artist-explorer William Bradford, and William Beard's *Mountain Stream and Deer* (1865), a quiet and contemplative view of an unspoiled wilderness forest. Among the few American paintings she purchased after 1870, mountain landscapes continued to prevail. These later purchases included a coastline view painted in 1867 by the Philadelphia artist James Hamilton (acquired by Elizabeth in 1875) and a

137. In the Yosemite Valley,
1866. Albert Bierstadt,
American, 1830–1902
(WA #05.22).

138. A Passing Storm in
the Adirondacks, 1866.
Sanford Robinson
Gifford, American,
1823–80
(WA #05.23).

139. Mount Washington
from the Conway Valley,
1867. John F. Kensett,
American, 1816–72
(WA #05.13).

140. Youth and Old Age
(Grandfather's Pet),
1867. Jared B. Flagg,
American, 1821–99
(WA #05.17).

141. The Mountain Pasture,
1877. Eugene Joseph
Verboeckhoven,
Belgian, 1799–1881
(WA #05.30).

somewhat gross and misshapen view of North Dome, Yosemite Valley (1884), which hung in the dining room at Armsmear, the work of William Ongley, a painter best known for his views of the Adirondacks.[54]

Her few remaining works by American painters were genre scenes. She acquired a pair of historical tableaux by Studio Building artist John Beaufain Irving and a small and charming childhood genre by James Crawford Thom titled *The Snow Slide* (1865). Her purchases also included an intriguing allegorical scene of familial continuity titled *Grandfather's Pet* (1867; fig. 140) by an artist turned Episcopal priest and former resident of Hartford, Jared B. Flagg. The only American painting by an artist not living or working at the time she acquired it is a modest arcadian view painted by Thomas Cole in the early 1830s and perhaps acquired through Frederic Church, who brokered pictures from the estate of his mentor.[55]

Aside from Hugues Merle's monumental *Angel's Offering* and a few others, there is no evidence that European pictures were well represented in Elizabeth Colt's picture gallery prior to 1870. The Merle was the only European picture to appear in the first views of the gallery, made about 1869. During the next couple of years she acquired half a dozen European pictures, notably *The Mountain Pasture* by Eugene Verboeckhoven (1866; fig. 141), the two stunning night scenes by Petrus van Schendel (1864; fig. 142), *At the Prison Door* by Louis Gallait (1864; fig. 143), which she loaned to an exhibition in New Haven in 1867, and a charming still life by Johann Preyer (ca. 1850), so representative of the taste of the 1840s and 1850s that it may have been acquired during Sam's lifetime.[56]

Several of her largest and most impressive European pictures represent artists who enjoyed wide followings in the United States during the 1870s, notably the orientalist works by Adolf Schreyer, Vincent Stiepevich (fig. 144), and Felix Ziem, the pastoral views of animals by Wouterous Verschurr and Eugene Verboeckhoven, and the sentimental portraits of maternal love and childhood innocence by Anton Dieffenbach, Michael Arnox, and William Bougeureau, the two latter bearing the labels of Hartford art dealer A.D. Vorce and unlikely to have been acquired by Mrs. Colt any earlier than 1878. Following her tour of Italy in 1882, Elizabeth Colt acquired at least three works by Italian artists celebrating the romance of old Europe, notably Hermann Corrodi's *Venetian Lovers* (ca. 1880), Antonio Paoletti's *Venetian Lady with Parrot* (ca. 1880), which she purchased at the Guggenheim Gallery in Venice in 1882, and a portrait of a monk tuning a harp (1888) by Arnaldo Tamburini, the last picture added to the gallery.

The picture gallery was a world of make-believe. While Elizabeth's brother and sister joined the parade of American tourists who swarmed through Europe in the late 1860s, she stayed home, visiting only once for a grand tour in the early 1880s. Pictures of Europe brought to mind the romance of her bridal tour and a chance to enjoy vicariously the experiences of her family and friends. Felix Ziem's view of the Grand Canal and the doge's palace in Venice (ca. 1870; fig. 145) was a form of pictorial tourism that complemented the literary tourism provided by books like *Exploration of the Nile* (1868), *The Gates of the East* (1877), *Parks, Promenades, and Gardens of Paris* (1869), *Norway and its Scenery* (1853), and the travel writings of her friend and neighbor Charles Dudley Warner, which lined the bookshelves at Armsmear. Scenes such as Adolf Schreyer's *Well in the Desert* and Vincent Stiepevich's *Bargain in Algiers* pay homage to Sam Colt's triumphs in and fascination with Constantinople, North Africa, and the Middle East. Flawlessly finished images of childhood, such as William Bougeureau's portrait *Manon Lescaut* (ca. 1878; fig. 146), and of rural life, such as Eugene Verboeckhoven's *Mountain Pasture,* provided a soothing and painterly image of a perfect world.[57]

At the time she began buying European pictures, American collectors were enduring mounting criticism for abandoning indigenous artists. The press continued to champion the work of American painters, complaining

142. Candlelight, *1864.*
   Petrus Van Schendel,
   Dutch, 1806–70
   (WA #05.47).

143. At the Prison Door,
   1864. Louis Gallait,
   Belgian, 1810–87
   (WA #05.42).

that "we have done honor enough to European artists";[58] nonetheless it observed that "buyers of pictures do not come forward as they did last season" due to the "high prices which our artists got for their works two years ago" and to the "many Americans... abroad last year" at the Paris Exposition.[59] When the American painter Worthington Whittredge declared that "a man is of no use who does not have faith in the heritage of his own country," he surely had in mind the decline of patronage from its high-water mark of 1866.[60] In 1884 the Hartford press editorialized against the "demoralization of art" and the tendency to "adore whatever is foreign," decrying the "stupidly unpatriotic,... purchases... made in an ignorant and contemptible spirit" by American millionaires "eager to buy foreign trash because it is foreign."[61] By then, the balance between Europe and America had become increasingly lopsided.

What these collectors had in mind were pictures that were easy to understand, morally elevating, devoid of controversy, and flawlessly executed. European salon art, with its conspicuous display of workmanship, was easily

144. The Bargain in Algiers. *Vincent G. Stiepevich, Russian, working ca. 1900 (WA #05.36).*

145. Venice, *ca. 1860. Felix Ziem, French, 1822–1911 (WA #05.29).*

146. Manon Lescaut, *ca. 1878. Oil on canvas, William Adolphe Bougeureau (1825–1905), Paris (WA #05.38).*

Manon Lescaut was the subject of a French pastoral novel (1753) by Abbé Prévost.

authenticated and provided incontrovertible evidence of its inherent "value," measured more by a picture's size and meticulous details than by its content or the artist's imagination.[62] History, however, has not been kind to the type of European salon art Elizabeth Colt collected. If investment value is any indication, she would have done better stuffing her money in a mattress.

In the rush to Europe that began in 1867 and continued throughout the 1880s, Americans became the best customers of every trinket-selling, name-peddling, fashion-mongering purveyor of envy in Paris. In 1876, with even farmers and machine tenders scraping up the car fare to partake in the nation's Centennial Exhibition in Philadelphia, the Hartford papers reported indifference on the part of New York's "moneyed class" who "would much rather take it over to Europe and trot around there for a few months," the same people who, "when questioned about things and places of interest in the United States... generally treated them as of no consequence."[63]

In 1868, while her sister and brother-in-law were gallivanting through Naples and Rome, Amsterdam, Cologne, London, Ireland, Holland, Paris, and Switzerland, rendezvousing with the likes of Civil War hero Adm. David Farragut and his wife and generally acting the part of upward-bound Americans abroad, Elizabeth, Caldwell, and friends visited the Shaker mother church in Mount Lebanon, New York, sauntered up to the White Mountains, and ended up in Newport, pretty much covering the ideological spectrum of vacationing Americans, from history and religion to nature and glittering wealth.[64] In 1873 Elizabeth and Caldwell toured Colorado and the West, and in 1876 she joined the millions who flocked to the Centennial in Philadelphia, where she appears to have caught the decorative arts collecting bug. She returned home with an assortment of British and Japanese art treasures (fig. 147). Elizabeth's picture gallery already revealed a tendency to hop on board the fashion trains of her time, not always to the advantage of posterity or what might today be described as investment considerations.

One of the great ironies of the Centennial is that it heightened the nation's anxiety about its status in the world of art and fashion. With cornucopias of decorative arts spilling forth from the exhibits of all the great nations of Europe, Asia, and the Middle East, the taste for exotic hand-crafted, custom-designed imported goods blossomed.[65] While a distinct subculture of Hartford's moneyed class used the event as a springboard into a fifty-year orgy of antiquarian and patriotic collecting, Elizabeth Colt resolved to build another collection, this time centered on the pottery, glass, and related decorative arts of Europe and Asia.[66]

Fifteen years after the first wave of Americans traveled to Europe, Elizabeth Colt, accompanied by Caldwell and her sister and brother-in-law, Hetty and Cyprian Nichols Beach, embarked on what would be her only extended absence from Hartford after the bridal tour of 1856. Leaving New York in November 1881, the travelers were in Italy by the following January and visited Spain, Germany, Turkey, Norway, and Sweden, and probably Paris, before returning home some twenty months later in 1883. As reported in *Harper's Monthly* in 1868, "the least acquisitive of people begin, before they have finished a European tour, to show some symptoms of the disposition of a 'collector,'" observing accurately that "it is not the intrinsic value... but the associations" that prompted

147. Pair of Covered Jars, 1875. Gilt and painted porcelain, Koransha Porcelain Works, decorated by Ichiryusai Uchimatsu, Arita, Japan (WA. #05.1279 and .1280). These jars were eventually installed in the Memorial Hall of the Memorial House.

such acquisitiveness.[67] For some Americans, such as the legendary Isabella Stewart Gardner of Boston, travel to Europe sparked a lifelong interest in collecting fine and decorative art. For Elizabeth Colt, it was relatively inexpensive cabinet fodder, formed more to preserve memories and heighten the worldly atmosphere at Armsmear than for any systematic purpose or jockeying for prestige. Indeed, the collections may have functioned in lieu of the standard travel chronicles used to document an experience most Americans found utterly transformative. If Elizabeth Colt kept a diary of the trip, it has not survived. Nor has extensive correspondence, leaving us with only fragments of an itinerary and a thin trail of purchase receipts for things bought along the way.

The travelers wrote from Nice before venturing on to Italy, where they toured Rome, Milan, Florence, Venice, and probably Naples, during the winter and spring of 1882, then heading to Spain. Crisscrossing the Mediterranean, reportedly on Caldwell's yacht, the party was in Constantinople (Istanbul) the following May, when "talk of going to Norway and Sweden," was the last word from abroad until a year later when "Mrs. Colt came home safe and well last week," leaving Caldwell and the Beaches in Europe through August 1883.[68] Purchase records indicate that they visited Vienna, Hamburg, Munich, and Seville; the collections suggest they also visited Hungary and Denmark.[69] Did they ever make it to Sweden and Norway? For the twelve months from June 1882 to June 1883, the paper trail is lost.

After Elizabeth Colt returned to Hartford, she embarked on a costly remodeling of the reception room at Armsmear. As her growing collection of decorative arts from around the world piled up in the reception room and cabinets of her library, Armsmear became even more exotic inside than out.[70] The collection is a sprawling bazaar of the kind of rich, eclectic, Victorian age stuff that, for the most part, has been cast into the dustbin of art history, almost never written about or appreciated, and rarely scrutinized nor displayed by museum cura-

148. Vase, 1876. Cloisonné enamel with gilt base and handles, Elkington and Co., England (WA #05.1369).

149. Stand, 1879. Brass and walnut, American (WA #05.1604).

This custom-made cast bronze stand was likely produced at one of Connecticut's brass factories.

tors. And yet, we have a remarkable and archaeologically intact document of what women—or at least one woman—from industrial America's moneyed class could do in the way of fashion and personal expression. The value lies mostly in the unveiling of a personality who willfully and, at times, skillfully negotiated the civic and artistic fashions of a time when the United States first emerged as a world power.

Elizabeth Colt's decorative arts collection consists of about two hundred pieces of pottery, sixty pieces of glass, samples of reproduction arms, armor, and furniture, more than a dozen pieces of mostly bronze sculpture, a handful of reproduction architectural fragments, several dozen bronze, brass, copper, and mixed material objets d'art, and about one hundred items best described as miscellaneous. It has everything from cloisonné treasures purchased at the Centennial (fig. 148), Russian coins, icons and malachite samples, Caldwell's mounted hunting trophies, and the brass stand Elizabeth had custom made as a gift for Caldwell's twenty-first birthday (fig. 149), to shadow box reproductions of the architecture of Alhambra (fig. 150), and a bewildering cross-section of things that may have belonged to Sam or Elizabeth. No information or explanation came along when they packed up a polished stone from the petrified forest of Arizona, a Turkish hookah pipe and related jambiya knives (fig. 151), or the small relic cannon made from muskets seized at the battle of Chepultepec in the Mexican War, which appears as a prop in Elliot's portrait of Elizabeth and Caldwell.

The collection consists mostly of contemporary pottery and glass representing the historical revival styles from England, France, Italy, Germany, Japan, and China. Like Mrs. Colt's European pictures, the collection has few aesthetic masterpieces. She ignored work by the leading French firms, such as Sévres and Limoges, and the most avant-garde in British art pottery, as well as the experimental in art glass, notably British cameo, French Gallé and Baccarat. American work, such as it was at the time, is also excluded. It mostly features high-end artistic sou-

150. *Shadow Box of Alhambra, ca. 1880. Painted and gilt plaster and alabaster in velvet-lined frame (WA #05.1112). Elizabeth Colt visited the Alhambra Palace (1354) in Granada, Spain, during her tour of Europe in 1882 and probably acquired this as a high-end souvenir. As the finest example of Moorish architecture in western Europe, the Alhambra epitomized the style of architecture suggested by the steel-and-glass conservatory at Armsmear.*

151. *Turkish Hookah Pipe and Jambiya Knife, ca. 1849. Brass, silver, turquoise, coral, gold inlay and steel, Turkey (WA #05.1087 and #05.1054). These are the best of a small collection of Turkish objects that Sam Colt acquired on his visit there in 1849, or at the Crystal Palace where similar works were displayed in 1851.*

152. *Italian Travel Souvenirs,*
ca. 1880.
Samples from a
collection of several
dozen objects and
pictures purchased
in Italy by Elizabeth
Colt. Left to right:
painting on copper,
copy of a work by
Lorenzo di Credi
(1459-1537),
purchased in Florence
(WA #05.1179);

miniature columns
from the Roman Forum
(WA #05.1115);
reproduction statue
of Caesar Augustus,
purchased from
B. Boschetti in
Rome in 1882
(WA #05.1132);
flagon, gold glass,
by Antonio Salviati,
purchased from the
artist in Venice in 1882
(WA #05.1189).

venirs representing the "national schools" of pottery and glass, and decorative baubles she picked up at the various "art" emporiums, now sprawling in number, in Hartford, Boston, and New York.[71] Not surprisingly, the collection closely parallels what was being shown at the international exhibitions and promoted in the torrent of books and articles dealing with international fine and industrial arts. Its self-conscious internationalism and historicism is, in fact, paramount. Although purchase records are scant, in all she probably spent less than thirty thousand dollars building the collection of decorative arts that found its way into the cabinets and reception room at Armsmear.

Elizabeth also bought reproductions of fine art masterworks from some of the museum and palace collections she and Caldwell visited on their tour of Europe. She acquired a bronze reproduction of a marble statue in the collection of the Vatican, reproductions of columns from the Roman Forum, a copy of a painting in the Borghese Gallery in Rome by the fifteenth-century painter Lorenzo di Credi, and "gold glass" from the Venice studio of Antonio Salviati (fig. 152). The delicacy of Venetian glass from the renowned studios of Murano (fig. 153) was offset by the grotesque baroque-style gilt pedestals, brackets, and mirrors, also bought in Venice, and costing the equivalent of about fifteen thousand dollars.[72] These gilt monstrosities eventually dominated the decor of Elizabeth's reception room, an aesthetic that truly defies the cyclical tendencies of fashion. She bought a pair of enormous, expertly carved, albeit grotesque, walnut chairs, upholstered with metallic threads and Venetian cut velvet, copies of originals in the doge's palace in Venice. So off-putting that in ninety years they have never been formally accepted or registered by the museum that owns them were the "Venetian Nubian figures"—two card receivers, a pier table, and a mirror (fig. 154)—from what Elizabeth described as "my dear son's smoking room." Gaudy and symbolically imperious though they are, they represent the most-expert and expensive Venetian workmanship and must have cost the

153. *Venetian Art Glass,*
ca. 1880;
Left to right:
vase with handles,
marbled glass
(WA #05.1218);
pitcher, opalescent glass
(WA #05.1212);

vase, gilt blown glass
(WA #05.1211);
vase with handles,
opaline glass
(WA #05.1210).

equivalent of fifty thousand dollars or more. There were reproductions of sixteenth-century repoussé work by Benvenuto Cellini from the Italian Court in Florence, purchased in Florence and apparently recognized by Elizabeth, who had seen and read about it at the American Centennial.[73] There was a copy of a Roman chariot in gold from the collection of the Vatican, a copy of Raphael's *Madonna della Sedia* from the collection of the Pitti Palace in Florence, and a cornucopia of lovely and curious things of contemporary design, including a marble portrait bust of the Egyptian queen and operatic diva Aida (fig. 155) purchased in Milan.

Ceramics predominated in Elizabeth's decorative arts collection. Among her ceramic cabinet treasures were the exhibition-quality French Barbotine painted vase from the pottery region of Gien (fig. 156), Italian Capo di Monte tankards, tazzas, and vases (fig. 157), Urbino-style Italian majolica modeled on fifteenth-century masterpieces (fig. 158), a diverse array of Japanesque and Arabesque forms by the French firm Longwy (fig. 159), a florid and artistic vase and ewer from the Hungarian firm of Zsolnay-Pecs (fig. 160), a mixture of Japanese artistic "export wares" (fig. 161), including a pair of enormous and expensive Arita-ware vases from Japan's Koransha Porcelain Works, which Elizabeth purchased at the Centennial. The collection is especially strong in British wares by Minton, Royal Worcester, and Royal Doulton, notably the Japanesque plates and vase (fig. 162), two of which were acquired at the Centennial, as well as Royal Doulton's revival seventeenth-century Rhenish stoneware and their hand-painted aesthetic vases (fig. 163), also acquired at the Centennial. Plates decorated with images of nurturing mothers and the tourist spots of Europe reveal a high level of sentimentality and association.

Elizabeth Colt's glass collection, although overwhelmed in quantity and diversity by her pottery, includes some notable highlights and is also highly representative of the historical revival tastes in Western art. The Bohemian painted cameo glass of the so-called Mary

154. *Venetian "Blackamoor" Mirror, ca. 1880. Polychrome and parcel-gilt on oak; two standing figures, one kneeling figure, and a pier table complete this set, which Elizabeth Colt described in her will as "the Moorish figures from my dear son Caldwell's smoking room."*

155. *Bust of Aida, ca. 1875, Pietro Calvi, 1833-1884. Bronze and marble (WA #05.1128). Bought in Milan in 1882. Aida was the Egyptian queen and the subject of Giuseppe Verdi's opera, written in response to the opening of the Suez Canal and first performed in 1871 at the Cairo Opera House.*

156. *Barbotine Painted*
Vase, ca. 1876.
Painted earthenware
with bronze mounts,
Gien, France.
May have been
purchased at the
Centennial where
Gien pottery was
prominently featured
(WA #05.1304).

157. *Capo di Monte Pottery,*
ca. 1880, Italy.
Left to right:
tankard, marbled
glass (WA #05.1214);
framed plaques,
opalescent glass
(WA #05.1310/11);
subjects are the
judgment of Paris and
feast of the gods;
tazza, (WA #05.1307);
covered vase
with handles
(WA #05.1312).

158. *Urbino-style Majolica*
Pottery, ca. 1880.
Tin-glazed earthenware,
Italy; Left to right:
pair of ewers
(WA #05.1480-1);
standing tureen
(WA #05.1292);
bottle, signed "Ginori"
(WA #05.1479);
pair of vases
(WA #05.1482-3).

159. *Japanesque Pottery.*
Left to right:
plate, 1877,
Worcester Royal
Porcelain Co.,
Worcester, England,
(WA #05.1421);
jardinaire, ca. 1880,
Worcester Royal
Porcelain Co.,
Worcester, England
(WA #05.1298);
plate, 1876,
Worcester Royal
Porcelain Co.;
purchased at the
Centennial Exhibition
in Philadelphia in 1876
(WA #05.1422).

160. Persian and Japanesque
Majolica Pottery,
ca. 1880.
Earthenware,
Longwy Pottery, Paris.
Left to right:
pair of vases
(WA #05.1342-3);
plate (WA #05.1348);
vase with cover,
(WA #05.1334);
pair of vases
(WA #05.1344-5).

161. Zsolnay Pottery, ca. 1880,
Pecs, Hungary.
Left to right:
vases (WA #05.1330);
funfkerchen, large
forms among the finest
pieces of art pottery
in the collection
(WA #05.1329).

162. Japanese Pottery,
ca. 1875. Porcelain,
Japan. Left to right:
satsuma vase
(WA #05.1354);
satsuma teapot
(WA #05.1355);
plaque
(WA #05.1286);
kutani bowl
(WA #05.1357);
kaga rose jar
(WA #05.1377).

163. *Pair of Faience Vases,*
1877. Doulton and Co.,
Lambeth, England,
underglaze painted
earthenware; decorated
by "Mrs. Johnson."
This is the style
of work taught in the
New York Decorative
Art Society china
painting class by
émigré-artist and
Doulton alumnus
John Bennet
(WA #05.1299
and .1300).

164. *Painted Cameo Glass,*
ca. 1890. Possibly
painted by the glass-
decorating firm of Julius
Mulhaus in Haida (now
Novy Bor), Bohemia.
Left to right:
vase (WA #05.1195a);
covered vase on stand
(WA #05.1194);
vase (WA #05.1195b);
vase on stand
(WA #05.1197).
Bohemia was the center
of decorative glass-
making for most of the
nineteenth century.
Elizabeth Colt may have
purchased these at the
Columbian Exposition
of 1891. The scenes are
based on Wagnerian
operas.

Gregory type, which peaked in popularity during the late 1880s and 1890s (fig. 164), indicates that she continued collecting until the end of the century. Popular in the oak-paneled dining rooms of the late 1880s and 1890s were Germany's tall glasses painted with heraldic emblems and crests, of which Mrs. Colt acquired several at some unspecified date (fig. 165).

Although the collection is a remarkable document, perhaps it is no less remarkable that it is still intact in an age that has witnessed the wholesale dismantling and dispersion of entire collecting areas by museums entrusted with care by well-intentioned and often generous donors. It is, perhaps, no surprise that the bulk of this collection has languished in museum storage for decades, and yet the Wadsworth Atheneum has retained it all during the past ninety years, when Elizabeth Colt's taste has been decidedly out of fashion. In some ways more radical than the outré contributions of our many politically motivated "contemporary artists," this quiet testimony on the part of an urban art museum struggling, as most are, to sustain public interest and support amid the overwhelming cacophony of televised popular culture is perhaps itself an extension of the "faithful memory" that inspired the woman behind it.

If there is an overarching theme to the art and architectural patronage of both Sam and Elizabeth, it is their determination to infuse their works with a sense of personal biography. Not only did she repeatedly intervene with architects, writers, artists, and sculptors, but an air of intention and personal will hover over every monument she created. At the 1896 dedication of the Caldwell Colt Memorial House adjoining the Colt memorial Church of the Good Shepherd, Elizabeth held court with journalists and guests, taking "great pleasure... in pointing out the designs and their significance," identifying the sources for columns and capitals, and explaining the associations behind its form and dozens of its ornamental details.[74] The picture gallery was no less personal in its associations, and it is easy to imagine Elizabeth in the roles of curator, docent, and interlocutor.

Internationalism was also a theme. Although Elizabeth Colt participated in the programs and exhibitions that embellished Hartford's reputation as a cultural center, she broke ranks with some of her peers by building a collection that was studiously internationalist in sensibility and devoid of regional associations or artists.[75] Together Sam and Elizabeth Colt played a dual role as Hartford's ambassadors to the world and the world's ambassadors to Hartford. As the pistols fanned out from their point of origin to the world beyond, the Colts installed worldly treasures on the grounds and in the cabinets at Armsmear, on the dome of Colt's Armory. Both provincial and cosmopolitan, the Colts successfully merged the two spheres of their lives, using collections and art as an instrument of personal identity.

165. Germanic Glass, ca. 1885. Decorated with enamel, probably made in Bohemia. Left to right: bottle (WA #05.1120); goblet with cover (WA #05.1193); goblet with cover (WA #05.1191); bottle (WA #05.1121). A fondness for heraldry may have persuaded Elizabeth Colt to purchase these.

BUILDING MEMORIALS

I feel that if he & Caldwell know how things are now they would approve of this action on my part.
I have tried to honor their memories always.

ELIZABETH HART COLT, 1901

Elizabeth Colt moved with a sense of purpose after Sam's death to launch the first of what became a remarkable and eclectic series of campaigns aimed at memorializing her husband and burnishing the reputation of the city she loved. By building monuments and memorials and by founding an array of organizations and institutions, she quickly emerged as a key civic leader, playing a major role in creating the apparatuses of charity and civic amenities that helped make Hartford one of the most respected "provincial cities" of America's industrial age. Just thirty-five at the time of Colt's death, Elizabeth drew inspiration from her faith and from the five and a half years of married life to a man she revered, admired, and loved.

Through all the years of collecting, philanthropy, and memorial building, Elizabeth Colt remained actively involved with Colt's Fire-Arms Manufacturing Company, hiring executive officers, grooming (unsuccessfully) her brother and son for executive responsibilities, and jeal-ously guarding its reputation and profitability. An 1887 encyclopedia of American biography describes Sam Colt's widow as a "manufacturer of arms."[1] To the end of the nineteenth century, she retained a controlling interest in its stock.

At the end of 1862, Elizabeth Hart Colt launched the first of her many campaigns of patronage, philanthropy, and works of faithful remembrance, fulfilling Sam's deathbed plea that she "carry out all his plans."[2] In one campaign after the next Elizabeth consciously broke new ground, seeking to "make anew" her chosen fields of endeavor, and, in so doing, creating a new pattern of possibility for American women. However they rank as innovative initiatives, her campaigns were almost singularly unprecedented in being the public acts of a woman. And while privately and outwardly deferential to the men she commissioned to carry out her plans, she could be a steely combatant, as in 1866 when she dumped one of America's leading architects reportedly for lacking bold-

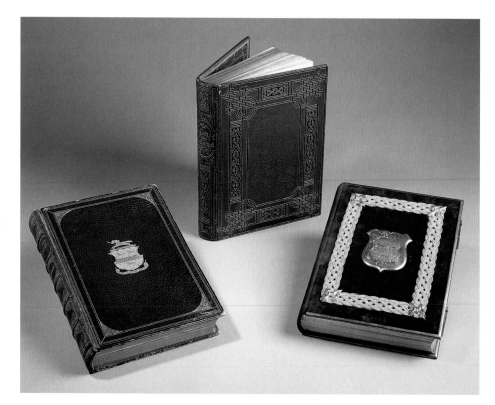

166. Armsmear: The Home, the Arm, and the Armory of Samuel Colt: A Memorial, 1866. By Henry Barnard, printed by C.A. Alvord Printers, New York, 1866. C.E. Matthews, one *of the finest book-binders of his day, was commissioned to execute a variety of deluxe bindings for* Armsmear. Left to right: *leather with gilt, embossed Colt coat of arms on the cover; tooled, dyed leather with a Celtic knot design, signed "Matthews" inside the front cover (courtesy of the Episcopal Diocesan House, Hartford); velvet with gilt brass mounts and a Colt coat of arms medallion (courtesy of Auerbach Art Library, Wadsworth Atheneum).*

ness and creativity; in 1870 when she battled the city fathers to preserve the integrity of her estate from the encroachments of the railroad; in 1878 when she hosted a congress of the male and female stewards of Hartford's diverse charities and virtually demanded they work cooperatively; and in 1901 when she drove the family and the armory's paid managers and investors to the bargaining table to sell the company.[3]

Elizabeth Colt's first project was commissioning *Armsmear: The Home, the Arm, and the Armory of Samuel Colt, a Memorial,* a lavishly illustrated, four-hundred-page profile of her husband's life and world. *Armsmear* was the first book-length treatment of a contemporary American mansion and its occupant, a work of art and literature in its own right. It was a stunning representation of the "house book," a genre of literature that reached its apogee in the breathtaking ostentation of Edward Strahan's *Mr. Vanderbilt's House and Collection* (1883). Hailed as the "most splendid Book... that was ever printed in this country," *Armsmear* was exhibited at the World's Fair in Paris in 1867, where it set a standard for American illustration and book-binding (fig. 166).[4]

Armsmear revealed Elizabeth Colt as a hands-on patron with high standards and the self-confidence to oversee the design and content of her ventures. She conceived of the book at the end of 1862. After rebuffing the supplications of Hartford author Lydia Sigourney, she hired Henry Barnard, author, educator, and president of the Connecticut Historical Society, to supply the text.[5] Barnard probably influenced the choice of Nathaniel Orr and Harry Fenn of New York as illustrators and almost certainly directed Elizabeth's use of his New York publisher C.A. Alvord. Orr, a prolific woodblock engraver, had worked for Colt in 1857, supplying illustrations of the inner workings of Colt's Armory for a puff piece that appeared in the *United States Magazine.*

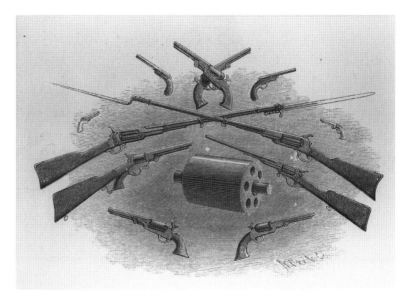

Elizabeth oversaw the photography. Amassing the illustrations involved constant supervision. Elizabeth directed photographers to specific scenes and locations around Coltsville, inside the house, and throughout the grounds. In January 1864 she wrote the illustrator to suggest artistic arrangements of Colt's revolvers (fig. 167) and to share her thoughts on "lettering on the frontispiece," recommending the "same arrangement" used in "Pages & Pictures... [by] Fenimore Cooper.... I have been trying to devise something a little different from Mr. Fenn's suggestion," she wrote. "I would suggest... 'Armsmear' be placed above the vignette... [and] below,... the full title to the book." In May she made the first of half a dozen payments to Orr for illustrations that eventually cost about twenty-five hundred dollars, the equivalent today of more than one hundred thousand dollars.[6]

Eager to complete the book for Christmas 1864 and increasingly uneasy about the cost overruns, she became "quite discouraged about it."[7] A year and a half later, with the book still incomplete, she concluded that it had been "a pretty expensive experiment" that taught her "a lesson which I shall not soon forget."[8] As the first of the five hundred copies finally arrived for Christmas 1866, the family estimated that the effort had cost the equiva-

167. Illustration from Armsmear, *engraved by Nathaniel Orr, depicting a variety of Colt firearms.*

lent today of four hundred thousand dollars, or about eight hundred dollars per copy for a book she distributed privately to friends, politicians, and presidents during the remaining forty years of her life.[9]

With four styles of binding—two grades with embossed covers, a deluxe cover of velvet with gilt brass trim, and two personal copies bound in Charter Oak (fig. 168), one decorated with the rampant colt and Colt coat of arms; the other, with Sam's portrait framed by oak leaves and acorns—clearly, despite frustrations, Elizabeth did not quit or stop spending until she got it right. With chapters on the house and gardens, the armory, Charter Oak Hall and the factory village, the cabinet of memorials, an account of Sam's early struggles as an inventor, and a chapter written by a Professor J.D. Butler on the development of the Colt revolver, *Armsmear* remains to this day a remarkable document, a work of art, and an authoritative source on the life of Sam Colt.[10]

Publishing books, and building and endowing libraries remained a consistent theme in Elizabeth Colt's philanthropy for the next forty years, ending with a provision that her will be printed and distributed after her death. In 1879 there was a Jarvis family genealogy to which she contributed and may have played an anonymous role in publishing.[11] In 1881 she published and largely wrote a ninety-page illustrated memorial biography of her mother.[12] In 1898 she commissioned her cousin the Reverend Samuel Hart to write an illustrated memorial tribute to her son Caldwell, which also profiled the Caldwell Colt Memorial House she built in his memory for her church.[13] The next year she published and cowrote *A Memorial of Caldwell Hart Colt,* and in 1902 she published *A Memorial to the Soldiers and Sailors... in the War of 1898* (fig. 200, p. 221) to document the "first monument ever built in the National Cemetery by a Society of women," the monument having been dedicated by President Theodore Roosevelt on May 21, 1902.[14]

In addition to the library at Armsmear, she created libraries for the parish house of her church, for the Union for Home Work shelter for the poor, and for her son's

168. Armsmear *with binding carved of wood from the Charter Oak, one with Samuel Colt's head in bas-relief (WA #05.1573);* *the other with the Colt coat of arms (WA #05.1574). These were Elizabeth Colt's personal copies of the book.*

Delta Psi fraternity at the Sheffield Scientific School at Yale.[15] She also contributed to, and almost certainly helped sponsor, articles that appeared in popular periodicals and books, both on Sam Colt and on Armsmear.[16] Elizabeth Colt believed in the power of words.

The frustrations of relying on a coterie of male experts who appear to have been reluctant to negotiate or take direction from a female client did not stop Elizabeth Colt from embarking on an even more ambitious memorial while the book was in progress.[17] When Sam Colt died in 1862, it was James Batterson who inscribed the words on his marble tomb. The death of their son Samuel Jarvis Colt prompted Sam and Elizabeth to create a burial plot on the grounds at Armsmear (fig. 169), despite having purchased a family plot in Hartford's Old North Cemetery only a few years earlier. Whether from a wish to be engaged in a high-profile civic improvement or because she felt uneasy about having her husband and children buried in the back yard, three years later, Elizabeth Colt became the first subscriber to Hartford's new rural cemetery, Cedar Hill, where she, Sam Colt, and their children are now buried.

James Batterson was one of the founders of Cedar Hill, designed on a new open field plan by Swiss-trained landscape designer Jacob Weidenmann. It was described to the public in 1865 as the "Greenwood Cemetery of New England," featuring three hundred acres of "smooth-shaven verdant lawns,... willowy vales," and "leafy slopes." Cedar Hill was Hartford's most ambitious public works project of the late 1860s, and Batterson moved quickly to secure Elizabeth Colt's patronage and support. During 1865 and 1866 up to one hundred men worked steadily to transform Cedar Hill's high ridge, moving tons of earth and stone, creating "smooth green lawns, verdant valleys, and gently-sloping hills," planting trees and shrubs, and rerouting a natural brook to create a "reservoir pond." At its opening in July 1866, Cedar Hill boasted a ten-thousand-foot-long picket fence, "elevated ridge crowned with cedars," "extensive views," lakes, islands, cascades, and handsome stone and rustic bridges.[18]

Illustrating the cover of Cedar Hill's first annual report was the Colt family monument (fig. 170), which Elizabeth commissioned from James Batterson in 1864, more than a year before the initial published reports about the new cemetery and two years before it opened.[19] That spring, Batterson returned from a business and art-buying trip in Europe with a pair of marble Venetian dogs, which he sold to Elizabeth for the grounds at Armsmear.[20] As early as 1861, Batterson advertised that he was "now importing" a "rare... rose granite" from "the celebrated quarries at Peterhead,... Scotland,... beautifully modelled with black hornblend and red feldspar," ten years before granite of any color made significant inroads into a market dominated by marble and Connecticut's own brownstone.[21] There is no evidence he found any takers before Elizabeth Colt, and certainly none on so grand a scale. Eventually rose granite became one of Batterson's most distinctive products, used in Civil War monuments for which he developed a national market by the end of the decade.[22] It appears, however, that the Colt monument, costing an estimated twenty-five thousand dollars, was the commission that, more than any other, established Batterson solidly in the business of making and marketing high-profile granite public statues and cemetery memorials.[23]

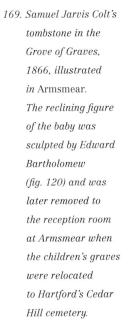

169. *Samuel Jarvis Colt's tombstone in the Grove of Graves, 1866, illustrated in* Armsmear. *The reclining figure of the baby was sculpted by Edward Bartholomew (fig. 120) and was later removed to the reception room at Armsmear when the children's graves were relocated to Hartford's Cedar Hill cemetery.*

With this commission, Elizabeth Colt set the pace for the generation of family monuments that epitomize Hartford's celebrated Gilded Age. Standing thirty-two feet high, with an Egyptian-style red granite bronze-mounted plinth, column, and capital, the monument was topped by a seven-foot-high bronze figure of Gabriel, the angel of the Resurrection, designed by American expatriate sculptor Randolph Rogers and cast at the Royal Foundry in Munich. The shaft, which Batterson reportedly "modelled from the examples... found among the ruins of the grand temple at Karnac," with secondary columns based on "an ancient temple at Luxor," was cut and finished to his specifications in Scotland.[24] By May it was "on board a vessel en route to New York."[25] Although Colt's Armory Band played at the consecration ceremony for Cedar Hill in September, the bronze statue was not installed until after it was featured in an exhibition of "memorial art" Batterson staged on the grounds of his monument company in conjunction with the annual fair of the Hartford County Agricultural Society in October.[26]

170. Colt Family Monument, 1866, Cedar Hill Cemetery, Hartford. The monument is marked "J.G. Batterson" on the base. Batterson, the owner of a monument works, designed and supplied the various components of the monument, including the bronze sculpture of the angel Gabriel by American sculptor Randolph Rogers (1825–92), which stands atop the imported rose granite column.

Towering alone over the freshly landscaped grounds, the Colt family monument, "its massive proportions" corresponding "with the power and energy of the man," captured the spirit of Colt's outsized reputation. On its base was affixed the Colt-Jarvis coat of arms in bronze, bearing emblems of marital union, surmounted by the rampant colt, and engraved with words Elizabeth chose to describe Samuel Colt: "a devoted husband... a steadfast and generous friend," whose "fame... at home and abroad" had "won honor... and contributed largely to the prosperity of his fellow citizens, and of his native city, which he loved," signed by "his wife in grateful affection." A tribute to Sam, the monument was also a bold affirmation of Elizabeth's growing reputation as a patron of art.

An even more fitting memorial to her industrialist husband was Elizabeth's heroic rebuilding of the armory after the disastrous fire of February 5, 1864. Described as the "severest calamity that ever visited the industry of Hartford," the "Great Fire at Colt's Armory" did an estimated $782,000 in damage (today equal to about $33 million). It threw as many as nine hundred men out of work and dangerously compromised the arsenal of a nation at war.

All hell broke loose in fiery fury on that cold winter morning as a community woke to the sound of the "steam gong" at Colt's Armory wailing, as if in pain, "so protracted [in] its noise that the community generally anticipated the ringing of the bells and the cries of fire which soon followed." The panic and excitement began at 8:15 in the morning, an hour and a quarter into the day's work at the plant, when "workmen discovered smoke issuing from the attic wing... over the polishing room... near the main driving pulley." The internal water hose that was first turned on the fire came up dry. As the "flames burst out from the drying room" in the attic, they "ignited with the belt on the pulley and shot with lightning rapidity to the roof" sweeping "across the building faster than a man could run," igniting the "floors of yellow pine... sat-

urated with oil.... Flames shot through the openings" in the windows "with terrible fury" as "black smoke curled in the air." Below it was pandemonium as the "seventeen or eighteen hundred workmen aroused by... the fire" discovered that they "were all locked in their respective departments, it being a rule of the company." Finally, with the "doors thrown open," the workmen poured out into the street, now crowded with thousands of spectators. Several of Colt's workmen and contractors with "great coolness,... courage and daring" rushed "through the flames" saving as much property as they could.[27]

At "about nine o'clock," in an event that would haunt Elizabeth Colt for the rest of her days "the dome... on which was the large gilded globe supporting a large colt... fell in with a tremendous crash." As a crowd now estimated at twenty thousand stood by and watched, the fire worked its way north, shooting across the "covered bridge-way... to the office." Elizabeth wept as she watched from her bedroom window. "To think that the

magnificent, noble structure is in ruins," her father wrote, and "so identified with the Colonel, it seems like burying him again."[28]

In the days that followed, photographers rushed to the scene, producing souvenir pictures of the ruins, a sight the most "awfully grand... any building on this continent ever presented" (figs. 171 and 172). The newspapers printed special editions of the story, selling more than one thousand copies nationwide.[29] Hartford's poet Lydia Sigourney dashed off "On the Burning of the Armory, Erected by the Late Colonel Colt," while speculators bought up whatever revolvers were on hand among arms dealers across the country.[30] Elizabeth Colt gathered relics for the cabinets at Armsmear (fig. 173). "Historical allusions" were frequent in the days that followed, not least for the "herculean efforts of the armorers themselves" who "rose in their strength to do battle... determined to save" the West Armory, laboring with "the might that only affection can give; they loved their chief,

172. *Colt's Armory after
the Great Fire, 1864.
N.A. and R.A. Moore,
Hartford, photographer.
(Courtesy of the
Connecticut Historical
Society.)*

173. *Molten Steel and Pistol
Components, 1864,
Colt Collection.
A bullet mold* (left) *and
the rare loading tool
for the 1862 model
police pistol* (right) *are
among the recognizable
components from what
may have been hundreds
of comparable fused
gun parts found in
the wreckage after
the armory fire of 1864.*

who had gone from them forever, and they willed that their hands should save his monument from destruction."[31] Their efforts reportedly saved the West Armory, enabling the company to resume work on its government contracts within three days of the fire.

For Elizabeth Colt, for whom the emotional toll of such a loss was incalculable, the financial loss was also enormous. For once, the press and entire community rallied behind Colt's memory, urging the "company to act as they believe *he* would have acted" and advising the city to "pledge its credit to them for... reconstruction."[32] Astonishingly, the only life lost was that of Edwin Fox, a workman crushed by the weight of falling timbers while he struggled to retrieve company property. But the loss of patterns, drawings, models, and machinery, including screw machines that "were the only specimens of their class on this side of the Atlantic," was tremendous.[33] Initial estimates of the loss of machinery and finished work tallied about $782,000, including $300,000 to rebuild the armory.[34] The cause of the fire was never determined, though the fact that "the fire broke out... during the only half-hour in the 24 when a watchman was not present" and that Colt's had recently hired "a deserter from the rebels" gave speculation of Confederate arson great currency.[35] Also blamed was the presence of "combustible... cotton waste... near the main driving pulley" and the possibility that "friction" near a "drying room heated by steam pipes... [and] filled with pistol stocks" was responsible.[36]

Although Sam Colt never insured his property, Elizabeth did, for an amount that covered about half the loss. Ten days after the fire, the press heralded as "very liberal and highly honorable" the decision of the stockholders, meaning principally Elizabeth Colt, "to pay the contractors all that was due them at the time of the destruction."[37] While the family privately maintained that "they intend rebuilding it," the decision to do so was not made public until 1865.[38] To compound matters, Elisha K. Root, who assumed the presidency of Colt's after Sam's death in 1862, died in September 1865, leaving Elizabeth's brother Richard temporarily in charge of the company.

In October, Colt's Fire-Arms was congratulated by the press for hiring Maj. Gen. William B. Franklin, a "capable and estimable man" and "one of the national celebrities of the war," as the new president. Franklin was paid $10,000 per year (today $384,000), a salary Elizabeth's father described as "not too dearly purchased" for a man of "his remarkable business talents and qualifications."[39] Franklin, a civil engineer and West Point classmate of Ulysses Grant, George McClellan, and Stonewall Jackson,

had impeccable military and technical credentials. Having served the U.S. Treasury Department as head of its engineering department before the war, Franklin was especially well suited to the task of rebuilding the armory and maintaining its military contracts.[40]

Two years after the clean-up began, the new armory was "substantially complete." Although almost identical, at first glance, with Sam Colt's original armory building, the new east armory substituted brick for brownstone, was one story taller (fig. 174), and with the "bold projections" of its central bay—"not unlike the lower story of the Pitti Palace in Florence"—"more ornamental than would otherwise have been."[41] The signature "Moorish" style, "blue, star-spangled" dome—a "precise copy of the dome of a church in Moscow," complete with a "gilded globe on which a colt is rampant, as if it were triumphant over the world"—was complemented by the addition of latches and hinges "fashioned in the shape of pistols" on all the exterior doors. With "over five hundred and fifty windows... over thirty-seven miles of heating, gas and water pipes," and a "well-lighted attic... fitted up as a shooting gallery for the testing of arms," the new Colt's Armory probably was, as described, "the most complete establishment of the kind in the world."[42] It may also have been the first industrial building in America of "fire-proof" construction, designed at considerable additional cost with floors constructed of "wrought iron beams, and brick arches."[43] Fire walls "three feet thick" were built at a cost estimated at $300,000, or the equivalent today of about $11.5 million.[44] The new armory—"the design of Mrs. Colt and the directors"—was "not only an unsurpassed workshop, but, also a monument to the memory

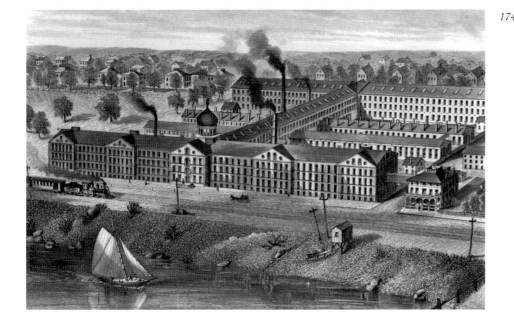

174. Colt's Armory and Machine Shops, ca. 1875. From Industrial America; or Manufacturers and Inventors of the United States (Atlantic Publishing and Engraving, 1876). Although the 1855 armory stood for fewer than nine years, it has been illustrated much more frequently than the larger and more elaborate armory Elizabeth Colt had rebuilt in 1867. This is the best view showing both East and West Armories, the office building restored after the fire, the worker house, and in the distant background, the steeple of the Church of the Good Shepherd. (Courtesy of the Connecticut State Library.)

of the late Colonel Colt" and was fully consistent with Elizabeth's determination to live a life of "faithful affection" and memory.[45]

Six months after assuming the presidency of Colt's Fire-Arms, General Franklin was in New York meeting the architect involved in Elizabeth Colt's most ambitious program of patronage and civic improvement, building the Church of the Good Shepherd. The church, which originated as a parish mission to the South Meadows in 1859, maintained a Sunday school in a room reserved for the South Meadows Sons of Temperance. By 1864 up to 145 parents and children attended weekly services, and a "sewing school" had been established for the children of the parish, which at that point "ceased to be a mission" and "assumed an independent shape."[46] For Elizabeth Colt, who counted numerous Episcopal ministers and an Episcopal bishop among her immediate family and ancestors, her move to build a church for the armory workmen, although ambitious, was hardly surprising.

By 1866 plans for a church, originally bearing the name of the Church of the Holy Innocents (fig. 175), were in hand when William Franklin met in New York with the architect Frederick C. Withers, a successful, British-trained designer, author of *Church Architecture* (1871), and partner with Calvert Vaux in one of the nationally successful architectural practices of the 1860s and 1870s. The following August, Franklin received "from Mrs. Colt the working drawings of the Church," which, for reasons unknown, dissatisfied her.[47] By October, a different architect, Edward Tuckerman Potter, was at work preparing details and elevations for a more inspired and ambitious design, for which he was paid two thousand dollars, in September 1867, the month Connecticut's Episcopal bishop laid the cornerstone for Elizabeth Colt's new church.[48]

Edward Potter's early retirement in the mid-1870s deprived him of a well-earned reputation among the celebrated American architects of the Victorian age. Prolifically experimental and renowned for his elaborate programs of sculptural ornament, Potter pushed the boundaries of design and materials technology in such masterpieces at the Nott Memorial at Union College (1858 and 1871) and the First Dutch Reformed Church (1862) in Schenectady, the George Hunter Brown house (1867) in Millbrook, New York, and the Hartford mansion of Mark Twain (1873).[49] Potter, whose father was Elizabeth Colt's father's classmate at Union College, belonged to a clan interwoven into the leadership of the Episcopal Church in the United States, making Elizabeth's choice of him as architect almost a family affair.[50]

On September 7, 1867, the cornerstone was laid and a construction crew broke ground for a new church at the north end of Coltsville, following a ceremony officiated by the Episcopal bishop.[51] News of Mrs. Colt's plans for a church "for the benefit of those living in the region of the Armory" circulated around town all year.[52] When the church was finally completed and consecrated in January 1869, after twenty months of continuous work by stonecutters, painters, stained glass artists, wood-carvers, masons, and upholsterers from Hartford and New York, its final cost was at least $175,000 (today almost $7 million), and its effect was dazzling (figs. 176 and 177).[53]

Regarded as a "gem of Ecclesiastical Architecture" and the "pride of the city" the Church of the Good Shepherd was an artistic and technological marvel.[54] "From turret to foundation-stone" the church was praised as "a work of art."[55] Built as a "free church,... entirely at the cost of Mrs. Colt as a memorial to her husband... and

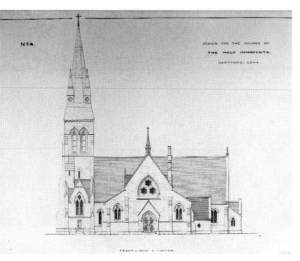

children," the body of the church was sixty feet tall and featured an innovative Sunday school wing with "magnificently carved screen and doors." From the iron columns and state-of-the-art gas lighting of its interior to its ambitious program of ornamentation and scriptural passage, the building introduced Hartford to the English modern Gothic style that would soon dominate its skyline.

Built of hammer-dressed Portland brownstone, trimmed with a smooth light yellow Ohio sandstone, the exterior of the church was rich in scriptural passage and ornament, including wreaths of ivy, polished red granite columns, a 150-foot spire, and a roof covered in multicolored, zigzagging bands of slate, trimmed along the crest with Gothic-style ironwork tracery. The plan included an elaborate hierarchy of doors and entrances, each with a symbolic program of ornament and sacred inscriptions, and each leading to a different precinct within the church: one for the vestry and priest, one for the Sunday school, a third used perhaps by the family, and the extravagantly decorated "armorer's door."

With its sacred motto, "Whatsoever Thou Doest, Do All to the Glory of God," and workmanlike hammer-dressed surface, the armorer's door is indeed "a work of art" (fig. 178). But what has most intrigued and perplexed generations of spectators is its unique program of ornament "fashioned out of various parts of the revolver."[56] Column capitals braced by Colt's navy revolvers carved in sandstone (fig. 179), crosses interwoven with bullet molds, pistol hammers, cylinders, pistol grips and barrels, and accents composed of revolver cylinders festooned with ivy represent one of the few juxtapositions of these sacred and secular icons in American art. What was the point? Had Colt's revolver now become a sacred icon? Certainly the revolver's role in bringing the light of Christian civilization to the dark corners of the earth was rarely questioned by the generation that colonized the American West, opened Japan and Asia to trade, and explored the interior of Africa. Whatever one thinks of the politics, the kinship of religion, industry, and empire has rarely been so boldly proclaimed.

As intriguing as it was outside, the interior of the Church of the Good Shepherd was a gas-lit spectacle. At the head of the basilica was an elevated chancel illuminated by thirteen windows representing Christ and the twelve apostles (fig. 180), chestnut and oak paneling beneath a richly frescoed border, an enormous organ by "Messrs. Hook of Boston," and an altar floor inlaid with "magnificent patterned tiles" imported from England. In front of the chancel, next to the baptistery, was a gilt brass lectern "equal" to one "exhibited at the late Exposition in Paris," featuring a gold-plated brass eagle standing on a globe ringed with "real carbuncles, which… gleam forth like balls of fire" (fig. 181).[57] In the baptistery, a remarkable baptismal font (fig. 182), a gift from Elizabeth's sister Hetty, featured three marble figures honoring the deceased infants of Sam and Elizabeth, Samuel, Elizabeth, and Henrietta. The church also featured the city's first set of chimes.[58]

The symbolic becomes personalized inside with the Church of the Good Shepherd's most innovative feature,

180. Apostle Windows,
     Church of the Good
     Shepherd, 1868.
     Stained glass, Henry E.
     Sharp and Son,
     New York.

181. Lectern,
     Church of the Good
     Shepherd, 1868. Brass
     and gilt brass,
     probably British;
     high-end liturgical
     furnishings like
     this were produced
     by Connecticut's brass
     industry beginning
     in the late 1870s.
     Whether work of
     this quality was ever
     produced in Connecticut,
     it was certainly
     not produced as
     early as 1868.

182. Baptismal Font,
     Church of the Good
     Shepherd, 1869.
     Marble, by
     John M. Moffitt
     (1837–87), New York;
     Moffitt was a London-
     trained sculptor whose
     practice emphasized
     work for cemeteries and
     churches. The
     culmination
     of his career was
     the commission for the
     Soldier's Monument
     on East Rock in
     New Haven in 1887.

183. Colt Memorial Window,
Church of the Good
Shepherd, 1868.
Stained glass, designed
by Elizabeth Colt,
made by Henry E. Sharp
and Son, New York;
fresco painting by
"Moore of New York"
(Steve Lakatos
Photography).
This window was the
emotional and artistic
focal point of the church.
The scriptural passage
in fresco that frames
the window reads,
"The Lord which is in
the midst of the Throne
shall Feed them and
Lead them unto Living
Fountains of Waters and
God shall wipe away
all Tears from their
Eyes." Elizabeth Colt's
authorship of this
feature is suggested by
her description of Sam
Colt as "my husband,"
language revealing an
active voice and hand
in the effort.

its stained glass windows. Elizabeth Colt, who approved and may have devised the concept for the armorer's door, clearly had a hand in selecting the works of art on which the stained glass windows were modeled. One of the earliest programs of fine art-based stained glass in the country, for which Henry E. Sharp and Son of New York received $5,325 (today about $200,000), it features several apostle windows, a memorial window, and a multifoil, or Catherine wheel, with a "protecting angel" bearing its infant to heaven.[59] Although the process by which Sharp and Mrs. Colt gathered and selected visual models for the windows is not known, a bas-relief sculpture by Ernst Rietschel, reproduced in a contemporary art periodical, was clearly the source for the protecting angel. The posture, gaze, and hairstyle of the baby being transported to heaven bear a remarkable resemblance to a portrait bust the Colts commissioned of their firstborn, Samuel Jarvis Colt, in 1858. Elizabeth cannot have failed to recognize her son's visage in this work of art.[60]

Next to the memorial window are two lancet windows featuring angels modeled after paintings by the fifteenth-century Dominican monk Fra Angelico, a gift from her brother Richard and brother-in-law C. Nichols Beach.[61] The chancel windows featuring Christ, his "face beaming with love and forgiveness," flanked by the twelve apostles were modeled after Johann Friedrich Overbeck's celebrated paintings of the saints.[62] The founder of the Nazarene Movement in Munich, which revived the tradition of religious painting from the Italian renaissance, Overbeck was one of the most popular religious painters of the mid-nineteenth century.

The memorial window (fig. 183), offered a personal lesson, one of many suggested by the stained glass, symbolic decoration, and almost two dozen sacred mottoes that burst from every nook, arch, and cranny, inside and out, at the Church of the Good Shepherd. The memorial window, itself framed by sacred mottoes, recalls Elizabeth Colt's devastating familial losses, three children and a husband in three and a half years. One side features the Good Shepherd—He Shall Gather the Lambs

into His Arms—and the names of the Colts' infant children. The other side features the Old Testament figure of Joseph of Egypt, shown at the height of prosperity. Colt's friends liked to compare his life with the story of Joseph, who "had been sold into slavery, who had been betrayed by his brethren, who had conquered evil fortune, and who, [having] risen to power and wealth,... turned with lavish generosity towards his old father and brothers."[63]

The consecration of the Church of the Good Shepherd in January 1869 was one of the "most imposing ceremonies that ever took place in any church in this State" and was officiated by Episcopal bishops from Connecticut, Maine, western New York, Pittsburgh, and Albany, at the head of a procession that included forty clergymen from all over the Northeast followed by the "Reverend Professors from Trinity College," the church vestry, Elizabeth Colt, her family and friends.[64] With the Church of the Good Shepherd, Elizabeth Colt made her entrance as a major patron of religion and art, united by memory and faith.

Even before the church was completed, supplications from churches and church causes had become so numerous that Elizabeth printed a form letter to explain how the "large number of applications which nearly every mail brings from all parts of the United States, for aid in some church work, renders it quite impossible to answer each one individually, and I am compelled to resort to this method of saying... that my own plans for church purposes, at present absorb all the means that I can devote to such objects."[65] And yet, seven years later she accepted a leadership role, with the attendant financial obligations, of the Bureau of Relief's fund to support "students preparing for the Episcopal Ministry in Connecticut" and the "education of the daughters of clergymen."[66] And in 1881, when she was offered the presidency of the Connecticut Branch of the Women's Auxiliary to the Board of Missions of the Episcopal Church, she accepted.

The Women's Auxiliary, as a philanthropy, was the perfect spiritual companion for an age of globalism,

184. Caldwell Colt
Memorial House,
1896.
Edward Tuckerman
Potter (1831–1904),
architect,
New York.

empire, and Western expansion. Imagining themselves as carrying "the light" to the dark corners of the earth, Elizabeth Colt spoke on behalf of the auxiliary at its founding in 1881 of a "new zeal and interest in mission work." Echoing her advocacy of "union as strength," on behalf of the Board of Associated Charities, Elizabeth spoke "in meekness and humility" of winning "all of our household of faith to work together... in Mexico and... China," sponsoring "a trained nurse... to work in our mission... and hospital... in Wuchang," and supporting "the school of Bishop Tuttle at Salt Lake City" and a "school for colored children in Georgia."[67] A year later, at the auxiliary's annual meeting, the need "to unite previously-existing societies" was reiterated.[68]

Under Elizabeth Colt's leadership as president, the Episcopal Women's Auxiliary supported mission churches in Oregon and Colorado, explored ways of "providing for the six millions of emancipated slaves," and operated Indian schools in the northwest. It raised a large fund by tapping her friends and acquaintances from New York, Newport, and Boston to support the Hill Memorial Christian School in Athens, Greece; not exactly dealing with the problems at your doorstep, but probably a soothing corrective for an age perplexed by its mixture of prosperity, dislocation, abundance, suffering, and empire.[69]

The untimely death of Caldwell Hart Colt in 1894 provided another test of Elizabeth Colt's personal faith. It all came to an end in Punta Gorda, Florida, where, at the age of thirty-five, Sam and Elizabeth's sole heir was found dead, allegedly due to complications from tonsillitis and almost certainly from natural causes. Rumors of suicide circulated quickly, prompting William Cowper Prime, a widely respected New York and Hartford friend who was with him in the last hours of his life, to write the editor of the *Hartford Daily Courant* urging him to print his firsthand account of Caldwell's last hours: "We have heard that a rumor has been started at the north that Colt killed himself. This is absurd and pure fabrication. If it be so published it will grieve Mrs. Colt very much and ought to be promptly denied.... When we went in to dinner at 6 o'clock Sunday evening I told Annie that Colt was dying & he died two hours later."[70]

Elizabeth had been through this drill many times, with parents, siblings, children, and spouse. Grief stricken, she arranged for the body to be transported in a special car attached to the regular 2:05 train. Caldwell "lay in state" in a casket draped with the United States flag, his commodore's yachting cap, and mounds of carefully arranged lilies and violets. In New York a contingent of Colt's friends embarked for the ride to Hartford where the car was greeted by the workmen at the armory who presented a "broken column of roses" to the stockholder and executive few had known or much seen. His pall bearers included Commodore E. D. Morgan of the New York Yacht Club, Vice-Commodore H. A. Sanderson of the Larchmont Yacht Club, Lt. William Henn of the British Navy, and a phalanx of cronies from New York, Newport, and Hartford society.[71]

For Elizabeth Colt, who years earlier had written of the "pathetic... thought of being the last of one's generation of a family," the loss of Caldwell would inevitably prompt another wave of memorialization.[72] Now, with no direct heirs to inherit the Colt fortune, and perhaps, with the honor and reputation of her beloved son in jeopardy, she embarked on her most audacious campaign of memorial building.

The Caldwell Colt Memorial House (fig. 184) ranks as one of the eccentric masterpieces of Victorian American architecture, a triumph of personal exhibitionism and one of the country's intriguing examples of the Victorian architectural form known as the memorial hall. The memorial hall typically accommodated a variety of civic functions, including galleries for art and history, town

offices, libraries, public meeting rooms, and the inevitable regimental flags and memorial tablets honoring the Civil War dead. They were almost always the most artful, expressive, and expensive buildings belonging to the towns, cities, and institutions that built them. Drenched in symbols of spirituality and remembrance, they embody the most deeply felt principles of Victorian art, an art that has been so flagrantly and repeatedly violated by the one-size-fits-all, top-down professionalism of modernism that it is hard to imagine a time when the poignant, the personal, and the situational were essential components of good design.

Completed in 1896, the Caldwell Colt Memorial House is a triumph of personal associations and contextual reference, conversing eloquently and sympathetically with its companion Church of the Good Shepherd and yet saturated with secular details and personal affectations. Elizabeth Colt called Edward Potter out of an eighteen-year retirement to design a building that would be compatible with her memorial church. The memorial house is Potter's greater (and more expensive) masterpiece and arguably one of the finest examples of memorial and associational architecture in the United States. It represents a scrapbook of memories enshrined in the same secular, early English Gothic style that architect Edward Potter had introduced in Hartford almost thirty years earlier. As a free-standing building beholden and subordinate to the design of an existing structure on the site, the memorial house was actually forward looking in the way it searched back and around to achieve aesthetic compatibility with its parent memorial (fig. 185). It is, however, a difficult building to absorb, much less to understand. References and symbols pile up like biographical confection on this remarkable, boat-shaped monument. Elizabeth Colt was determined that, at least in the city of his birth, Caldwell Colt would never be forgotten.

185. Church of the Good Shepherd and Caldwell Colt Memorial House, ca. 1896. (Courtesy of the Connecticut Historical Society.)

186. *Memorial Hall inside the Caldwell Colt Memorial House, 1896. Note the Eastman Johnson portrait situated within the tribune.*

No correspondence about the building survives. This is regrettable. Even more than with her pictures, books, and related memorials, this building is so loaded with personal and biographical details that it might not be inaccurate to describe Elizabeth Colt as a full partner and collaborator in its design. She certainly dictated its uses and form, its ornament and materials. The building is suffused with nautical references, from the portholes that pierce its roof to the ship's bridge that originally ran along the ridge pole and supported a mast-shaped iron spire carrying a cross aloft 110 feet above ground (fig. 186). Inside the memorial hall on the second floor is a ship's catwalk that runs under the ceiling. Stone-carved replicas of ancient ships' prows, including warships, yachts, and ships of commerce, are complemented by anchors, rigging, capstans, quadrants, and yacht signals carved in stone at every turn. Front and center, on the ground level facing north, is a bas-relief of the *Dauntless* (fig. 187). Inside the grand hall was the tribune, a recessed space resembling an altar, framing the portrait of Commodore Caldwell Colt that Elizabeth commissioned, at great expense, from the acclaimed American painter Eastman Johnson (fig. 188). To the side are bas-relief allegorical portraits of the *Dauntless*, outward bound, along with "the voyage ending," united beneath an arch carved with the words from the Canticles, "Many Waters Cannot Quench Love." As imaginative as Potter may have been, no architect ever designed anything this intensely personal without the client's input and direction (figs. 189 and 190). Sacred mottoes burst forth with victorious conviction, springing loose a torrent of words, words, and more words, enveloping the viewer in a frenzy of language, both inside and out.

Facing south toward Colt's Armory was the Colt-Jarvis crest in mosaic, flanked on one side by the stone-carved words "Man Conquers Life by Labor/God Conquers Man by Love," and on the other by Sam Colt's motto of victory through suffering, "Vincit quit patitur." Inside the tribune, carved in stone, were Elizabeth's words, the words by which she expressed her fondest hopes for her son's reputation and remembrance.

He Carried his flag with credit to himself, and honour to his country in many seas. The master of his own vessel, he never feared to bear hunger. Always mindful of the comfort and pleasures of others, he won and kept the affectionate regard of many friends who mourn his early loss. He was a true and loyal friend, strong in his affections, honest in his dealings, courageous by nature, considerate of those under his command, never giving an order which he himself would not dare execute, liberal in his gifts, noble in his hospitality, courteous in his attentions, tenderly devoted to his Mother.

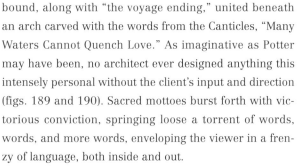

187. *Details of the Caldwell Colt Memorial House, 1896.*

At the dedication of the memorial house in September 1896, witnesses were amazed by the power of mother love exhibited in this eccentric shrine to the colorful personality of Caldwell Colt. Newspapers described it as "an outward and visible sign of an inward and spiritual history," the result of a "campaign of devotion to God and affection for a dear son." While acknowledging that it was a "beautiful structure," the speaker affirmed that "the mother-love which finds expression here is far more beautiful... sharing the immortality of divinity."[73] All told, the building cost Elizabeth at least $300,000 (today about $10.5 million), a little less than she spent rebuilding Colt's Armory, and half again what was spent building the church in 1868.[74]

As a civic and religious monument, the Colt Memorial House is an irrepressible presence. From its stained glass to the marble, onyx, and green spar columns gathered "from parts of the world where he most loved to travel," the building draws from a deep reservoir of architectural materials and technologies to celebrate not only a life of yachting but also hunting, fishing, outdoor activities, and other "hobbies," both real and imagined, from "his boyhood." In addition to its capacious memorial hall, a function room with stage and "minstrel gallery" capable of seating three hundred, there was a library, kitchen, sewing room, and club rooms for the parish and community on the first floor, an elegant study with fireplace for the rector on the second floor, and rooms "devoted to physical education, bowling, billiards, and dumb-bells" on the ground floor.[75]

Altogether, Elizabeth Colt spent the equivalent of $30 million building three of what today rank among Hartford's ten most significant architectural treasures, a rare legacy of civic embellishment. Later she left the city a park, a museum wing, and a retirement home for clergy widows. By thinking and acting out her personal dramas on a grand scale, Elizabeth Colt sought to ensure that the people and institutions she loved would be hard to forget and almost impossible to destroy.

188. Patriotic and Family Emblems inside the Caldwell Colt Memorial House, 1896.

189. American Bison Head, exterior detail, Caldwell Colt Memorial House, 1896. Several of Caldwell's hunting trophies are accurately rendered in stone in the ornament of the building. These trophies were among the collections bequeathed to the museum by Elizabeth Colt.

190. Rampant Colt, exterior detail, Caldwell Colt Memorial House, 1896.

FIRST LADY OF CONNECTICUT

When a disagreeable philanthropist said to the lady at the head of a charity fair, "You never undertake anything for the poor unless it yields amusement for yourselves," she very properly replied: "You mistake. We never amuse ourselves unless it benefits the poor."

HARTFORD DAILY TIMES, DECEMBER 3, 1883

Elizabeth Colt's contributions to the civic culture of Victorian Hartford were not all material or memorial. Indeed, she earned the title First Lady of Connecticut in her full-page, front-page obituary in Hartford's leading newspaper as much for her efforts as a champion of impoverished women and children as for her work as an institution builder. And for a woman negotiating a leadership role in a culture of civic leadership predominately segregated by gender, Elizabeth Colt faced challenges that transcended wealth and vision, rendering her achievements all the more intriguing.

The Victorian culture of noblesse oblige, which has been ably chronicled by Kathleen McCarthy, Gertrude Himmelfarb, Robert Bremner, and others, does not always paint a pretty picture of the motivations and results achieved when the women, primarily, of the Victorian age emerged as champions of charitable and social service causes. And yet, from the vantage point of the late twentieth century, when socially cooperative behavior and the whole concept of civic and even personal responsibility has declined, the spectacle of the Victorian age, with its intense regional pride and boundless confidence in the power of human will to uplift and uphold, is, to me, breathtaking. As Kathleen McCarthy has noted, "Each generation has interpreted the notion of civic stewardship to fit the special contours of its world.... Goaded by pulpit, press, and popular expectations, antebellum urbanites flooded into the side streets of their cities, personally aiding the sinner and the poor, surveying developing needs, and creating new associations to deal with them."[1]

While Hartford was a much smaller city and had far fewer than the "hundreds of individual charities and cultural institutions" of McCarthy's Chicago, its tradition of civic improvement and its varying apparatuses of support and protection for the weak and needy are perhaps even more remarkable because the city was, at the time Sam and Elizabeth arrived, little more than a glorified farm town and river port with a modest past and a very uncertain future. Several of Hartford's charitable institutions, notably its school for the deaf (1817), its mental hospital (1822), and its free public high school (1847), were the first or among the first institutions of their kind in the United States. By the end of the century, the range of its private nonprofit institutions was astonishing.

Elizabeth Colt was one of Hartford's most conspicuous leaders in developing these institutions. She helped found and presided over five clusters of organizations, each with its own complex genealogy and mission. In addition to the Union for Home Work, the Hartford Decorative Art Society, and the Women's Auxiliary to the Board of Missions of the Episcopal church, she was actively involved in the Connecticut Society of Colonial Dames of America and the Ruth Wyllys Chapter of the Daughters of the American Revolution (DAR). She was not involved with the Connecticut Woman's Suffrage Association, although several of her closest friends were, and its presiding officer, Hartford's Isabella Beecher Hooker, was a national figure in the women's suffrage movement.[2]

Elizabeth Colt entered, quite literally, the public stage of civic leadership in 1863 with the first of her many fund-raising theatrical performances, for the Soldier's Aid Society. On a cold November evening, more than thirteen hundred people paid the equivalent of about fifty dollars each to attend the first tableaux vivants, the costumed, theatrical exhibitions of living pictures that emerged as a popular theatrical genre during the Victorian era. Raising, in one night, the equivalent of about forty-two thousand dollars for Hartford's recently founded Soldier's Aid Society, the event involved "members of some of the 'first families'" in scenes of frozen action including the "Coronation of Alexander II," the "Volunteer's Departure," "Famous Old People," and a "sad and thrilling" depiction of "Bluebeard," presumably with his head in a noose. Elizabeth Colt was lauded for her "natural nobility of carriage, ladylike grace, handsome features, and queenly ease of manner" in her representation of "Catharine of Aragon... as she appealed to Heaven." The first tableaux were so successful that the society staged a second in February, enacting "Scenes from Mother Goose," "Eve of Waterloo," "William Tell," "Art Gallery," and a scene described as the "artist's dream," itself an interesting document of the intellectual preoccupations of a war-torn nation.[3]

A month earlier, in January 1864, Elizabeth was drafted as president of the Soldier's Aid Society's "fiscal campaign" that culminated at the Metropolitan Fair in New York the following spring.[4] The Soldier's Aid Societies, especially the U.S. Sanitary Commission that gave rise to the Metropolitan Fair, gave new impetus to women's philanthropy, in what historian George Fredrickson has described as the "inner Civil War" waged on the home front by an increasingly sovereign and idealistic class of affluent women.[5] The war greatly increased the ranks of Hartford's widows and orphans. "No foolish extravagance, or lack of energy has brought them to want," the newspapers opined, the "families of deceased soldiers... are entitled."[6]

Hinting at what is now the most universal criticism of this age of private charity, the female champions of Soldier's Aid reserved special sympathy for "persons... who have enjoyed refining advantages... [and] good social positions."[7] Many among Hartford's recently affluent female stewards of the poor could doubtless imagine themselves, a generation past or a generation forward, in the same compromised straits. By 1864 most had a

brother, husband, son, or neighbor in the war or among the war dead. In what Horace Bushnell described as a "fiscal campaign to match the sublimity and true majesty of the spectacle," Hartford's women rose to the occasion, hosting fund-raising events, donating time, talent, and possessions to the cause of Soldier's Aid.[8]

As the richest person in town, Elizabeth Colt was, not surprisingly, regularly tapped to lead efforts for the relief of soldiers and the poor. That she did so with verve, style, and what her peers described as "sound, conservative judgement" and "wisdom, thorough conscientiousness, and perfect self-control" was not so inevitable.[9] Writing of her mother, and perhaps unconsciously, of herself, Elizabeth described a woman who was "taught... the privileges and duties wealth bestows" and who "went forward in her Christian life, strengthened by its holy sacraments to withstand the perils of prosperity, as well as the trials of faith."[10] This, and what appears to have been a deep well of human sympathy shaped by her own losses and disappointments, encouraged Elizabeth Colt to leverage her considerable wealth to achieve a position of effective leadership in a variety of civic causes.

The Hartford Soldier's Aid Society, like its national counterpart, the U.S. Sanitary Commission, provided nursing and bandages, hospitals and ambulance services during the war. Historians have described these organizations as representing "the largest and most powerful, and most highly organized philanthropic activity that had ever been seen in America." The wartime experience fostered an emphasis on "professionalism... in the field of philanthropy" and resulted in an approach to the problems of an urban environment being tackled with warlike determination, "with vice and poverty as the enemy."[11] The Hartford chapter hosted a soldier's Thanksgiving and fund-raising events aimed at "founding a Home for discharged and disabled Connecticut soldiers."[12]

Mimicking an enormously successful fund-raising tactic pioneered at the Metropolitan Fair, in 1865, Elizabeth Colt donated a "pair of engraved Colt's Army pistols... to be competed for between the friends of Generals Hawley and Terry," with the proceeds earmarked for a "home for disabled Connecticut soldiers."[13]

After the war the society appears to have disbanded until 1872, when the same cast and crew reemerged to found the Union for Home Work. The Union was, in many ways, the organization that most clearly bore the imprint of Elizabeth Colt's personality and prestige. Founded with the help of a diversified group of wives of manufacturers, lawyers, politicians, and clergymen, the Union for Home Work was created for "the relief of all kinds of suffering, and the physical, intellectual, and spiritual elevation of the women and children of... Hartford," whose needs were to be addressed through "visitation," the provision of a reading room, and the "establishment of a Nursery, where working women may leave their young."[14] The union was nondenominational, offering consolidation and coordination in lieu of the duplication of effort too often associated with Hartford's many churches. Described as "a pioneer in developing... [a] modern system of charity," the union became a source of civic pride and a milestone in the organizational efforts of Hartford's women.[15]

The success of the Union for Home Work owed much to the partnership between Elizabeth Colt and its professional superintendent, Elizabeth S. Sluyter. One of the few historians to survey this intriguing episode in the city's history suggests that women like Elizabeth Sluyter were "poorly paid, for they worked for terms devised by a wealthy group often insensitive to the manager's need."[16] In fact, the hundred dollars a month (today about forty-six thousand dollars annually) plus board she was eventually paid was in addition to whatever psychic value was derived from being lionized by a community she served with distinction for almost forty years. The work was undoubtedly grueling, but as remunerative as any professional woman could imagine in the 1870s and far more than women were paid in the textile mills, as domestics,

or in the dangerous work of packing gun cartridges. For Elizabeth Colt, the union's strongest and steadiest benefactor and its president for twenty-two years, the gratification of overseeing a charitable institution no less systematic, goal-oriented, and prolific in its achievements than Colt's Armory was palpable.

For her twenty-fifth anniversary as the Union's superintendent Elizabeth Sluyter was showered with praise and testimonials, and Elizabeth Colt "presented to Mrs. Sluyter a handsome cake basket containing a draft for $1,227," the equivalent of more than a year's salary. It was Elizabeth, always the preacher's daughter with a powerful capacity to weave scripture and religious sentiment into her speech and writing, who spoke of the "the deep and true affection we feel for you... [and] your wise, loving, self-denying toil for others." Surveying the accomplishments of a quarter century, she wrote: "The Union for Home Work was organized in faith and in the conviction that those who love God should seek to help His children less favored than themselves.... How many homes have been transformed by your care and counsels; how many young lives rescued from misery;... how many turned from the paths of sin to the way of holiness?"[17]

Elizabeth Sluyter was the consummate professional in an age that offered few professional opportunities to women. Settling in Hartford in 1871, from Binghamton, New York, where she and her husband Capt. E.L. Sluyter ran a social services missionary, she gained a reputation for system, management, and tactical maneuvering. Curiously, it was actually her husband who, in 1871, "had been called to take charge of a coffee-house enterprise" in which the Union originated.[18] Elizabeth superseded him almost immediately, although he remained actively involved. Over and over, the accounts of Mrs. Sluyter point consistently to a gritty determination to transcend the stigma of disorder too often leveled against women's charitable efforts. "It requires rare judgment to lend a helping hand when it is needed, and to know when to withdraw it, when it is best to do so. It is hard to make alms-giving orderly and justifiable," wrote Frank

Cheney in his testimonial.[19] The Union's patrons commended it for successfully guarding "against the fatal mistake of pampering and pauperizing its beneficiaries" and described its superintendent as "a woman as remarkable for executive force as for tender-heartedness, as judicious as zealous, as unmoved by the whine of fraud as quick and eager in compassion for real suffering,... a student with the soundest practical sense of proportion in things, insight into human nature, [and as] understanding of the feelings and modes of thought of the poor as... of their material necessities."[20]

A culminating event in the working relationship between Elizabeth Sluyter and Elizabeth Colt occurred in October 1878 when Colt called a meeting at Armsmear, attended by "40-50 ladies and gentlemen" representing most of the city's several dozen charities, church missions, and agents of relief for the indigent. The subject was organized charity. With the mayor, the chief of police, priests from the Roman Catholic Church, and representatives from the YMCA, the Women's Aid Society, the Women's Temperance Society, and the Widow's Fund Association collected around Elizabeth Colt's reception room, Sarah Cowen and Elizabeth Sluyter spoke of the "necessity... for... union [among] all charities... to suppress imposture." Citing the example of London's Charity Organization Society, Elizabeth Sluyter called for "consolidation" in "one grand organization," as the best method "to systematize charity" and the value of "requiring labor on the part of those assisted" to "suppress and detect impostors" who abused the system by "getting aid from several different sources."[21]

After a five-year recession, cities across the country were reeling from the pressures and turmoil of relief for the poor. Within weeks of the meeting at Armsmear the leadership of the new Hartford Board of Associated Charities urged citizens "not to give anything [to] those who come to the back door for food, clothing, fuel and money, but to direct all such persons to the headquarters of the society," noting that the "police commissioners have directed the policemen to cooperate with the board."[22]

Depending on one's point of view, this rising indignation by the city's elites was either unconscionably selfish or what might today be regarded as a refreshingly firm deployment of discipline. During the rest of 1878 and throughout 1879 the newspapers were filled with descriptions of "professional tramps," who were "often surly and insulting" and "mean to live upon those who do labor." They instituted a system by which men, primarily, "cannot be fed or lodged here unless he pays his fare by... labor. This test clears up the question as to the moral obligation of our town... to lodge and feed these roving persons."[23] A wood-yard was "established, where the tramps can work... to pay for their meals... or be committed for vagrancy." Making an increasingly sharp distinction between the "worthy" and "unworthy" poor, the goal was reputedly to assure that "willful laziness... be left to its own... punishments," though, of course, "savings of expense to the town" were also cited.[24] Once the new Board of Organized Charity was incorporated it immediately set about registering all those receiving charity, in order to "obtain full information of the numbers and actual condition of our poor,... to detect the unworthy," and "to identify the worthy."[25]

The Union for Home Work was credited with revolutionizing Hartford's approach to relief for the indigent and praised for its "utter absence of cant" and its relentless quest "to understand thoroughly the causes of poverty." Its services expanded throughout the 1870s to include a sewing school, a day nursery, a flower mission, soup kitchen, summer excursions, and a generalized effort to "teach poor people... how to live better and be better."[26] This culminated in 1884 with the dedication of its Market Street headquarters and facility (fig. 191). This large brick structure, designed by architect George Keller, contained a committee room and employment room, a department of secondhand clothing, a "commodious kitchen" where "pupils were taught the art of cooking and keeping house," a laundry where "poor children and adults come for bathing," a day nursery, bathroom, playroom, and wide piazza "for the... innocent babes of hard-

191. *Union for Home Work Building, 1883. George Keller (1842–1935), architect. Illustration from the* 34th Annual Report on the Union for Home Work, *1906. (Courtesy of the Connecticut State Library.)*

192. *Souvenirs from Martha Washington Reception and Tea Party, April 1875. The cups and saucers were manufactured in New Jersey with inscriptions customized for the various cities around the country that participated in this Centennial-based charitable enterprise (WA #05.1394 and .1398-99).*

working women," a "hall with a seating capacity of 300" where "sewing classes [are] taught," and no doubt reflecting Elizabeth Colt's passionate belief in the redeeming power of reading and the written word, a library. For Elizabeth Colt, by then the Union's president, it was another opportunity to build—for permanence, love, and memory.[27]

In addition to making use of her executive ability, the Union for Home Work provided an outlet for Elizabeth Colt's love of theater, drama, and social events. Space does not permit a full account of the escalating variety of tableaux vivants, Union bazaars, and exhibitions that became eagerly awaited festivities. Combining theater, music, pageantry, exotic gifts, and displays of art and historical treasures, they provided opportunities for the socially aspiring to mingle with the established elite. Internationally known authors, all Hartford neighbors and friends—Mark Twain, Harriet Beecher Stowe, Charles Dudley Warner, and William Cowper Prime—contributed original poems or essays for the Union bazaar's tabloid newspaper, and in 1880 Mark Twain conducted an impromptu auction.[28] From the first Union bazaar in 1874, events like the Oriental Tea Party (1886) and Carnival of Authors (1883) regularly raised the equivalent of two to four hundred thousand dollars each, constituting a major source of the Union's revenue.

The Martha Washington Reception and Tea Party was Elizabeth Colt's greatest theatrical achievement, involving months of preparation, specially printed programs, catalogs, and invitations, and custom-designed tea cups and saucers (fig. 192).[29] Conceived as a fundraiser for the Union for Home Work and scheduled to coincide with the national celebration of the centennial of the Battle of Lexington and Concord, the event celebrated New England's illustrious role in the Revolution and a century of economic and cultural expansion. Colt's Armory Band provided the music for an event, staged as a series of tableaux vivants, with a retail bazaar, exhibition of historical relics, and reception that packed luxurious Allyn Hall for three nights in April. As the crowds paying one dollar each filed into the public hall to the sound of Colt's Armory Band playing "Washington's Grand March," the curtain rose and revealed Hartford's leading lights, in ersatz colonial dress, enacting the artist Daniel Huntington's *Republican Court in the Time of Washington, or Mrs. Washington's Reception Day* (1861; fig. 193), a scene of an imaginary reception of patriots and political leaders hosted by Mrs. Washington during Washington's presidency.[30] Surrounded by family, friends, and Union cronies was Elizabeth Colt as Mrs. James Madison, wearing a "black-satin dress" with "lace and bows of yellow ribbon," a "damask petticoat," earrings "once worn by Mrs. Madison," and a bracelet owned by "a lady ancestor of the Hart family" (fig. 194). Her brother-in-law C.N. Beach posed as Vice-President John Adams, her sister Hetty as Martha Washington (fig. 195), Mark Twain as Gov. William Rivingston, and

seventeen-year-old Caldwell Colt as Monsieur William Louis Otto, secretary of the French legation (fig. 196).[31] Following a second tableau in which they represented "Trumbull's famous picture in the Historical rooms," presumably either the Battle of Bunker Hill or the signing of the Declaration of Independence, the performers and audience of two thousand adjourned to the reception, exhibitions, and bazaar.[32]

The guests mingled among the thirteen tables of refreshments, one for each of the colonies, in the grand reception room at Allyn Hall, which was decorated with "streams of bunting, flags, flowers, evergreens, [and] candles." Massachusetts featured a "miniature ship Dartmouth, from which the tea was thrown," and a "Sugar candy model of Bunker Hill monument." Connecticut was "decorated with flowers... articles from the Charter Oak... Connecticut 'election cake,'... and

193. The Republican Court, 1861. Daniel Huntington (1816–1906), New York. "Society then and now" was a popular topic in Centennial-era *literature and art, the implication being that while Victorian society was clearly "superior" in its technological achieve-* *ments, it was no less so in its social graces. (Courtesy of the Brooklyn Museum.)*

194. Elizabeth Colt as
Mrs. James "Dolley"
Madison, 1875.
R.S. DeLamater,
Hartford, photographer
(WA #05.1572).

195. Hetty Beach as Martha
Washington, 1875.
R.S. DeLamater,
Hartford, photographer
(WA #05.1572).

196. Caldwell Colt as
Monsieur William
Louis Otto, 1875.
R.S. DeLamater,
Hartford, photographer
(WA #05.1572).
Caldwell is dressed in
the coat (WA #05.1593)
his father wore when
he was appointed
attaché to the U.S.
legation at Berlin
in 1854. On his lapel
is the Turkish medal of
the Order of Abdülmecid
(WA #05.1532; fig. 196).

stewed oysters." Mrs. Samuel Colt presided over the table representing "Virginia... the 'Mother of States and Presidents,'" decorated with a "large... temple of liberty, made of flowers" and a centerpiece featuring a bronze bust of Washington, with John Smith and Pocahontas watching "over piles of tobacco and cigars... sold as relics."[33] In an adjoining room the guests, many of whom "wore robes in which their grandmothers had been married," admired an exhibit of early American antiquities featuring such treasures as a chair owned by Gen. Israel Putnam, Governor Trumbull's sword, and a "copy of the farewell sermon preached at Northampton by Jonathan Edwards." Elizabeth Colt loaned a "medallion locket containing a lock of General Washington's hair" (fig. 197) that she bought in 1871 for three hundred dollars (today more than eleven thousand dollars).[34] The ladies sold "Martha Washington tea cups,... articles of Charter Oak," and souvenir replicas of "the miniature weapon with which the juvenile Washington destroyed his father's cherry tree," inscribed with the words "I cannot tell a lie."[35] Indeed. When it was over the Union had raised enough money to provide 16,400 "good meals" for those in need.[36] The exhibition was repeated in expanded form as the Centennial Loan Exhibition the following fall, with the proceeds donated "to aid... in the erection of a building at the Philadelphia Exhibition and display the work of American women."[37]

Described by her friends as "one, who gave with hidden hand the material help, and with large heart a full measure of sympathy" and whose "private charities reached all classes, and were constant and unfailing," Elizabeth Colt's philanthropy is well documented and yet may have been even greater than the record shows.[38] When the investigators from R.G. Dun's credit rating service described the Coburn Soap Manufacturing Co., operated in Coltsville by Sam Colt's loyal friend William H. Green, as "not in best standing," running down, and yet able to "continue for some unknown reason," it is easy to imagine Elizabeth Colt as its mystery angel.[39] Return on capital was not foremost in mind in 1889 when Elizabeth Colt hired the architect George Keller to design and build a house "she intends to rent to old Mrs. Russell," another friend.[40]

For Elizabeth Colt, whose husband was as much artist as entrepreneur, and who, by 1875, had devoted almost twenty years to the patronage of sculpture, architecture, and paintings, the watershed of the 1870s provided a chance to test the boundaries of what an American woman could do in the way of public work. This was the decade that gave birth to the Metropolitan Museum of Art, the Museum of Fine Arts, Boston, the Philadelphia Museum of Art, and more. There has never been a time in U.S. life when art mattered more or was more on the minds of more classes and types of people than in the decade of the 1870s. Art was their television, fine arts, and film, their computer games, sports spectacles, and politics. Art has never been more front and cen-

197. Elizabeth Colt's
Colonial Memorabilia.
Left to right:
*George Washington
inaugural button, 1789
(WA #05.76);
George Washington's
hair bracelet clasp, 1871
(WA #05.1529);
badge of the DAR, 1893,
engraved "Elizabeth
Hart Colt 2246"
(WA #05.1531);
Connecticut script, 1776
(WA #05.1622/3).*

ter nor enjoyed a greater sense of urgency and purpose than it did about 1875.

It was in this environment that Elizabeth Colt and several of her Union for Home Work cronies embarked on a program that ultimately rehabilitated the moribund art program at Hartford's Wadsworth Atheneum and led to the founding of Hartford's most successful school of art. The Hartford Decorative Art Society (1877) and its parent and sister institutions in New York, Boston, and elsewhere were founded to address both the humiliating comparisons made between the United States and Europe throughout the Centennial Exhibition of 1876 and the national "uprising of women… agitating for… civil rights… better education… [and] professions."[41] The first of the nation's decorative art societies was founded in New York in 1877 to "establish a place for the exhibition and sale of sculpture, paintings, wood-carvings, lace-work, art needle-work, and decorative work… done by women" and, more generally, to "encourage profitable industries among women."[42] Having received "circulars from the newly organized Society of Decorative Arts in New York urging the organisation of a similar… society" in Hartford, Elizabeth Colt presided over an august committee of civic movers and shakers to found the Hartford Decorative Art Society, which opened its studio and art classes in the Cheney Building the following January (fig. 198). This was an important step toward founding a free public art gallery and permanent school of art in Hartford.

Elizabeth Colt, assisted by Union president Sarah Cowen, presided over a ladies committee composed of Harriet Beecher Stowe, Susan Warner (Mrs. Charles Dudley), Mrs. Frank Cheney, Olivia (Mrs. Samuel) Clemens, and Annie Trumbull Slosson. Hartford never witnessed so large a concentration of powerful female intellectuals devoted to a civic mission. Deep pockets, deep thoughts, and a men's Friends of Council that included Gen. Joseph Hawley, Rev. Francis Goodwin, Frank Burr, James Batterson, Charles Dudley Warner, Marshall Jewell, and the influential New York critic and internationally renowned author and collector William

Cowper Prime, ensured that this effort was a triumph waiting to happen, bringing together an assemblage of politicians, journalists, writers, intellectuals, and private fortunes equal to any ever devoted to a public cause in Connecticut.[43]

The society engaged "Miss Wheelwright… a pupil of [Richard Morris] Hunt, of Boston" and John Wells Champney as art instructors and flung open their studio and classrooms to crowds that quickly exceeded capacity.[44] Describing their efforts as "experimental and involving a large outlay," they hoped the public would "support the society in its efforts to secure for Hartford the best and highest art instruction," which involved instruction in "drawing and oil painting," lectures on such topics as "the chemistry of artists' colors," classes in china painting and art needlework, and an annual exhibition and sale of "embroideries, paintings, and plaques,… for the benefit of poor women."[45]

The society flourished for about seven years, then appears to have foundered on matters requiring larger outlays of money and perhaps conflict in priorities and authority between the male and female advocates of civic art and art education. For a society in which roles and responsibilities were so notoriously typecast by gender, art offered one of the first instances in which men and women worked to achieve parallel, if not unified, goals. Kathleen McCarthy and other scholars have largely concluded that the price of unification was the absorption and appropriation of women's goals and money by male stewards of civic culture. In this, Hartford was no different from the larger cities that McCarthy chronicles. But the fact that it took twenty years instead of five for the men to gain unequivocal control over their community's programs and institutions of civic art, is, I suspect, due in no small measure to the organization, intelligence, and independent wealth of Elizabeth Colt and her Decorative Art Society cronies.[46]

In 1886 Elizabeth Colt and the Decorative Arts Society helped break a fifteen-year stalemate in the campaign to create a free public art gallery when they united

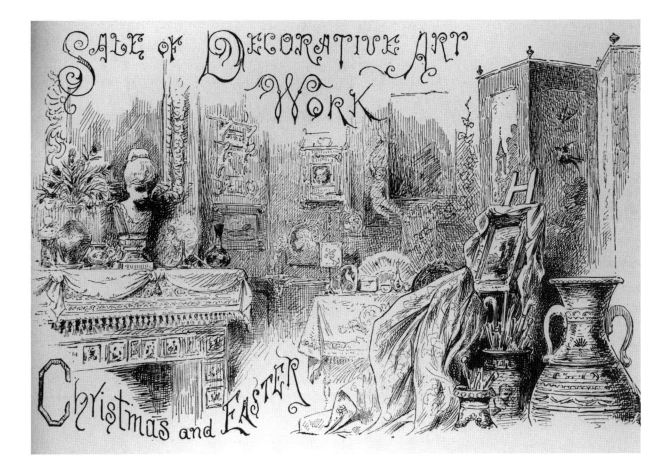

with the three competing men's factions to form the Art Society of Hartford. Incorporated by "a body politic," composed of Elizabeth Colt, Sarah Cowan, Joseph R. Hawley, Francis Goodwin, James Batterson, Charles Dudley Warner, and others, its purpose was "to promote and encourage the knowledge and practice of art," to operate "classes and schools... of art for ornamental, industrial, or other purposes," to hold annual "sales of decorative and artistic work," and with a collective gasp twenty years after Batterson's initial overtures, "to take charge of, add to, and manage the art gallery now in the Athenaeum."[47] In 1886 Hartford gained a public art museum.

Opening "the Atheneum Gallery for Public Use" was "brought about by the managers of the Hartford Art Society," an offspring of the Hartford Decorative Art Society, which Elizabeth Colt and Hartford's women had founded nine years earlier. Agreeing "to pay for the whole care of the room," they got off to a good start by hiring Dwight Tryon, back after more than ten years in Europe, "to teach drawing and painting." Noting that the art gallery of "the old Atheneum" had "been practically deserted or unvisited for many years" and that the institution was "too poor to keep its doors open to the public," the press speculated that "now that it is free... it may become more of an attraction."[48]

Despite the fact that the leadership of the Hartford Art Society was composed mostly of women and men connected with the Decorative Art Society and Union for Home Work, within five years of its founding, the Goodwins and their cousins Junius Spencer Morgan and

198. *"The Sale of Decorative Art Work, Christmas and Easter, 1883."* From Report of the Society of Decorative Arts of the City of Hartford, Connecticut *(New York: Studio Press, 1883). Classroom instruction in painting and drawing was complemented by courses in decorative painting and artistic needlework.* *(Courtesy of the Harriet Beecher Stowe Center Library, Hartford.)*

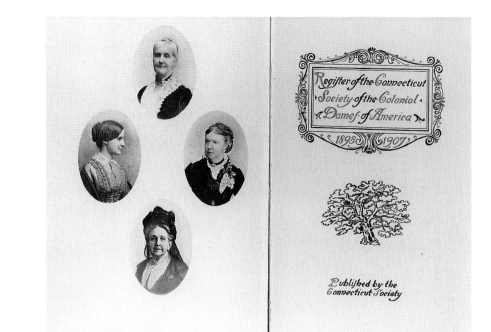

199. *Elizabeth Colt and the*
Founders of the Colonial
Dames. Frontispiece
from Register of the
Connecticut Society
of the Colonial Dames,
Hartford 1907.
Clockwise from top:
Elizabeth Hart Colt,
Elizabeth J. Hamersley,
Evelyn MacCurdy
Salisbury,
Mary D. Hoppin.

J. Pierpont Morgan were firmly in charge, commencing a dynasty of arts patronage and leadership that would last more than eighty years. Backed by a pledge of $150,000 (today $5.4 million) from the Morgans, Rev. Francis Goodwin, his mother, brother, and friends launched the most ambitious expansion in the Atheneum's history, resulting in the "opening of new libraries and art galleries" and culminating, ten years later, in the addition of the Morgan and Colt Memorial Wings to the sixty-year-old Wadsworth Atheneum and its thirty-year-old Watkinson library wing. Elizabeth Colt was one of several Hartford women who pledged $5,000 (today $175,000) to the campaign. But the Goodwin-Morgan alliance contributed forty times as much, gaining supremacy and control, which they maintained through the 1970s.[49]

Following the death of her beloved sister Hetty in 1898, Elizabeth, now in her seventies, began to withdraw from the limelight, leaving the work of cultural and civic stewardship to a new generation of rising matriarchs to whom she was a beloved and idealized role model. In 1899 she wrote to her friend Emily Holcombe about "how interested I have been in your work," a campaign to restore Hartford's ancient burying ground, but

lamented that the "losses have come to me so crushingly in the past few years that I have little heart for anything of a public character."[50] As she passed the baton to a new generation of charitable entrepreneurs, family history and religion were the two remaining pillars of devotion that occupied Elizabeth Colt's time and resources in the last decade of her life.

In 1892 Elizabeth Colt became the seventh member of Hartford's Ruth Wyllys Chapter of the DAR, and on November 22, 1893, she "invited to Armsmear a group of representative Connecticut women to discuss the formation of a branch" of the National Society of the Colonial Dames of America.[51] The Connecticut Society of the Colonial Dames of America was subsequently founded, with Elizabeth Colt elected its first president (fig. 199). With the last years of her life dedicated to the cause, she rose to a position as vice-president of the national society in Washington.[52]

She undertook this work with the same executive ability and seriousness she had applied to her many earlier campaigns. The Colonial Dames was founded "to further a clearer knowledge not only of Colonial events and persons but also of the ideas and principles which underlay them, to encourage in contemporary life a true patrio-

200. Spanish American War Monument and Publication. Illustration from A Memorial to the Soldiers and Sailors... in the War of 1898, *1902. (Courtesy of Shepherd M. Holcombe.)*

tism built on such understanding," and "to preserve the history and, to some extent, the ideas of a time when there was no central government."[53] No central government? Giddyap! Sam Colt and his libertarian friends in the Democratic Party of the 1850s would have loved it. Among the men's clubs of the time were such organizations as the Society of the Sons of Colonial Wars and the Connecticut Society of the Order of the Founders and Patriots of America, which was founded "to bring together and associate congenial men whose ancestors struggled... for life and liberty, home and happiness.... To teach reverent regard for the names and history, character and perseverance, deeds and heroism of the founders.... To inculcate patriotism in the associates and their descendants. To discover, collect, and preserve records, documents, manuscripts, monuments, and history relating to the genealogy and history of the first colonists." The 1890s was a decade of rampant colonialism.[54]

Elizabeth Colt served on the national society's Constitutional Committee in 1894, and in 1896 the Connecticut society organized an "exhibit of Colonial Relics at the Atheneum" and raised funds by sponsoring a Colonial tableau, with music performed by Colt's Armory Band, still active forty years after its founding in Coltsville.[55] In 1899 she played an active role in the national society's "first public project," building a "monument at Arlington [Cemetery] to the soldiers and sailors who died in the Spanish-American War" (1902; fig. 200), described as the "first monument ever built in the National Cemetery by a Society of women."[56] Ever ready with the pen or to send a work off to the printers, Colt printed *A Memorial to the Soldiers and Sailors* to commemorate the Dames' efforts and memories of the monument's unveiling by President Theodore Roosevelt, on May 21, 1902. Seated on the dais with fellow dignitaries while a U.S. president lauded the work of American women, Elizabeth Colt could not have failed to reflect on her forty years of charitable entrepreneurship with a sense of having helped shape the contours of so auspicious a moment.

In the years that followed, the Colonial Dames championed efforts to "extend to foreigners the real spirit of welcome" by distributing "helpful literature... printed in Yiddish, Polish, Italian, and Bohemian" and producing films with titles such as *The Making of an American* and *Hats Off* "to be shown to immigrants on incoming boats from Europe."[57] Today, a century later, the society that came together in the reception room at Armsmear operates the Webb-Deane-Stevens Museum, a collection of painstakingly preserved and maintained house museums in nearby Wethersfield, Connecticut. Within days of Elizabeth's death, the Hartford press, describing Armsmear as having "all the requirements for an ideal museum," circulated a rumor that she had left "the property… in the custody of the Colonial Dames, with an endowment," a fate that was not to be.[58] The Colonial Dames, nonetheless, revered Elizabeth Colt as their "beloved friend and first president," printing and circulating the most affective account of her life and character:

In the dark hour of trial she was ever present to help and comfort. There are those who remember how in the anguish of… the loss of a mother, she spoke words of peace to the mourners. In her sweet low voice she said, "Peace will *come."… The trials she… suffered, the loss of her husband and infant children, and in more recent years the parting with her son and a beloved sister and brother—such trials might have crushed many a spirit, but out of all she rose triumphant, bearing on her face each time… a more serene… and beautiful expression. "The Peace of God which passeth all understanding" she knew, and in that strength she lived and walked. In a noble character of remarkable symmetry and perfectness a wonderful element was* love, *tender love and devotion. In all that she did this is shown. In the memorials she erected, the church offered in memory of her husband and little children, in the parish house in memory of her son, what careful, loving thought is revealed!… Perhaps we cannot hope to see her like again, but those of us who have seen the light which shone in her beautiful and sanctified life and example, may well "Thank God and take courage."*[59]

Elizabeth's final campaigns involved rallying the family shareholders to negotiate the sale of Colt's Fire-Arms, commissioning one last monumental work of art, the Colt Memorial Statue, and working with her attorney Charles E. Gross to draft a will involving the final disposition of assets and the formation of three new civic amenities, Colt Memorial Park, The Colt Trust and its "home for the widows of clergymen and impoverished but refined and educated gentlewomen," and the Colt Memorial wing of the Wadsworth Atheneum.

Elizabeth Colt's decision to sell Colt's Fire-Arms Manufacturing Company was prompted by the fact that, with no direct heirs and a need for more liquid assets, it was, simply, time. In 1901 she wrote of having rebuffed "proposals… to buy the Armory," stating that, "I could in no sense say I *wished* to sell, but that under the circumstances… feel that if he [Sam] & Caldwell know how things are now they would approve of this action on my part. I have tried to honor their memories always." Having "asked that the name should be retained," as well as "the men, who have stood so loyally by us," she orchestrated the sale of the family's majority shares to outside investors.[60]

The trigger was undoubtedly the series of strikes and walkouts—among Hartford's first—that rippled through the machinists' union and industry nationwide in 1901. Although Colt's workers stayed on the job, Elizabeth sought a private buyer.[61] What survives of correspondence and documentation reveals a woman eager and able to negotiate subtle details. She ended up with a deal that retained the Colt name over the door, assured the employment of loyal, long-time Colt's employees, and placed the company with a Boston and New York holding company "who would not run up the price in speculation & then sell out."[62]

In the last week before the transfer was completed she wrote of "feeling badly to have the Armory go out of the family. No one can feel as sorry as I do," lamenting that had there "only been some one [in the family] to take the business… I should not have had to part with what I have been so proud of. I was at the office yesterday morning & it was very hard to think that in a few days I should have no right there…. [It was] like bidding a last good bye to a dear friend."[63] In the end, the armorer's widow had become an armorer herself.

Elizabeth Colt apparently traveled to St. Louis in 1904 to witness one of the last of the great World's Fairs, a nineteenth-century idea that bore its sweetest fruit at the opening of the twentieth.[64] Elizabeth Colt's DAR chapter, led by her friend and protégée Emily Seymour Goodwin Holcombe, took charge of the antiques exhibition in the Connecticut House.[65] Described as bigger than the World's Columbian Exhibition in Chicago in 1893 and almost twice the size of the Centennial Exhibition in Philadelphia, the St. Louis Fair drew exhibits from forty-four states and fifty countries. With its simultaneous "revival of Olympic Games," the Louisiana Purchase Exposition celebrated a century of empire and western expansion that Elizabeth Colt could not have failed to identify with.[66]

At the fair, she would have encountered the work of John Massey Rhind, whom she selected in 1902 to carry out her final work of artistic patronage. John Massey Rhind was at the peak of his fame in 1904, in a prolific career that spanned the age of civic statuary and bronze. The fact that he was younger and began later has prevented Rhind from joining the pantheon of America's best-known sculptors in bronze, such as Augustus St. Gaudens, Daniel Chester French, and John Quincy Adams Ward. Rhind had studied under Jean Jacques Dalou in Paris and by the mid-1890s was enjoying a steady patronage in civic monuments, most notably in Pittsburgh, Youngstown, Ohio, Providence, Rhode Island, and New York. The Astor Memorial Doors (1896) of Trinity Church in New York, the Henry King Fountain (1893) in Washington Park, Albany, his contribution to the embellishment of Grant's Tomb (1897) in New York, and the Corning Fountain (1899) in Hartford were among his best-known works prior to 1903, the year Elizabeth Colt engaged him to design the Colt Memorial Statue.[67]

Elizabeth Colt began pondering the idea of a grand public statue to honor her husband in the early 1890s. In 1896 her attorney Charles E. Gross approached her with a proposal to build "a memorial to Col. Colt situated" in what he called Barnard Park, a small pocket park "at the very gateway to that section of the city which [Colt's] energies developed." Reportedly, "Mrs. Colt was willing.... Two years later... [she] sent for Mr. Gross and told him that she had not given up on the idea" and had resolved

to go it one better. She would "erect it on the grounds of Armsmear" and "give to the city for park purposes" one hundred acres belonging to the estate.[68] Gross and Warner, then working together with Rhind on the Hartford Terrace and Corning Fountain projects, probably suggested the artist, whose work Elizabeth Colt would have known anyway. The small bronze miniatures, or artist's maquettes (fig. 201), used as models for the Gorham Foundry's large castings, were finished in 1903. The monument was under construction and very much on her mind during the summer of 1905 when she died, leaving her attorney with the authority and a fund of $60,000 (today about $2 million) to finish the work.[69]

The completed Colt Memorial Statue (fig. 202), unveiled in August of 1906, a year after its patron's death, features larger-than-life bronze castings of Sam Colt, "the boy genius" and inventor, and Colonel Colt, the "captain of industry" and master of Coltsville. The statue guards the entrance to Colt Park, which, at one hundred

201. Maquettes for Colt Memorial Statue, 1902. J. Massey Rhind, designer. Roman Bronze Works, New York (WA #05.1554 and 05.1120).

202. *Colt Memorial Statue,*
1905–6, Colt Park,
Hartford. Granite
and bronze; J. Massey
Rhind, sculptor, Gorham
Manufacturing Co.,
Providence, R.I.

acres, was the third largest of the half dozen parks created in a jagged ring around Hartford during the 1890s.[70] The monument, commissioned "in faithful affection by his wife," is a classical formulation featuring two figures and two bas-relief tableaux mounted in polished granite, carved with symbols of prosperity and empire. The tableaux illustrate two of the high points in Sam Colt's life, his presentation at the court of Czar Nicholas in 1854 and his testimony before the British House of Commons (fig. 203) on the use of machines in the manufacture of firearms. Inscribed in granite are the words "On the grounds which his taste beautified/By the home he loved/ This memorial stands to speak of his genius his enterprise and his success / And of his great and loyal heart."

At the time of Elizabeth's death, Charles E. Gross, her attorney and the man responsible for completing the project, cited the statue as another example of her "all absorbing and always predominating loving loyalty to the memory of her husband and children."[71] "Faithful affection" had prevailed.

Forty years of memorial devotions, and Elizabeth Colt had not yet shaped her own legacy. How did she want to be remembered? What could she build or create that would assert her autonomy without disrupting her deep-felt obligations to the memory of her husband and family? She answered by building a museum wing for the city's increasingly art-oriented Wadsworth Atheneum.

Elizabeth contributed significantly to the campaign of 1893, the most ambitious since the Atheneum's founding, and it is easy to imagine her associates in the Hartford Art Society and the Reverend Francis Goodwin, the Atheneum's president and leader of the Goodwin-Morgan faction, seeing green in 1894, after the death of Elizabeth Colt's sole heir. The Goodwins, who would have mixed with Sam Colt and his Democratic cronies like oil with water, increasingly insinuated themselves into the life of Elizabeth Colt through the Episcopal Church, especially through the Reverend Francis Goodwin. Goodwin preached and officiated at ceremonies and funerals at Elizabeth's Church of the Good Shepherd and was the reader at her funeral in 1905. Clearly, the final act of entrusting art and the sacred mementoes of Sam Colt's career to the Wadsworth Atheneum for safekeeping was an expedient that brought closure to a lifelong interest in art, art education, and civic embellishment.

Following considerable negotiations that almost ended in default over issues of the collection's sovereignty and permanence, a deal was completed in which seventy-six itemized objects, artworks, and firearms,

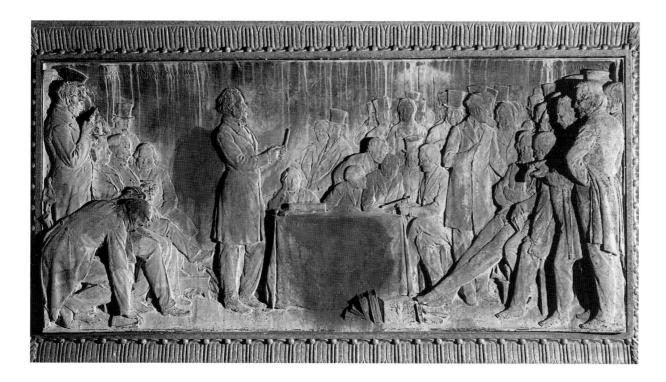

including "the Cabinet of arms... now in my second-story hall," plus "all of my other bronzes; statuary; curios; medals; articles of virtu, bric-a-brac; arms; armor and equipment, together with the cases in which they are or may be displayed by me" were bequeathed to the Wadsworth Atheneum, "where they shall be kept separate and apart and designated... as 'the Elizabeth Hart Jarvis Colt Collection' and 'the Elizabeth Hart Jarvis Colt Gallery.'"[72] In addition to the collection of objects, artworks, and documents, numbering almost one thousand pieces, Elizabeth Colt gave $50,000 (today $1.75 million) "for the erection of a suitable addition to the present buildings of said Athenaeum."[73] Make no mistake about it, the collection may have featured Sam Colt's guns and legacy, but it was Elizabeth Colt's name over the door, and it was her life and vision on display when the Colt Memorial wing of the Wadsworth Atheneum opened its doors on Main Street in 1910. Occurring just a few years after Isabella Stewart Gardner's Fenway Court (1903) opened its doors as a public museum in Boston, the Colt Memorial was the first art museum wing in America to bear the name of a female patron.

The Colt name had passed into legend by the time Sam's widow's memorial museum opened its doors, eighty years after the revolver king conceived the invention that made him rich. Struggling to summarize the unique slice of American history now transferred to the public domain, a journalist at the opening day ceremonies in 1910 cited "an old saying that civilization follows the revolver and the axe" and, without a blush of guilt or hesitation, proudly pronounced that the "Colt revolver was one of the great forces in the nineteenth-century spread of the white man's civilization,... one of the great forces in the nineteenth century," an age of empire.[74] In a statement that would foretell years of conflict and bewilderment in stewarding so eclectic and personal a collection, one writer cited its "many articles of great interest... [but] little value," including, lest we forget, a hundred-pound tarpon autographed by the friends that accompanied Sam and Elizabeth on their yachting excursion along the coast of New England in the summer of 1860.[75] Such objects, stripped of their histories and associations, might look like junk. They are not. Another writer stretched the boundaries of hyperbole by describing "articles collected from many centers of art and science and education, and depicting... notable triumphs of the human mind... an education and an encouragement to lovers of art and to the students of mechanics," including "treasures and statuary and works of art," a "museum of repeating firearms," veritable "lessons in

203. *Sam Colt's Demonstration before the House of Commons, 1854. Detail from bas-relief sculpture in bronze, Colt Memorial Statue, Colt Park, Hartford; 1904, J. Massey Rhind, sculptor, Gorham Manufacturing Co., Providence, R.I.*

patriotism and American history... taught in the cases." All that plus a "picture gallery, on the second floor," a "reproduction of the Colt Library" on the first floor, and a "family room" in the basement, containing "souvenirs directly connected with Colonel Colt & Mrs. Colt."[76]

The Colt Memorial (fig. 204) was designed by Benjamin Wistar Morris of New York, the same architect hired to design the much-larger Morgan Memorial, which the Goodwin-Morgan axis, now running the Atheneum, opened in conjunction with the Colt wing. Described as "strictly Tudor architecture" designed to make an elegant "transition from the castellated Gothics of the Wadsworth [building]... to the Italian renaissance of the Morgan," its interior was finished with oak paneling, oak doors, granite walls, and details decorated with the "owl of Minerva," a globe, "symbols of learning," and medieval armour.[77] Somewhat overwhelmed by, and set back from, the Wadsworth and Morgan wings of the now greatly expanded Atheneum complex, the Colt Memorial was one testy and tasty piece of meat, sandwiched between two monuments of the Standing Order. Daniel Wadsworth and Sam Colt would have been horrified to have been so conjoined. Union prevailed.

On August 23, 1905, six weeks short of her seventy-ninth birthday, Elizabeth Colt died at the home of her niece in Newport, Rhode Island. The headlines described "a life full of good works." "She was nobody's enemy, she was everybody's friend... democratic and dignified.... There are but few men or women," wrote her old friend Mary Bushnell Cheney, "whose departure from earth leaves in the community they have lived in a sense of emptiness, an aching void which no other presence can fill."[78] "Nobility of character," "an aristocracy of goodness," a city "blessed from her living in its midst."[79] "It would hardly be possible to overestimate the importance of Mrs. Colt's position in the life of this city."[80]

Elizabeth Colt's will is a remarkable document that, if read between the lines, has a narrative quality that is poignant, spiritual, and even heroic. The published version, at thirty-four pages, provides a snapshot of sovereign womanhood. Although her attorneys, Charles E. Gross and his son Charles W. Gross, detailed the published version, Elizabeth Colt, who oversaw the smallest details of her many commissions, played a significant role not only in the contents of her will but also its expression.

She left Elizabeth Hart Jarvis Robinson, Elizabeth Hart Jarvis Beach, and her sole surviving sibling, Mary, each $200,000 (today, $7 million). She gave "my nephew" (Sam's illegitimate son) money and the Colt family portraits, and her grand-nephew Caldwell Colt Robinson $10,000 (today about $350,000). She did well by her siblings' heirs, especially those that bore hers or Caldwell's names. The bequests requiring specific action or maintenance were well suited to the tasks, and while the $250 designated to the city "in trust" for "repair, preservation or renewal of the monuments... erected to the memory of the parents of my husband... in the Old North Cemetery" may sound inadequate, the income from its current value of $8,750, year after year after year, was just about what it should have taken to "repair, preserve, and renew."

After leaving generous but not extravagant gifts to her domestic servants, Elizabeth Colt turned to the matter of the Colt trust and the founding of a "home" for "widows or orphans of deceased clergymen of the Protestant Episcopal Church," and "as many impoverished but refined and educated gentlewomen, as the buildings" could support. The Colt trust was established with a bequest of $800,000 (today $28 million) to "keep in complete repair" the Church of the Good Shepherd, the Caldwell Hart Colt Memorial House, the church rectory on Wethersfield Avenue, and Armsmear, now a retirement home to be operated with such funds as were left after the needs of the various buildings and memorials were met. With the stroke of a pen, Hartford gained a museum, a park, and a retirement home, and one of the first proactive initiatives in historic preservation.

Elizabeth Colt's death marked the end of an era. As the flowers were banked up on the altar of the Church of

the Good Shepherd, few could deny that the young lady who had danced with her beau, the sea mist rolling across the twilight shores of old Newport in the summer of 1852, had herself become a legend.

In one campaign after the next Elizabeth consciously broke new ground, seeking to "make anew" her chosen fields of endeavor. In the end, a life devoted to faithful affection and memory had been a life filled with rock-solid accomplishment. Whether nineteenth-century Hartford was a better place for the Colts having lived in it, it surely was a bigger and very different place, a place that shaped and was shaped by two restless icons of the industrial age named Sam and Elizabeth.

204. Elizabeth Hart Jarvis
Colt Memorial Wing,
Wadsworth Atheneum,
1910. Benjamin Wistar
Morris, architect.

Sources frequently cited have been identified by the following abbreviations:

CHS   Connecticut Historical Society, Hartford, Connecticut.
CSL   Connecticut State Library, Hartford, Connecticut.
Colt   PFAMCColt's Patent Fire-Arms Manufacturing Co.

Throughout the book I have translated money into current values based on three tables included in *Historical Statistics of the United States, Colonial to 1970* (Washington, D.C.: U.S. Bureau of the Census, 1975): table D–722–727: "Average Annual Earnings of Employees: 1900 to 1970," p. 164; table D–718–721 "Daily Wage Rates on the Erie Canal: 1828 to 1881," p. 164; and table D–715–717 "Average Daily Wage Rates of Artisans, Laborers, and Agricultural Workers, in the Philadelphia Area: 1785 to 1830," p. 163. For purposes of comparison, I assume that moderately skilled laborers—carpenters and masons or those described as "artisans"—find a contemporary analogue in counterparts earning $12 per hour, $96 per day, and $24,000 per year, based on a fifty-week year.

PREFACE

1. Memo, Jean Cadogan to Bill Hosley, July 9, 1992, Colt Archives, Wadsworth Atheneum.
2. In addition to the several works by R. L. Wilson, Ellsworth Grant, Jack Rohan, Bern Keating, William B. Edwards, and others cited elsewhere in this book, the bibliography on Sam Colt contains numerous books published with increasing frequency beginning in the 1930s and peaking during the 1950s. Among these are Gertrude Hecker Winders, *Sam Colt and His Gun* (New York: John Day, 1959); James E. Serven, *Colt Firearms, 1836-1854* (Santa Anna, Calif.: 1954); and the most monumental and best-documented of the early works, Charles T. Haven and Frank A. Belden's *History of the Colt Revolver* (New York: William Morrow, 1940). Haven and Belden's book, like most of the published literature on Colt, describes its audience as "collectors and students of weapons and firearms history." Although collectors have contributed enormously to the scholarship and continuing interest in Colt, their distinct point of view has favored some aspects of the story over others.

MEET SAM AND ELIZABETH

1. Quoted anonymously in the *Hartford Daily Times*, January 10, 1862.
2. *Hartford Evening Press*, January 15, 1862.
3. *Hartford Courant*, August 25, 1905; admirer's quote is in the *Hartford Courant*, August 24, 1905.
4. Ibid., August 24, 1905.
5. Alexis de Tocqueville, *Democracy in America*, ed. J. Mayer (1848; reprint, Garden City, N.Y.: Anchor Books, 1969), 513.
6. Henry Howe, *Adventures and Achievements of Americans* (Cincinnati, 1859), 149–51.
7. Elizabeth Hart Colt, "Memoir," in Henry Barnard, *Armsmear: The Home, the Arm, and the Armory of Samuel Colt, a Memorial* (New York: C. A. Alvord, printer, 1866), 295–330.
8. Quotation is from Howe, *Adventures and Achievements*, 149–51; alternate explanation of Colt's inspiration is from Robert M. Patterson, *Samuel Colt vs. the Mass. Arms Company: Report of the Trial... June 30, 1851* (Boston: White and Potter, 1851), 211.
9. "Repeating Fire-Arms: A Day at the Armory of 'Colt's Patent Fire-Arms Manufacturing Co.,'" *United States Magazine*, vol. 4 (March 1857), 229.
10. James D. McCabe Jr., *Great Fortunes* and *How They Were Made; or the Struggles and Triumphs of Our Self-Made Men* (Cincinnati: E. Hannaford and Co., 1871), 5–6.
11. Testimony on House Bill No. 59, "for the relief of Samuel Colt," *The Congressional Globe* (33d Congress, 1st Session, July 8, 1854) (Washington, D.C.: John C. Rives, 1854), 1645; Sam Colt, Hartford, to Dr. C. B. Zabriskee, New York, December 11, 1847, Colt Collection, CHS; *Hartford Daily Times*, March 13, 1874. The quotation about genius was taken from a biographical sketch of the inventor Christopher Sharps, a kindred spirit and Hartford competitor with Colt; more kin than kind, he never fully succeeded and was apparently almost indifferent to the pursuit of fame and fortune.
12. Patterson, *Colt vs. the Mass. Arms Company*, 4. The copy of this book in the Colt Collection was intended as a presentation gift for President Millard Fillmore.
13. Edward N. Dickerson, New York, to *New York Herald*, January 19,

1854, as cited in "Reports of... committee... to inquire... whether money... offered members... to extend Colt's patent," *Reports of Committees of House of Representatives*, 353 (1854) (33d Congress, 1st Session, August 3, 1854), 40; Patterson, *Colt vs. the Mass. Arms Co.*, 6.
14. Wyn Wachhorst, *Thomas Alva Edison: An American Myth* (Cambridge: MIT Press, 1981), 74.
15. In spite of several published accounts of Christopher Colt's business ventures and personality—the most extensive running several pages and making numerous specific claims [Jack Rohan, *Yankee Arms Maker: The Story of Sam Colt and His Six-Shot Peacemaker* (1935; reprint, New York: Harper and Bros., 1948), 2–6]—very little documented evidence has ever been published and the published accounts are largely speculative. Christopher Colt appears to have moved to Hartford from Hadley, Massachusetts, in 1802 when he first advertised himself as purchasing "Francis Brown's stock in trade" [*Connecticut Courant*, May 10, 1802]. He appears to have succeeded to some degree in the retail and importation of dry goods, crockery, and glassware. In 1816 he advertised as a partnership with Horace Hayes in Horace Hayes and Co., importers of cutlery and hardware [*Connecticut Courant*, August 6, 1816], and may have been the partner in Caldwell and Colt's distillery until 1819, when he went bust. The facts get very fuzzy in 1823 and 1824. Christopher Colt remarried in 1823 and was allegedly involved as the sales agent of the Hampshire Manufacturing Company in Ware, Massachusetts, as early as 1824. He was employed and living in Ware from about 1828 to about 1836, when he returned to Hartford. The company was not in business until 1829 [*Hampshire Gazette*, June 17, 1829; Arthur Chase, *History of Ware, Massachusetts* (Cambridge: Cambridge University Press, 1911), 220], and it remains unclear where Christopher was during the mid-1820s or whether Sam Colt and his siblings may have been living with relatives during that period. There is also no documentation to support the account of his saving Sarah Caldwell's life in a driving accident about 1803, but they did marry in 1805, and Rohan's account of Colt's rapid rise and fall in the Embargo of 1808 is plausible. Sarah bore six children at somewhat regular intervals before dying in 1821, leaving her husband and children to their own resources.

Maj. John Caldwell (1755–1838) was a prominent West Indies merchant who settled in Hartford from Belfast, Ireland, arriving via Barbados about 1773 [*Connecticut Courant*, September 7, 1773]. During the War of Independence, he regularly advertised his growing trade in horses, rum, molasses, and other West Indies goods [*Connecticut Courant*, fifteen notices December 16, 1777, to June 5, 1781, inclusive] and in 1788 built one of the prominent retail dry goods establishments in Hartford [*Connecticut Courant*, December 29, 1788]. The next year he was appointed manager of a lottery to capitalize Hartford's first bank [*Connecticut Courant*, July 20, 1789] and in 1792 was appointed a member of the prestigious building committee for the Connecticut State House [*Connecticut Courant*, August 27, 1792] and the building committee for Hartford's first jail [*Connecticut Courant*, October 8, 1792]. In 1797 he was elected alderman and appointed president of the Hartford Bank [*Connecticut Courant*, April 3, June 12, 1797]. Between 1781 and 1810 his mercantile empire, John Caldwell and Co., with Jeremiah Wadsworth and Michael Olcott as partners, was Hartford's most prominent trading house. In 1815 he appears to have formed a store and distillery, under the name of Caldwell and Colt, in partnership with his son-in-law Christopher Colt [*Connecticut Courant*, July 12, 1815, through May 11, 1819].

In recollecting Sam Colt's childhood, Elizabeth Colt wrote of how Colt's mother died not long after "his father's business affairs became embarrassed, and... he lost the bulk of his fortune" [George A. Jarvis, George Murray Jarvis, William Jarvis Wetmore, *The Jarvis Family* (Hartford: Case, Lockwood and Brainard, 1879), 92–93]. One of the "embarrassments" appears to have been the failure of Caldwell and Colt. As one of only two heirs to Maj. John Caldwell, Sarah stood to inherit a considerable fortune. By predeceasing her father, it is not clear if that inheritance reverted to the family. There is no evidence of Caldwell family wealth being transferred to Christopher Colt or the Caldwell-Colt grandchildren after Sarah Caldwell Colt's death.
16. William N. Hosley Jr. and Gerald W. R. Ward, eds., *The Great River: Art and Society of the Connecticut Valley, 1635–1820* (Hartford: Wadsworth Atheneum, 1985); R. B. Jones, "Descendants of John Coult" (manuscript, CHS), 27; "Colonel Sam Colt's Ancestry," *Hartford*

Daily Times, January 15, 1862; Estate Inventory, Lt. Benjamin Colt, Northampton Probate Court District Records, vol. 13, 585–86, taken January 7, 1783. At the time of his death in 1781, Benjamin Colt's estate was sizable, containing assets valued at more than five hundred pounds not including the "lands and personal estate disposed of by will." In addition to blacksmithing he was a landowner and farmer possessing numerous plots of land around Hadley and a farm in nearby Colrain. When Colt died at the age of forty-three, the inventory of his blacksmithing shop reveals an active business, including two anvils, two pairs of bellows, a variety of hammers, tongs, shoeing tools, swages, grindstones, a large supply of iron and German steel, and a number of finished and unfinished scythes.

17. In a somewhat inexplicable "division of dower" of Lt. Benjamin Colt's estate in 1826, forty-five years after his death, assets were dispersed to two daughters, Lucretia Price and "Rhuamah" (Amy?) Porter, and to "Christopher Colt" being the "south front room" and a "privilege to pass through the kitchen... so long as the present house stands," suggesting, if this is, in fact, Lt. Benjamin's son Christopher, that the family actually lived in Hadley during the mid-1820s [Division of Dower, Lt. Benjamin Colt, Hadley, Northampton Probate Court District, box no. 35, file no. 27, April 4, 1826].

18. Although Elizabeth Colt's memoir of Sam's childhood suggests that Sam and his siblings were "under the care of Mrs. Price, a sister of Mr. [Christopher] Colt," the facts presented are vague [Colt, "Memoir," *Armsmear*, 296]. Not long after Christopher's second marriage in 1823 to Olivia Sargeant, Sam was reportedly sent to Glastonbury for a year to work on a farm [*Armsmear*, 298]. The "Aunt Price," was Christopher's older sister, Lucretia Colt Price (1776 to ca. 1859) of Hadley, Massachusetts [Roderick Bissell Jones, "Descendants of John Coult" (manuscript, CHS), 29]. Special thanks to Judy Johnson, genealogist and reference librarian at the CHS, for assisting in sorting through the tangled confusion of the Colt family genealogy. Sam Colt was in and around Amherst and Ware, Massachusetts, in 1829, and a recently discovered letterbook belonging to Colt places him at Hopkins Academy in Hadley at that time [private collection, cited, Wadsworth Atheneum, Colt Archives, p. 1805]. With three of their father's siblings still living in Hadley, it is plausible that Sam Colt and his siblings spent much if not all of their lives between 1824 and 1829 in or around the Hadley area.

19. Testimony on House Bill No. 59, "for the relief of Samuel Colt," *The Congressional Globe* (33d Congress, 1st Session, July 8, 1854) (Washington, D.C.: John C. Rives, 1854), 1651. The matter of whether Sam Colt's father was sufficiently supportive may never be fully resolved, but of the letters sent from Christopher Colt to his son, most offer moral support, money, loans, useful contacts, or a combination of them. Never, on record, did Sam's father openly disparage his son's enterprise.

20. Barnard, *Armsmear*, 150–56; Rohan, *Yankee Arms Maker*, 36–37. Colt lectured in the United States and Canada, in the North, South, and Midwest, thus gaining a much wider knowledge of people and places than was usual among Americans of that generation.

21. Joseph E. Walker, Richmond, to Sam Colt, Lynchburg, Va., April 5, 1835, Colt Collection, the CHS. Walker, who was a fellow lyceum circuit showman, appears to have been Colt's eyes and ears while out of town. Already on the run in April, Walker wrote Colt that "Secresy *[sic]* I require of you.... When I arrive in New York I will inform you precisely where I am... in Richmond.... [Say you] do not know anything about me.... They might find me out before I would be ready to meet them or their demands.... [Am] in much debt."

22. Joseph E. Walker, Baltimore, to Sam Colt, Richmond, December 28, 1834, Colt Collection, CHS; John Pearson, Baltimore, to Sam Colt, Baltimore, February 1, 1835, Colt Collection, CHS.

23. John Pearson, Baltimore, to Sam Colt, March 9, 1835, Colt Collection, CHS; John Pearson, Baltimore, to Sam Colt, New York, April 23, 1836, Colt Collection, CHS.

24. From a story in the *St. Louis Globe-Democrat*, as reported in the *Hartford Daily Times*, June 8, 1883. Pearson, described as an English gunsmith, was then seventy-nine years old and living in St. Louis.

25. The best-documented and most detailed account of the story is William B. Edwards, *The Story of Colt's Revolver: The Biography of Col. Samuel Colt* (Harrisburg, Pa.: Stackpole, 1953), which draws largely on the company archives at the CSL and the family papers preserved by Elizabeth Hart Colt, now at the CHS. Also see R. L. Wilson, *The Book of Colt Firearms* (1971; reprint, Minneapolis: Blue Book Publications, 1993); Rohan, *Yankee Arms Maker*.

26. Ellsworth Strong Grant, *Yankee Dreamers and Doers* (Chester, Conn.: Pequot Press, 1973), 12. Colt's great-uncle, Peter Colt of Hartford, was a prominent and affluent civic leader, state treasurer during the first administration after the Revolution; in 1793 he moved to Paterson to superintend a cotton mill, one of the first in the United States.

Charter and Prospectus, Patent Arms Manufacturing Company, March 9, 1836, reproduced in Philip R. Phillips and R. L. Wilson, *Paterson Colt Pistol Variations* (Dallas, Tex.: Jackson Arms for Woolaroc Museum, 1979), 126–27.

A. J. Yates, of Texas, writing from New York to James Colt, October 1, 1836, Colt Collection, CHS; A. J. Yates, New Orleans, to Sam Colt, Paterson, January 20, 1837, Colt Collection, CHS, describes "sending an expedition through Texas by the headwaters of the Colorado River"; Agent Jewett of the Brazilian Navy to Sam Colt, October 3, 1836, Colt Collection, CHS.

27. Arthur Chase, *History of Ware, Massachusetts* (Cambridge: Harvard University Press, 1911), 220. Christopher Colt was the company agent from its founding in 1829 until about 1834.

28. Testimony of Congressman Thurston of Ohio on House Bill No. 59, "for the relief of Samuel Colt," *The Congressional Globe* (33d Congress, 2d Session, February 3, 1855) (Washington, D.C.: John C. Rives, 1855), 554; Edward N. Dickerson to *New York Herald*, 41.

29. Broadside Catalogue of Machinery and Tools in Paterson, Colt Collection, May, 14, 15, 1845, CHS, illustrated in Phillips and Wilson, *Paterson Colt Pistol Variations*, 163. Aside from a few "drilling machines," "polishing machines," and "engine lathes" none of the machinery on hand when the final contents of the Paterson facility were sold at auction suggests more than a marginal capacity at uniformity or volume production.

30. Sam Colt, Paterson, to Dudley Selden, New York, April 15, 1837, Colt Collection, CHS. It is unknown how many workmen were eventually employed at Paterson. By midyear 1837 Dudley Selden wrote of wishing they could employ the one hundred to two hundred men Colt apparently had in mind and then itemized the ways Colt had compromised their ability to do so [Dudley Selden, New York, to Sam Colt, Paterson, April 18, 1837, Colt Collection, CHS].

31. *Houston Telegraph*, ca. January 1841, undated newspaper clipping in Colt PFAMC, Scrapbooks, February 1836–June 1852, np. CSL, box 63.

Sam Colt, New York, to Secretary of Navy James K. Paulding, Washington, D.C., May 12, 1840, contained in "Report from the Secretary of the Navy... in relation to the adoption of the improved boarding-pistols and rifles invented by Samuel Colt," *Public Documents of the Senate of the United States*, 360, no. 503, May 6, 1844, 6.

32. Information about movable parts is from "Committee on Military Affairs... relative to 'Colt's repeating fire arms,'" *Reports of Committees of the Senate of the United States*, 512, no. 136 (30th Congress, 1st Session, April 25, 1848), 2; Wilson, *Book of Colt Firearms*, 21; information about potential for unintended discharge is from Col. Samuel Colt, "On the Application of Machinery to the Manufacture of Rotating Chambered-Breech Fire-Arms, and the Peculiarities of Those Arms," *Minutes of Proceedings of the Institution of Civil Engineers*, vol. 11 (Session 1851–52) (London: Institution of Civil Engineers, 1852), 51; the overall list of problems is from the *Hartford Daily Times*, August 10, 1869.

33. Testimony by Congressman Stevens of Michigan on House Bill No. 59, "for the relief of Samuel Colt," *The Congressional Globe* (33d Congress, 1st Session, July 8, 1854) (Washington, D.C.: John C. Rives, 1854), 1645.

34. *New York Courier and Enquirer*, June 29, 1837, cited in Colt PFAMC, Scrapbooks, February 1836–June 1852, np. CSL, box 63, reviews an examination conducted at the U.S. Military Academy at West Point.

35. Sam Colt, Paterson, to Dudley Selden, New York, April 15, 1837, Colt Collection, CHS; Dudley Selden, New York, to Sam Colt, Paterson, April 18, 1837, Colt Papers, CHS.

36. Edwards, *The Story of Colt's Revolver*, 116, lowballs the estimate, while Wilson, *Book of Colt Firearms*, 19, 37, suggests higher numbers. Neither estimate is documented.

37. Edwards, *The Story of Colt's Revolver*, 139–48, 195–203. Beginning in 1839, Colt experimented with the design for waterproof casings to obviate the need for the flask-charged powder and ball system of loading guns. Colt developed a waterproof cartridge lined with tinfoil that attracted considerable government patronage and was manufactured on Colt's behalf by the Hazard Powder Company in Enfield,

Connecticut, beginning in 1843, the outset of a twenty-year friendship and business relationship between Sam Colt and Col. A. G. Hazard. The business carried on for a couple years, when it was abandoned so that Colt could launch the New York and Offing Magnetic Telegraph Company, a company that sold subscription rights to its communications network. The company failed.

38. Barnard, *Armsmear*, 276.

39. "Experiments with Mr. Colt's Submarine Battery," *Supplement to the Hartford Daily Courant*, April 27, 1844.

40. Barnard, *Armsmear*, 279–91.

41. Edward N. Dickerson to *New York Herald*, 40; Patterson, *Colt vs. the Mass. Arms Co.*, 42.

42. Charles Daly, *Discourse Delivered at the 35th Anniversary of the American Institute, November 11, 1863* (Albany: Van Benthuysens, 1864). The American Institute was founded in 1828 when it organized its first "exhibition of the products of American industry." The Massachusetts Charitable Mechanics Association commenced annual exhibitions along the same lines in 1837.

43. Philip K. Lundeberg, *Samuel Colt's Submarine Battery* (Washington, D.C.: Smithsonian Institution Press, 1974). Years later, Professor Draper was cited as the true inventor of Colt's "secret" technique for igniting submarine explosions ["Letter from the Secretary of War Relative to the Secret of Colt's Submarine Battery," *Index to the Executive Documents of the House of Representatives of the United States*, 465, no. 127 (May 15, 1844), 12]. In 1842 Colt worked directly with Samuel F. B. Morse, assisting with experiments in telegraphy and the fabrication and laying of underwater cable [Edwards, *The Story of Colt's Revolver*, 197].

   "Ledger of Rents, Fees, and Scholarships, 1838–1853," New York University Archives, record group 10, box 4, np.

44. Sam Colt, New York, to Christopher Colt, Hartford, January 18, 1846, Colt Collection, CHS.

45. Sam Colt, New Haven, to Edwin Wesson, Northboro, Massachusetts, February 5, 1847, Wesson Collection, Harriet Beecher Stowe Center.

   Patterson, *Colt vs. the Mass. Arms Co.*, 161. Perhaps in retaliation for the pressure, chaos, and complaints that surrounded Colt's collaboration with Whitney, Warner later participated in the patent infringement by the Massachusetts Arms Company.

46. Edwards, *The Story of Colt's Revolver*, 222–23, 226.

47. Sam Colt, Hartford, to Capt. Samuel Walker, New York, January 18, 1847, Colt Collection, CHS.

48. Sam Colt, New York, to Eli Whitney Jr., New Haven, December 20, 1846, Colt Collection, CHS.

49. Receipts and Memos, New Haven, September 27–28, 1847, Colt Collection, CHS; Slate and Brown, Windsor Locks, to Sam Colt, Hartford, August 28, 30, 1847. Having demanded credit references "in consequence of some remarks we have heard very lately," Slate and Brown threatened to withhold the patterns developed to produce the Walker-Whitneyville revolvers and were subsequently snubbed by Colt in their pursuit of additional employment.

50. In August 1847 Sam Colt signed a contract with his uncle Elisha Colt for a line of credit "not to exceed… fourteen thousand dollars," today the equivalent of about $1 million. Elisha Colt would receive "all money until he is indemnified against all liabilities… the said Samuel Colt… to have Insured… his machinery, tools, work in progress… sufficient to cover all and every liability" ["Memorandum of Agreement," Sam Colt and Elisha Colt, July 16, 1847, Colt Collection, CHS]. Trawling unsuccessfully for credit with his uncle Dudley Selden the month before, Colt assured him that "New York is of all others the place for such an establishment" [Sam Colt, New Haven, to Dudley Selden, New York, July 15, 1847]. Working on his behalf at the same time, Colt's brother James investigated relative labor costs, access to materials, and the availability of investment capital in St. Louis [James B. Colt, New York, to Mr. Samuel Hawken, St. Louis, May 3, 1847, Colt Fire-Arms Collection, CSL, box 6].

51. *Hartford Daily Times*, September 29, August 25, 1847; Slate and Brown, Windsor Locks, to Sam Colt, Hartford, August 27, 1847, Colt Collection, CHS.

52. *Hartford Daily Times*, January 21, 1849. Colt and Wesson both exhibited their firearms at the Hartford Agricultural Society fair in October 1847. The first mention of Wesson's plans to move his operations from the rural town of Northboro in Worcester County, Massachusetts, was reported that month in a story that provides a noteworthy anecdote about Hartford's appeal at the beginning of the industrial era:

"We understand that Mr. E. Wesson, the celebrated manufacturer of 'Wesson's Patent Muzzle Rifle,' has concluded to locate in this city…. He finds Hartford… [has] superior advantages…. [and] remarked [that] our 'Free Public High School' was a grand thing for the city" [*Hartford Daily Times*, October 25, 1847]. After months of wrangling with a Hartford booster with investments in the new industrial district on the canal in Windsor Locks, Connecticut, Wesson was approached by Elisha B. Pratt about building on Pearl Street [Elisha B. Pratt, Boston, to Edwin Wesson, Northboro, November 5, 1847, Wesson Collection, Harriet Beecher Stowe Center]. The facility was occupied in the spring of 1848. He was not there more than a few months before he began secret negotiations "to get around Colt's patent" [Joseph Childs, Woonsocket, R.I., to Edwin Wesson, Hartford, September 18, 1848, Wesson Collection, Harriet Beecher Stowe Center]. Just before his death in January 1849, Wesson reported that he and "Mr. Colt had not spoken in a long time & further that Mr. C. did not like him for fear that he would get the order from Government for his pistols" [M. H. Stevens, Washington, D.C., to Edwin G. Ripley, Hartford, March 15, 1849, Ripley and Talcott Collection, Harriet Beecher Stowe Center].

53. *Hartford Daily Times*, November 5, 1847.

54. "Committee on… petition of Samuel Colt for an extension of letters patent, granted… 25, February 1836," *Reports of the Committees of the Senate* 707, no. 279 (33d Congress, First Session, May 23, 1854), 3–4.

55. J. Deane Alden, *Proceedings at the Dedication of Charter Oak Hall* (Hartford: Case, Tiffany, 1856), 39.

   Estate Inventory, Elisha K. Root, 1865, Hartford Probate District Court Records, CSL. Among the ten machine patents bought from the estate after Root's death were patents no. 9941 (8/10/54) for "improvement in drop hammer"; no. 12002 (11/28/54) for "improvement in machine for boring… cylinders"; no. 12874 (5/15/55) for "improvement in slide lathe"; no. 12285 (1/23/55) for "improvement in compound rifling machine"; and various patents for cartridge design and manufacturing dated right up through 1864 [Colt PFAMC, Cash Book B, 1864–71 (February 6, 1866) 87, private collection, Colt Archives, Wadsworth Atheneum].

56. *Hartford Daily Times*, May 28, 1858, December 28, 1857.

57. Rohan, *Yankee Arms Maker*, 14.

58. Paul Uselding, "Elisha K. Root, Forging, and the 'American System,'" *Technology and Culture*, 15, no. 4 (October 1974), 543–68. This article explains Root's contribution in adapting die-forging technology to the production of firearms on an unprecedented scale at a time when the technology of milling or metal cutting had matured.

59. Grant, *Yankee Dreamers and Doers*, 49–55.

60. Edwards, *The Story of Colt's Revolver*, 267.

61. Sam Colt to Elisha K. Root, Employment Contract, October 13, 1853, Colt Collection, CHS. Payments of $250 per month (today equal to $176,687 per year; raised to $350 per month in 1861) posted in company ledgers [Colt PFAMC, Ledger, 1856–62, private collection, Colt Archives, Wadsworth Atheneum, 166].

62. Uselding, "Elisha K. Root," 557–64.

63. Barnard, *Armsmear*, 260–61; Uselding, "Elisha K. Root," 549.

64. *Hartford Daily Courant*, January 20, 1849.

65. *Hartford Daily Times*, January 5, 1849. These figures are very close to the production and employment figures projected in 1847 [*Hartford Daily Times*, August 25, 1847]. Typically with Colt, today's forecast became tomorrow's reality.

66. R. G. Dun and Co., Report, Hartford, August 14, 1849, 15:346, R. G. Dun and Co. Collections, Baker Library, Harvard University Graduate School of Business Administration.

67. *Hartford Daily Times*, June 5, August 7, 1851; R. G. Dun and Co., Report, Hartford, August 1851, 15: 346, R. G. Dun and Co. Collection, Baker Library, Harvard University Graduate School of Business Administration. Production was reported at sixty revolvers per day [Edwards, *The Story of Colt's Revolver*, 306].

68. Barnard, *Armsmear*, 199.

69. William Jarvis, Hartford, to William Jarvis Jr., Eyria, Ohio, December 4, 30, 1860, Jarvis Collection, CHS.

70. *Hartford Daily Times*, October 12, 1961. In 1883 the company was pleased to report a "Large Increase" in business, "over 40%," with employment up to "1,000 men busy," from just 700 two years earlier. Having liberated themselves somewhat from the cyclical dilemma of the arms industry, Colt's finally achieved a respectable level of diversi-

fication in the 1880s, manufacturing, among other things, guns, machines, sewing machines, printing presses, the Baxter patented steam engines, engines for propellers, electric motors, cigarette-making machines, and among its most profitable items, the famous Gatling gun [*Hartford Daily Times*, April 23, 1883].

71. R. L. Wilson, *Colt: An American Legend* (New York: Artabras Publishers, 1985), 205. Similarly, it has been estimated that Colt manufactured about 1 million weapons before his death in 1862 [R. L. Wilson, *The Peacemaker: Arms and Adventure in the American West* (New York: Random House, 1992), 133]. Figuring an average profit at about $3 per gun is consistent with Colt's eventual net worth of between $3 and $4 million.

72. *Testimonials of the Usefulness of Colt's Repeating Fire-Arms from European Journals* (Hartford: Hartford Times, 1852), as cited in Haven and Belden, *A History of the Colt Revolver*, 327.

73. Colt, "On the Application of Machinery."

74. In frustration at the inability of Britain's private arms industry to accommodate the demands of a nation at war, the British government launched a campaign to build the nation's first government-owned and government-run armory in Enfield. The Birmingham Small Arms Trade, a trade group, was formed to lobby on behalf of the industry [Nathan Rosenberg, *The American System of Manufacturers: The Report of the Committee on the Machinery of the United States* (Edinburgh: Edinburgh University Press, 1969), 40–42].

75. *Hartford Daily Times*, May 18, 1854.

76. "Extracts from the Minutes... Taken before the Select Committee on Small Arms, House of Commons, March, 1854," cited in Barnard, *Armsmear*, 371–72. In Charles Dickens's account of Colt's London armory, special note is made of the humble backgrounds of Colt's workforce, "the men... with scarcely an exception... hitherto ignorant of gunmaking," and composed of "carpenters, cabinet-makers, expolicemen, butchers, cabmen, hatters, [and] gas-fitters, porters" [1855; reproduced in Haven and Belden, *A History of the Colt Revolver*, 345–49].

77. "Extracts from the Minutes... Taken before the Select Committee on Small Arms, House of Commons, March, 1854," cited in Barnard, *Armsmear*, 365.

78. *Hartford Daily Times*, May 18, 1854.

79. Sam Colt, Hartford, to Maj. Lewis Carr, New York, August 9, 1852, as cited in Edwards, *The Story of Colt's Revolver*, 309.

80. Elizabeth Hart Colt, *A Memorial of Mrs. Elizabeth Miller Jarvis* (Hartford: Case, Lockwood and Brainard, 1881), 80. Elizabeth's grandfather's estate, estimated at half a million dollars "was divided between his widow and two daughters."

81. *Newport Daily News*, April 15, 1850, August 8, 1851.

82. Ibid., August 28, 1850; George William Curtis, *Lotus-Eating: A Summer Book* (New York: Harper and Bros., 1852), 175–76; *Newport Mercury*, July 26, 1851; *Newport Daily News*, July 30, 1851

83. Curtis, *Lotus-Eating*, 164–66.

84. *Newport Mercury*, August 15, 1857. The Colts had been guests at Peabody's Fourth of July banquet in London the summer before [*Newport Daily News*, August 2, 1860].

85. Curtis, *Lotus-Eating*, 175–76, 179.

86. *Hartford Daily Times*, June 6, 1856.

87. Olive H. Ulrich, Hartford, to William H. Ulrich, June 8, 1856, CSL. Special thanks to Dean Nelson for discovering and sharing this account.

88. *Hartford Daily Times*, June 6, 1856.

89. Barnard, *Armsmear*, 309.

90. Ron Chernow, *The House of Morgan* (New York: Atlantic Monthly Press, 1990), 7.

91. Hon. Thomas Seymour, as cited in *Hartford Daily Times*, March 1, 1862.

92. A letter, now lost, described this in considerable detail, as relayed in an interview with the great-granddaughter of Hetty Jarvis.

93. *Description du Sacre et du Couronement de leurs Majestes Imperiales l'Empereur Alexandre II, et l'Impératrice Marie Alexandresna, 1856* (Moscow: Privately Printed, ca. 1857). A copy of this book was received by Thomas H. Seymour of Hartford. The copy I examined is at the New York Public Library; the book has been described by scholars as perhaps "the largest book ever issued from the printing press," with special large type being cast for it, and fifty-two illustrations, fifteen as enormous color lithographs. Only four hundred copies were printed [Edward Kasinec and Richard Wortman, "The Mythology of

Empire: Imperial Russian Coronation Albums," *Biblion: The Bulletin of the New York Public Library*, 1, no. 1 (fall 1992), 77–100].

94. Sam Colt, Baltic Sea, to Richard Hubbard, Hartford, October 15, 1856, Colt PFAMC, Outgoing General, 1856, CSL, box 6.

95. Colt, "Memoir," *Armsmear*, 299, 302, 305–7, 316–17, 320.

96. Ibid., 318.

PRACTICALLY PERFECT

1. *Hartford Daily Times*, July 12, 1855.

2. Grant, *Yankee Dreamers and Doers*, 23, 239.

3. William Prescott Smith, "Opening Address delivered before the Maryland Institute... Second Annual Exhibition" (Baltimore: Maryland Institute, 1849), 20.

4. Grant, *Yankee Doers and Dreamers*, 115.

5. Horace Bushnell, *An Address before the Hartford County Agricultural Society, Delivered October 2, 1846* (Hartford: Brown and Parsons, 1847), 2–4, 7, 9.

6. Horace Bushnell, *Prosperity Our Duty: A Discourse Delivered... January 31, 1847* (Hartford: Case, Tiffany and Burnham, 1847), 3–5, 7, 9, 11–14, 18, 24.

7. *Hartford Daily Times*, November 25, 1878. For a recent biography, see Robert L. Edwards, *Of Singular Genius, Of Singular Grace: A Biography of Horace Bushnell* (Cleveland: Pilgrim Press, 1992).

8. *Hartford Daily Courant*, May 30, 1864.

9. Grant, *Yankee Dreamers and Doers*, 12.

10. *Hartford Daily Times*, August 18, 1853.

11. Ibid., November 2, 1875, April 19, 1884.

12. Ibid., June 7, 1847.

13. Alden, *Proceedings at the Dedication of Charter Oak Hall*, 20–22.

14. William Greener, *Gunnery in 1858* (London: Smith, Elder, 1858), iii.

15. "Mechanics, the Agents of Power," *Scientific American*, October 4, 1856, 29.

16. Rosenberg, *The American System*, 46, 44.

17. *Hartford Daily Times*, August 20, 1853. Two weeks later, Secretary Davis himself visited arms factories in Hartford and A. G. Hazard's gunpowder works in nearby Enfield.

18. Rosenberg, *The American System*, 65, 201.

19. Guy Hubbard, "Development of Machine Tools in New England, Part 10," *American Machinist*, January 1924, 129–30; Rosenberg, *The American System*, 201.

20. Rosenberg, *The American System*, 201–2.

21. Ibid., 65, 194, 388.

22. Ibid., 203–4.

23. Merritt Roe Smith, introduction to *Military Enterprise and Technological Change*, ed. Merritt Roe Smith (Cambridge: MIT Press, 1985), 4–38.

24. Merritt Roe Smith, "Military Entrepreneurship," *Yankee Enterprise: The Rise of the American System of Manufactures*, ed. Otto Mayr and Robert C. Post (Washington, D.C.: Smithsonian Institution Press, 1981), 63–102; Smith, "Army Ordnance and the 'American System' of Manufacturing, 1815–1861," in *Military Enterprise and Technological Change*, 39–86; Smith, *Harpers Ferry Armory and the New Technology: The Challenge of Change* (Ithaca: Cornell University Press, 1977); Smith, "The American Precision Museum," *Technology and Culture*, 15, no. 3 (July 1974), 413–37.

25. Rosenberg, *The American System*, 76–77.

26. Colt, "On the Application of Machinery," 61–62.

27. Charles H. Fitch, *"Report on the Manufacturers of Interchangeable Mechanism: The Manufacture of Fire-Arms," Report on the Manufacturers of the United States at the Tenth Census, June 1, 1880* (Washington, D.C.: Government Printing Office, 1883), 618.

28. Smith, "Army Ordnance and the 'American System,'" 54; William H. Hallahan, *Misfire: The History of How America's Small Arms Have Failed Our Military* (New York: Charles Scribner's Sons, 1994), 192.

29. Merritt Roe Smith, "The Military Roots of Mass Production: Firearms and American Industrialization, 1815–1913," manuscript (1994), 6–7.

30. "Extracts from the Minutes... Taken before the Select Committee on Small Arms, House of Commons, March, 1854," cited in Barnard, Armsmear, 370.

31. Felicia Johnson Deyrup, *Arms Makers of the Connecticut Valley: A Regional Study of the Economic Development of the Small Arms Industry*, 1798-1870, ed. Vera Brown Holmes and Hans Kohn, Smith College Studies in History, vol. 33 (Northampton, Mass., 1948), 20–21; David A. Zonderman, *Aspirations and Anxieties: New England*

Workers and the Mechanized Factory System, 1815–1850 (New York: Oxford University Press, 1992), 55.

32. Visitors were astonished by "the order, the system, the neatness, and almost military exactness and decorum which pervade every department of the works." Descriptions convey the impression of Shaker-like cleanliness and order with "grounds kept... [in] perfect condition... walls and floors... neat and clean... machinery and tools... symmetrically and admirably arranged... workmen... well dressed... thrift... no newspapers... no tobacco or intoxicating drinks... no unnecessary conversation... responsibility... definite and strict" [Jacob Abbott, "The Armory at Springfield," *Harper's New Monthly Magazine*, 5, no. 26 (July 1852), 17].

33. Estate Inventory, Roswell Lee, taken October 10, 1833, Hampden County Probate Court Records, doc. no. 6827. Although religious books predominate, Lee also owned works of history and science, a forty-seven-volume encyclopedia, a book on industrial accounting, and "2 copies Millwright Guide."

34. Samuel Colt, Diary, April 1836–April 1837, April 25, May 1–22, 1836, Colt Collection, CHS.

35. Hallahan, *Misfire*, 124.

36. Nathan Rosenberg, "Technological Change in the Machine Tool Industry, 1840–1910," *Journal of Economic History* 28, no. 3 (September 1963), 414–43.

37. Fitch, "Report on the Manufacturers," 2:26.

38. Hubbard, "Development of Machine Tools in New England," *American Machinist*, January 1924, 171.

39. *Hartford Daily Times*, May 16, 1860. Representatives from Smith and Wesson, and Eli Whitney Jr. were among the bidders as more than two hundred lots of equipment were auctioned off.

40. J. D. Van Slyck, *Representatives of New England Manufacturers* (Boston: Van Slyck, 1879), 23–25; John D. Hamilton, "The Ames Century," *Man at Arms*, 2, no. 6 (November–December 1980), 24–30; Rosenberg, *The American System*, 59. The Ames Manufacturing Company also manufactured firearms, first about 1843 with a contract to produce the Jenks patent carbine and later as the Massachusetts Arms Company, which spun off from Ames in 1849 and became the parent company of Smith and Wesson.

41. R. G. Dun and Co., Report, Springfield, December 16, 1850, May 1853, August 11, 1857, 41:86, R. G. Dun and Co. Collections, Baker Library, Harvard University Graduate School of Business Administration.

42. *Hartford Daily Times*, January 5, 1852.

43. J. Leander Bishop, *A History of American Manufacturers*, 2 vols. (Philadelphia: Edward Young, 1864), 2:691.

44. Edward H. Knight, "The First Century of the Republic," part 4, *Harper's New Monthly Magazine*, 50, no. 297 (February 1875), 377; Hallahan, *Misfire*, 75.

45. Maj. T. A. Mordecai, "Report of the Military Commission to Europe, 1855–1856," *Reports of Committees of the House of Representatives*, 172.

46. John D. McAulay, *Civil War Breech Loading Rifles* (Lincoln, R.I.: Andrew Mowbray, 1987), 39–40.

47. Joseph Childs, Woonsocket, R.I., to Edwin Wesson, Hartford, September 18, 1848; Edwin Wesson, Hartford, to Dr. Gale, Washington, D.C., October 14, 1848; Daniel Leavitt, Chicopee, to Mr. D. B. Wesson, Hartford, March 21, 1849, Wesson Collection, Harriet Beecher Stowe Center.

48. *Population Schedules of the 7th Census of the United States, 1850* (Washington, D.C.: National Archives, 1964), microfilm no. 41, "Hartford," 442.

49. *Hartford Daily Times*, January 5, 1850; Wayne R. Austerman, "Rebel Steel on the Rio Grande," *Man at Arms* 17, no. 6 (December 1995), 17; Guy Hubbard, "Pioneer Machine Builders: The Mechanics of the Windsor Region and Their Products," manuscript (1922), 9–10, 83–85.

50. "The Volcanic Repeating Rifle," *Frank Leslie's Illustrated Newspaper*, October 9, 1858, 291.

51. McAulay, *Civil War Breech Loading Rifles*, 93–97.

52. Hallahan, *Misfire*, 181–85.

53. J. Schon, "A Brief Description of the Modern System of Small Arms" (Dresden, 1855), translated for inclusion in Mordecai, "Report of the Military Commission," 197–98. Among the advantages over Colt's, the Adams and Deane revolver was of solid-frame construction, was loaded more rapidly, and weighed considerably less than Colt's comparable gun. Most important, the Adams and Deane revolvers had a double-action mechanism, which means it cocked and fired with a single pull of the trigger, whereas Colt's had to be cocked with one action and fired with another. The Adams and Deane revolver also had more easily removable cylinders for cleaning. Colt's advantage was its greater relative interchangeability and greater long-range accuracy. Colt's was decisively superior for frontier use where repairs could not be easily made. The adoption of the Adams and Deane as the official arm of the British military in 1855 played an important role in the closing of Colt's London armory the next year.

54. *Hartford Daily Times*, February 20, 1855.

55. Colt, "On the Application of Machinery," 44, 61.

56. "Committee on... petition of Samuel Colt for an extension of letters patent granted... 25 February 1836," *Reports of Committees of the Senate of the United States*, 707, no. 279 (33d Congress, 1st Session, May 23, 1854), 2.

57. "Select Committee on Small Arms, House of Commons, March, 1854," cited in Barnard, *Armsmear*, 366.

58. George Kubler, *The Shape of Time* (New Haven: Yale University Press, 1962), 7.

59. J. E. Coombes, "Notes on United States Military Small Arms" (manuscript, CSL, 1906), 89.

60. Wilson, *Book of Colt Firearms*, 75.

61. Amos Colt, cited in Wilson, *Book of Colt Firearms*, 181.

62. Wilson, *Book of Colt Firearms*, 194.

63. Ibid., 152.

64. "Repeating Fire-Arms: A Day at the Armory," *United States Magazine*, 236–37.

65. "Steel-Making—Most Valuable Improvement," *New York Times*, May 22, 1860, as cited in Advertising Broadside, Colt PFAMC, private collection, Colt Archives, Wadsworth Atheneum, 2038e.

66. Norman J. Whisler, "What Were They Made Of?" *Man at Arms*, 12, no. 1 (January–February 1990), 24–30.

67. *Hartford Daily Times*, February 20, 1856.

68. Ibid., January 20, 1852, November 5, 1847.

69. Ibid., August 1, 1851, January 20, 1849, November 7, 1853.

70. Ibid., February 19, 1884. Tuller and Copeland, and Philander Hotchkiss were among Hanks's competitors.

71. R. G. Dun and Co., Report, Hartford, August 1846, April 13, 1848, 15:368, R. G. Dun and Co. Collections, Baker Library, Harvard University Graduate School of Business Administration; J. Hammond Trumbull, *The Memorial History of Hartford County* 2 vols. (Boston: Edward I. Osgood, 1886), 1:570.

72. *Hartford Daily Times*, October 8, 1858, June 4, 1859; *Hartford Daily Courant*, June 17, 1864, June 22, 1869; Bishop, *A History of American Manufacturers*, 747–49.

73. R. G. Dun and Co., Report, Hartford, October 17, 1850, May 1, 1857, November 1866, 15:288, 379, R. G. Dun and Co. Collections, Baker Library, Harvard University Graduate School of Business Administration.

74. Hubbard, "Development of Machine Tools in New England," *American Machinist*, January 1924, 171–72.

75. *Hartford Daily Times*, December 13, 1876.

76. Rosenberg, "Technological Change in the Machine Tool Industry," 423. Rosenberg describes "technological convergence" as the interplay of technology and innovation that occurs where a network of industries producing seemingly unrelated commodities benefit from their mutual reliance on commonly shared technology.

77. *Hartford Daily Courant*, May 12, 1866; *Hartford Daily Times*, December 6, 1873.

78. "The Charter Oak City," *Scribner's Monthly*, 13, no. 1 (November 1867), 3. With a population of about forty thousand, Hartford was then thirty-fourth in size among U.S. cities.

79. Trumbull, *History of Hartford County*, 1:570.

80. *Hartford Daily Times*, January 15, 1873, December 13, 1876; *Hartford Daily Courant*, April 1, 1864; Gerard L. Studley, *Connecticut, the Industrial Incubator* (New York: American Society of Mechanical Engineers, 1981), 78–79.

81. Hubbard, "Development of Machine Tools in New England," *American Machinist*, April 1924, 617–18.

82. *Hartford Daily Courant*, December 9, 1864.

83. Hubbard, "Development of Machine Tools in New England," *American Machinist*, April 1924, 617–20.

84. *Hartford Daily Times*, June 2, 1874.

85. Uselding, "Elisha K, Root."
86. *Hartford Daily Times*, March 22, 1880.
87. Hosley and Ward, eds., *The Great River*, 208–9, 214–17, 349–51. The Spencer, Cheney, Colt, and Loomis families are represented among the region's most innovative eighteenth-century tradesmen.
88. Hallahan, *Misfire*, 140–42.
89. *Hartford Daily Times*, January 13, 1882.
90. Hubbard, "Development of Machine Tools in New England," *American Machinist*, August 1924, 313–14; Rosenberg, "Technological Change in the Machine Tool Industry," 429.
91. Fitch, "Report on the Manufacturers," 644.
92. Edward N. Dickerson, New York, to *New York Herald*, January 19, 1854, as cited in "Reports of... committee...to inquire... whether money... offered members... to extend Colt's patent," *Reports of Committees of House of Representatives*, 353 (1854) (33d Congress, 1st Session, August 3, 1854), 37.
93. Testimony on House Bill No. 59, "for the relief of Samuel Colt," *The Congressional Globe* (33d Congress, 2d Session, February 3, 1855) (Washington, D.C.: John C. Rives, 1855), 548.
94. Edwards, *The Story of Colt's Revolver*, 73; Uselding, "Elisha K, Root," 547.
95. Colt PFAMC, Cash Book A, January 1856–March 1864 (December 18, 1856), 27; ibid., Journal A, 1856–63 (March 1, 1862), 484, private collection, Colt Archives, Wadsworth Atheneum.
96. Ibid., Ledger, 1856–62 (March 12, 20, 1862), 485, 488, private collection, Colt Archives, Wadsworth Atheneum. These are only a sampling of the machine purchases documented in the ledger.
97. Ibid., Ledger, 1856–62 (March 20, 1862), 487, private collection, Colt Archives, Wadsworth Atheneum; ibid., Journal A, 1856–62 (December 24, 1862), 451.
98. Ibid., Cash Book A, January 1856–March 1864 (February 24, 1858), 91, private collection, Colt Archives, Wadsworth Atheneum.
99. Ibid., Ledger, 1856–62 (March 17, 1862), 487, private collection, Colt Archives, Wadsworth Atheneum.
100. Ibid., Journal A, 1856–63 (7/28/56, 9/18/56, 2/28/62, 2/24/62, 7/19/61, 8/20/61, 12/31/61) 35, 44, 481, 235, 239, 455; ibid., Cash Book A, January 1856–March 1864 (9/14/61), 243, private collection, Colt Archives, Wadsworth Atheneum.
101. John Ross Dix, "A Visit to Hartford in 1859," *Boston Olive Branch and Atlantic Weekly*, October 8, 15, 1859, cited in Wilson, *Book of Colt Firearms*, 70–71.
102. Colt Advertisement, March 1857, reproduced in Haven and Belden, *A History of the Colt Revolver*, 368. A lone advertisement describing "Colt's Patent Fire-Arms Manufacturing Company" as "prepared to furnish, to any extent, machinery, of the most approved patterns *[sic]* and designs, for the manufacture of various kinds of fire-arms" represents an enterprise that never fully got off the ground.
103. *Hartford Daily Times*, December 28, 1857. Colt's brother-in-law John Jarvis departed Hartford for Russia in April 1858 to supervise the installation of machinery. In addition to gun-making machinery, Colt sent steam engines to Russia, built from castings made by the George S. Lincoln Company. The first mention of Colt's building machinery for commercial purposes is in 1857 in "Repeating Fire-Arms: A Day at the Armory," *United States Magazine*, 239.
104. Colt PFAMC, Journal A, 1856–63 (August, September 1856, 3/31/57, 4/30/57), 25, 41, private collection, Colt Archives, Wadsworth Atheneum.
105. Testimony on House Bill No. 59, "for the relief of Samuel Colt," *The Congressional Globe* (33d Congress, 1st Session, July 8, 1854) (Washington, D.C.: John C. Rives, 1854), 1643, 1426, 1651. There is every reason to believe that Colt's profits by mid-1854 had totaled at least $.75 million. R. G. Dun estimated Colt's profits at $5 per gun in 1851, and between $200,000 and $400,000 in 1853. Colt's personal assets and equity were reported at more than $1 million by 1856, and at about $3.5 million (today almost $200 million) when he died in 1862 [R. G. Dun and Co., Report, Hartford, August 1851, August 1853, November 27, 1855, May 30, 1857, 15:346, 348, R. G. Dun and Co. Collections, Baker Library, Harvard University Graduate School of Business Administration].
106. Dix, "A Visit to Hartford in 1859," 70–71.
107. Colt, "On the Application of Machinery," 46. It is important to note that Colt's reference to "women and children" was made to demonstrate how little strength or skill was required to operate the machines, not to suggest that women and children were actually operating the machines. They did not.
108. Colt, "On the Application of Machinery," 62.
109. Testimony on House Bill No. 59, "for the relief of Samuel Colt," *The Congressional Globe* (33d Congress, 1st Session, July 8, 1854) (Washington, D.C.: John C. Rives, 1854), 1643. The most famous account of Colt's London armory was written and published by Charles Dickens in 1855 (*Household Words*, vol. 9), as transcribed in Haven and Belden, *A History of the Colt Revolver*, 345–49.
110. Most accounts are relatively short and nondescript. When then Secretary of War Jefferson Davis visited in 1853, he toasted Colt at an evening banquet and announced, "These machines, sir,... which we have seen today, are wonderful to us" [*Hartford Daily Times*, September 6, 1853]. In 1848 Sam Houston, of Texas fame, visited Colt's and Wesson's [*Hartford Daily Times*, March 18, 1848], as did Adm. David Farragut in 1866.
111. *Hartford Daily Times*, April 20, 1868; from a letter written to the newspaper following a recent visit.
112. *Manufacturer's Gazette* (Philadelphia), June 1861, as cited in Colt PFAMC, Scrapbooks, 1860–70, np. CSL, box 56. In correspondence, architectural historian Richard Candee [10/16/93, Wadsworth Atheneum, Colt Archives, file no. 26)] cited several New England factory buildings as large or larger than Colt's, notably the Tremont Mill in Lowell (1862; 442 feet long, with five stories); Wauregan Mills in Danielson, Connecticut (1858; 500 feet long); and the Androscoggin Mill in Lewiston, Maine (1860; 542 feet long x 72 wide, with five stories). However, no larger factory has been established as predating Colt's.
113. *Hartford Daily Times*, February 27, 1863.
114. Bishop, *A History of American Manufacturers*, 2:739.
115. Dix, "A Visit to Hartford in 1859," 70–71.
116. Fitch, "Report on the Manufacturers," 636; Rosenberg, *The American System*, 194.
117. Barnard, *Armsmear*, 236–37.
118. Fitch, "Report on the Manufacturers," 643.
119. Barnard, *Armsmear*, 213, 220, 224. The book of the prophet Ezekiel provides one of the most utopian visions in the Old Testament and was clearly the inspiration for Barnard's utopian assessment of Colt's enterprise.
120. Fitch, "Report on the Manufacturers," 643–44.

GUNS AND GUN CULTURE

1. *Hartford Daily Times*, June 30, 1885.
2. George Bernard Shaw, *Major Barbara*, in *Selected Plays* (New York: Dodd, Mead, 1948), 417.
3. Basil Collier, *Arms and the Men: The Arms Trade and Governments* (London: Hamish Hamilton, 1980), 1, 5.
4. Sam Colt, Hartford, to Charles Manby, London, May 18, 1852, Colt Collection, CHS.
5. "The Volcanic Repeating Rifle," *Frank Leslie's Illustrated Newspaper*, October 9, 1858, 291.
6. *Hartford Daily Times*, September 9, 1874.
7. *Dictionary of American Biography* (New York: Charles Scribner Sons, 1934), 14:89.
8. *Hartford Daily Times*, January 1, 1850.
9. James M. McCaffrey, *Army of Manifest Destiny: The American Soldier in the Mexican War, 1846–1848* (New York: New York University Press, 1992), 31.
10. *Hartford Daily Times*, January 2, 1860.
11. Shearer Davis Bowman, *Masters and Lords: Mid-Nineteenth Century United States Planters and Prussian Junkers* (New York: Oxford University Press, 1993), 114.
12. Donald S. Spencer, *Louis Kossuth and Young America: A Study in Sectionalism and Foreign Policy, 1848–1852* (Columbia: University of Missouri Press, 1977), 33, 40.
13. *Hartford Daily Courant*, January 14, 1851.
14. Summarizing contemporary scientific literature on matters of race, Lionel Tiger recently wrote that "all of contemporary population genetics and molecular biology underscores the notion of races as discrete and different entities as false," and urged that we "stop using 19th century biological concepts as a platform for 21st century social policies" ["Trump the Race Card," *Wall Street Journal*, February 23, 1996].
15. Spencer, *Louis Kossuth and Young America*, 12–14.

16. Neil Arnott, *A Survey of Human Progress from the Savage State to the Highest Civilization Yet Attained* (London: Longman, Green, Longman and Roberts, 1861), 5, 24.
17. *Hartford Daily Times*, December 10, 1866, May 25, 1882.
18. Reginald Horsman, *Race and Manifest Destiny: The Origins of American Racial Anglo-Saxonism* (Cambridge: Harvard University Press, 1981), 1–3. Quotation is from Charles DeWolf Brownell, *The Indian Races of North and South America* (Hartford: Hulbut, Scranton, 1864), 159.
19. Brownell, *Indian Races of North and South America*, 709, 718, 720.
20. Marcus Cunliffe, *Soldiers and Civilians: The Martial Spirit in America, 1775–1865* (Boston: Little, Brown, 1868), 339–40.
21. "Our assumptions about deterrence… for the past 20 years… [have been the] basic idea that if each side were able to… threaten retaliation against any attack and thereby impose on an aggressor costs that were clearly out of balance with any potential gains, this would suffice to prevent conflict" ["The Strategic Defense Initiative," *Special Report No. 129* (Washington, D.C.: U.S. Department of State, 1985), 1].
22. Cunliffe, *Soldiers and Civilians*, 22. Given the intensity of the debate around this issue—it is at the core of pro– and anti–gun control arguments—it is surprising that the history of the "peacemaker" defense has not been studied. Although there is little evidence of its currency in the United States before 1840, the French penal reform theorist Cesare Beccaria wrote in 1764 that "laws that forbid the carrying of arms… disarm those only who are neither inclined nor determined to commit crimes" [Gary Kleck, *Point Blank: Guns and Violence in America* (New York: Aldine De Gruyter, 1991), 144].
23. Gen. James Tallmadge, *Address before the American Institute… 26th of October 1841* (New York: Hopkins and Jennings, 1841), 14.
24. Bernard Brodie and Fawn M. Brodie, *From Crossbow to H-Bomb* (Bloomington: Indiana University Press, 1973), 140.
25. *Hartford Daily Times*, January 5, 1852.
26. Patterson, *Colt vs. the Massachusetts Arms Co.*, 11.
27. Howe, *Adventures and Achievements*, 149.
28. Frank T. Morn, "Firearms Use and the Police: A Historic Evaluation of American Values," *Firearms and Violence: Issues of Public Policy*, ed. Don B. Kates Jr. (San Francisco: Pacific Institute for Public Policy Research, 1984), 494.
29. Stephen Halbrook, "The Second Amendment as a Phenomenon of Classical Political Philosophy," ibid., 372–73.
30. Cunliffe, *Soldiers and Civilians*, 52.
31. *Hartford Daily Times*, November 8, 1860.
32. Eric H. Monkkonen, *Police in Urban America, 1860–1920* (Cambridge: Cambridge University Press, 1981), 42, 45.
33. Sidney L. Harring, *Policing a Class Society: The Experience of American Cities, 1865–1915* (New Brunswick, N.J.: Rutgers University Press, 1983), 30–31, 107; Morn, "Firearms Use and the Police," 501.
34. Price List, Colt PFAMC, January 1, 1861, illustrated in R. L. Wilson, *The Arms Collection of Colonel Colt* (Hartford: Wadsworth Atheneum and Herb Glass, 1964), xxii.
35. Morn, "Firearms Use and the Police," 491, 503.
36. Harring, *Policing a Class Society*, 107.
37. Morn, "Firearms Use and the Police," 500.
38. *Newport Mercury*, July 26, 1851; *Newport Daily News*, August 19, 1851.
39. *Hartford Daily Times*, July 4, 1866.
40. Rosenberg, *The American System*, 127–28.
41. Russell S. Gilmore, "'Another Branch of Manly Sport': American Rifle Games, 1840–1890," *Hard at Play: Leisure in America, 1840–1940*, ed. Kathryn Grover (Amherst: University of Massachusetts Press, 1992), 93–95.
42. Osha Gray Davidson, *Under Fire: The NRA and the Battle for Gun Control* (New York: Henry Holt, 1993), 22–23, 30–32. The NRA's turn away from its role as a representative body for hunters and target shooters began in 1868 with the passage of the first gun-control legislation and culminated in what is known within the organization as the Cincinnati Revolt of 1977.
43. Morn, "Firearms Use and the Police," 495; Buffalo Bill's Wild West Show first toured Hartford in 1883 and in 1884 attracted an attendance of eight thousand to its "show at Charter Oak Park" [*Hartford Daily Times*, July 31, 1884].
44. Testimony on House Bill No. 59, "for the relief of Samuel Colt," *The Congressional Globe* (33d Congress, 1st Session, July 8, 1854) (Washington, D.C.: John C. Rives, 1854), 1651.

45. Barnard, *Armsmear*, 160.
46. *Hartford Daily Times*, October 4, 1854.
47. McCabe, *Great Fortunes*, 343.
48. Patterson, *Colt vs. the Mass. Arms Co.*, 287–88.
49. Walt Whitman, *Walt Whitman (Song of Myself)*, in *The Walt Whitman Reader* (1855; reprint, Philadelphia: Courage Books, 1993), 29–105.
50. Colt PFAMC, Cash Book A, January 1856–March 1864 (May 29, 1861), 225. The schooner, possibly built at S. Belden and Son's shipyard near Colt's Armory, may not have been built much before its first documented reference in these account books.
51. Sam Colt, Vienna, to Elisha Colt, Hartford, July 18, 1849, as cited in R. L. Wilson, *Colt Engraving* (1974; reprint, Bienfield Publishing, 1982) 22.
52. *London Evening Sun*, August 11, 1852, as cited in Colt PFAMC, Scrapbooks, nd., np. CSL, box 56. The medal is in the Wadsworth Atheneum's Colt Collection and is illustrated in Barnard, *Armsmear*, 121.
53. *Hartford Daily Times*, October 20, 1849.
54. In 1852, the year this organization was chartered by the general assembly, Sam Colt offered the use of a piece of land on the South Meadows where he was building his armory [*Transactions of the Connecticut State Agricultural Society* (Hartford 1852), 16].
55. *Hartford Daily Times*, October 13, 1853.
56. Colt PFAMC, Cash Book A, January 1856–March 1864 (October 29, 1858), 129, private collection, Colt Archives, Wadsworth Atheneum.
57. *Hartford Daily Times*, November 21, 1849. Although the activities of this organization are not well documented, its founding leadership was dominated by several of Colt's closest business associates, including James H. Ashmead, Samuel Woodruff, E. C. Kellogg, and James G. Batterson.
58. Daly, *Discourse Delivered at the 35th Anniversary of the American Institute*, 16–30. In spite of the greater attention given to Britain for its Great Exhibition of 1851, the French are credited with originating industrial exhibitions in which manufacturers competed for awards, having mounted thirteen such displays between 1798 and 1855. This report notes that the French industrial fair of 1844 was "such as no other nation but France could have made" and the inspiration behind Britain's exhibition, which attracted a record-setting attendance of about 6 million in 1851. The first U.S. industrial exhibition took place in Pittsfield, Massachusetts, in 1811, but it was not until the 1830s that organizations like the American Institute and the Massachusetts Charitable Mechanics Association began holding annual exhibitions on a considerably larger scale. Although claiming that the Hartford County Agricultural Society fair of 1853 "far exceeds that of any previous year," Hartford boosters noted that the 1853 Massachusetts Mechanics fair had been attended by more than 100,000 people, due in part, to its emphasis on industrial more than agricultural matters [*Hartford Daily Times*, October 15, 1853].
59. *Charleston Courier*, February 19, 1836, as cited in Colt PFAMC, Scrapbooks, "Records of Colt Company Property," February 1836–June 1852, np. CSL, box 63. No medal for 1836 has been found, but the medals for later years are included in the Wadsworth Atheneum's Colt Collection; Colt's appointment to a committee of the American Institute is recorded in a letter [American Institute, New York, to Sam Colt, Paterson, N.J., 1837, Colt Collection, CHS].
60. New-York Historical Society to Sam Colt, New York, December 12, 1838, Colt Manuscripts, Wadsworth Atheneum; National Institution for the Promotion of Science, to Sam Colt, Washington, D.C., August 22, 1840, Colt Collection, CHS; Connecticut Historical Society to Sam Colt, Hartford, June 7, 1854, Colt Manuscripts, Wadsworth Atheneum.
61. Appointment to the Local Committee for the State of Connecticut at the World's Fair in London, 1851; Appointment to the Connecticut Commission for the Exhibition in Paris, 1855, Colt Manuscripts, Wadsworth Atheneum.
62. *Hartford Daily Times*, February 20, 1855.
63. Certificate of Honor, Universal Society for the Encouragement of Arts and Industry, London, June 30, 1856, Colt Manuscripts, Wadsworth Atheneum. The medals cited are contained in the Colt Collection at Wadsworth Atheneum; several are illustrated and described in Barnard, *Armsmear*, 121–27,
64. "Committee on… petition of Samuel Colt for an extension of letters patent," *Reports of Committees of the Senate of the United States*, 707, no. 279 (33d Congress, 1st Session, May 23, 1854), 2.
65. Robert Friedel, "Perspiration in Perspective: Changing Perceptions of Genius and Expertise in American Invention," in *Inventive Minds:*

Creativity in Technology, ed. Robert J. Weber and David N. Perkins (New York: Oxford University Press, 1992), 11. The annals of American invention are filled with legends of heroic struggle and divine inspiration. From Charles Goodyear, for whom rubber was "a divine mission… his passion, mistress and Holy Grail" [Grant, *Yankee Dreamers and Doers*, 210–11], to Thomas Edison, the Wizard of Menlo Park, with his reliance on random probing and his contempt for "the-or-retical science" [Wachhorst, *Edison: An American Myth*, 35], the story of invention has been told as a heroic quest composed of a set structure leading from heroic persistence to an unexpected chance encounter or clue, culminating with the gift of insight [David N. Perkins, "The Topography of Invention," *Inventive Minds*, 239].

66. Gustavus Myers, *History of the Great American Fortunes* (1907; reprint, New York: Modern Library, 1936), 399–400.

67. *Hartford Daily Times*, August 8, 1851.

68. Ibid., January 12, 1850; *Report of Commissioner of Patents, for the Year 1850*, part 1 (Washington, D.C.: House of Representatives Printers, 1851), 6.

69. Patterson, *Colt vs. the Mass. Arms Co.*, 34.

70. Sam Colt, Hartford, to Gen. Thomas Rusk, Washington, July 19, 1848, Colt Collection, CHS.

71. Broadside, Edward N. Dickerson to Arms Dealers, November 12, 1852, as cited in Edwards, *The Story of Colt's Revolver*, 1001.

72. V. B. Palmer, "Hints for Business Men," *Hartford Daily Times*, October 22, 1851.

73. "How Colt Got His Pistols Adopted by the British Government," *Scientific American*, January 16, 1864, 34.

74. *New York Times and Commercial Intelligencer*, October 15, 1839, as cited in Colt PFAMC, Scrapbooks, "Records of Colt Company Property," February 1836–June 1852, np. CSL, box 63.

75. *New York Evening Post*, October 18, 1839, as cited in Colt PFAMC, Scrapbooks, "Records of Colt Company Property," February 1836–June 1852, np. CSL, box 63.

76. George Catlin, *Episodes from Life among the Indians and Last Rambles*, ed. Marvin C. Ross (Norman: University of Oklahoma Press, 1959); Wilson, *Book of Colt Firearms*, 94, 179. Catlin's departure for South America was reported by Colt's friends at the *Hartford Daily Times*, who described his mission in seeking "a race who have not yet been changed by the march of civilization" [*Hartford Daily Times*, March 6, 1855]. Nine of the twelve paintings were included in the inventory of Sam Colt's estate in 1862. By the time of Elizabeth Colt's death in 1905, she had dispersed the pictures to family members; today they are scattered among museums and private collections [Estate Inventory, Col. Samuel Colt, 1862, Hartford Probate District Court Records, CSL]. Nine Catlin pictures, appraised at $27 (today's equivalent of about $1,500), were in the billiard room at Armsmear.

77. As cited in Bern Keating, *The Flamboyant Mr. Colt and His Deadly Six-Shooter* (Garden City, N.Y.: Doubleday, 1978), 164.

78. Sam Colt, Hartford, to J. Deane Alden, Arizona, February 23, 1860, Colt Collection, CHS.

79. Colt, "Memoir," *Armsmear*, 307; Testimony on House Bill No. 59, "for the relief of Samuel Colt," *The Congressional Globe* (33d Congress, 2d Session, February 3, 1855) (Washington, D.C.: John C. Rives, 1855), 550; Bill of Sale, J. Morgan Hall, Washington, July 18–31, 1837, Colt Collection, CHS; Bills of Sale, Edward Simms, Washington, January 8, 11, 18, 25, February 9, 17, March 5, 8, 1939—during the three months Colt spent in Washington lobbying to secure a patent, existing bills record the purchase of about 85 bottles of brandy and sherry costing $75.37 (today's equivalent of almost $5,000). At the time of his death Colt's wine cellar at Armsmear contained $2,715 (today's equivalent of almost $150,000) in liquor including 100 gallons of whiskey, 90 gallons of brandy, 1,428 bottles of port wine, 939 bottles of sherry, and "15 Boxes Champaign *[sic]*," plus 5,400 cigars, appraised at the equivalent of almost $2,000 [Estate Inventory, Col. Samuel Colt, 1862]. Astonishingly, the "wine cellar at [the] Armory Office" contained almost two and a half times as much. Its contents were appraised at $6,379 (today, almost $350,000) and included 109 dozen-bottle cases of champagne, 1,102 cased-bottles of "old pale Brandy," claret, cider, madeira, gin, scotch, Jamaica rum, cherry brandy, and 650 bottles of "25 year old brandy," which, at $2.66 per bottle (almost $150 today), was the most expensive item on the shelves. It is hard to imagine a larger private inventory in the country at that time. Colt was no collector. His liquor was for gifts and for entertaining on an enormous scale.

80. Barnard, *Armsmear*, 366.

81. *West Point Courier and Enquirer*, June 29, 1837, as cited in Colt PFAMC, Scrapbooks, "Records of Colt Company Property," February 1836–June 1852, np. CSL, box 63.

82. Sam Colt, Paterson, to William A. Brown, West Point, June 29, 1837, Colt Collection, CHS.

83. Rohan, *Yankee Arms Maker*, 93.

84. *Report on Patent Arms* (Paterson, N.J.: Colt Patent Arms Manufacturing Co., 1838), CSL, box 63. Harney wrote Colt in February 1839, providing the most optimistic prognosis received during the Paterson years. The lieutenant colonel's comment that "the Rifles of your invention… have surpassed my expectations (which were great) in every particular… [and in] my honest opinion… no other guns… will be used in a few years," proved premature [Lt. Col. William S. Harney to Sam Colt, Washington, D.C., February 6, 1839, Colt Collection, CHS].

85. *New York Times*, January 18, 1841, as cited in Colt PFAMC, Scrapbooks, "Records of Colt Company Property," February 1836–June 1852, np. CSL, box 63.

86. William S. Harney to Sam Colt, Hartford, January 14, 1848, Colt Manuscripts, Wadsworth Atheneum.

87. J. K. Paulding, "Report from the Secretary of the Navy," *Reports of Committees of the Senate of the United States* (May 18, 1840), as cited in Colt PFAMC, Scrapbooks, "Newspaper Clippings," February 1836–June 1852, np. CSL, box 63.

88. Cited in "Repeating Fire-Arms: A Day at the Armory," *United States Magazine*, 229.

89. *Hartford Daily Times*, January 5, 1852.

90. *Minutes of Proceedings of the Institution of Civil Engineers*, vol. 11, session 1851–51 (London: Institution of Civil Engineers, 1852), 53–54.

91. Barnard, *Armsmear*, 160.

92. P. W. Henry to Sam Colt, February 22, 1840, cited in "Report from the Secretary of the Navy… in relation to the adoption of the improved boarding-pistols and rifles invented by Samuel Colt," *Public Documents of the Senate of the United States*, 360, no. 503, 6, 11–12.

93. C. Downing to Gen. Waddy Thompson, [ca. February 1840], cited in "Report from the Secretary of the Navy… in relation to the adoption of the improved boarding-pistols and rifles invented by Samuel Colt," *Public Documents of the Senate of the United States*, 360, no. 503, 13.

94. This document, from a private collection, is reprinted in R. L. Wilson, *The Colt Heritage* (New York: Simon and Schuster, 1979), 36.

95. *Hartford Daily Times*, November 11, 1846.

96. Ibid., May 3, 1847. For a readership hungry for news of this exciting new fighting force, the *Times* ran a story describing how "at San Antonio once, when Hays wished to impress the Comanches," he staged an exhibition in which "man after man rode round a hat at full speed… shooting into it five bullets in succession from his revolving pistol." The story concluded by describing the "Texas Ranger," as "a picked man" who had "left the old States," and whose "reckless and undaunting courage,… genial and hospitable" disposition, and "natural love of wild independent life," signified the possibility of rebirth through western migration.

97. John L. Davis, *The Texas Rangers* (San Antonio: University of Texas Institute of Texan Cultures, 1991), 21–22. Important to Colt, this was also one of the primary stories in the legend of the Texas Rangers.

98. Capt. Samuel A. Walker to Sam Colt, New York, November 30, 1846, Colt Collection, CHS.

99. *Hartford Daily Times*, February 19, 1847. Although Colt's Walker revolver was first manufactured in New Haven, Colt was in constant motion between New York, New Haven, and Hartford during this formative period and was apparently in Hartford, perhaps at his father's home, when Walker visited.

100. Ibid., November 15, 1847.

101. Ibid., February 24, 1849. One of the most hair-raising of these stories appeared in the Hartford newspapers in 1849, in the story of "A Texian *[sic]* Ranger, Dan Henrie," who was pursued by Comanches through the desert for days on horseback. The details read like an episode of Indiana Jones, complete with wild narrow escapes, quick wit, dashing "through fire," being "chased by wolves and Indians who'd killed his companions," having a "horse [which he "shot and ate"] torn to bits by wolves," and after "a week of almost incredible suffering," dashing across the country to "Hartford… to procure some of Colt's Repeaters."

102. Edwards, *The Story of Colt's Revolver*, 96.
103. Sen. Thomas J. Rusk, Washington, D.C., to Sam Colt, Hartford, December 27, 1848, Colt Collection, CHS.
104. Samuel Colt, New Haven, to Gen. Samuel Houston, Washington, D.C., February 24, 1847, Colt Collection, CHS.
105. Ibid. Many of the approximately one hundred letters selected for special care and bequeathed as part of the Colt Collection to the Wadsworth Atheneum in 1905 are testimonials that may once have been bound and maintained as a documentary portfolio.
106. G. Garibaldi to Samuel Colt, January 15, 1860; Victor Emanuel to Samuel Colt, July 24, 1860; Louis Kossuth to Col. Samuel Colt, March 27, 1853; Thomas Addis Emmet to Samuel Colt, May 28, 1852; T. Butler King to Gen. James Wilson, February 26, 1850; William M. Givin to Andrew Ewing, February 26, 1850; Maj. G. T. Howard and Capt. I. S. Sutton to Samuel Colt, February 26, 1850; Maj. O. Cross to Samuel Colt, February 26, 1850, Colt Manuscripts, Wadsworth Atheneum.
107. "Committee on Military Affairs... relative to 'Colt's repeating fire arms,'" *Reports of Committees of the Senate of the United States*, 512, no. 136 (30th Congress, 1st Session, April 25, 1848), 3. Serving on this committee was Thomas J. Rusk, senator from Texas and the secretary of war of the Republic of Texas, who was one of Colt's staunchest allies.

    "Report... relative to 'Colt's repeating fire arms,'" *Reports of Committees of the Senate of the United States*, no. 296 (30th Congress, 1st Session, February 12, 1849).
108. Lord Palmerston to Colonel Colt, December 21, 1853, Colt Manuscripts, Wadsworth Atheneum.
109. *Hartford Daily Times*, May 6, 1853.
110. Jabez Alvord, London, to George Alvord, Winsted, Connecticut, August 1, 1853, Alvord Papers, CSL.
111. Colt PFAMC, Cash Book A, January 1856–March 1864 (May 1, 1857), 47, private collection, Colt Archives, Wadsworth Atheneum.
112. Dix, "A Visit to Hartford in 1859," 70–71.
113. Among Colt's more prominent retail distributors were A. W. Spies and Co., "importers of guns, pistols, rifles," John Moore and Son, and Smith Young and Co., New York; B. Kittredge and Co., "importers of guns & sporting apparatus," Cincinnati and New Orleans; Tryon and Bro., Philadelphia; Palmers and Bachelders, Boston; Jasper A. Maltby, Galena, Illinois; James Canning, Mobile, Alabama; and Jerome B. Gilmore, Shreveport, Louisiana [Colt PFAMC, Ledger, 1856–62 (February 28, 1861), 180–520, private collection, Colt Archives, Wadsworth Atheneum].
114. A. E. Burr, Bill of Sale, September 2, 1854, Colt PFAMC, Bills and Receipts, CSL, box 63.
115. Wilson, *Book of Colt Firearms*, 76.
116. Ibid., 68.
117. John Sedgwick, "The Complexity Problem," *Atlantic Monthly*, March 1993, 96–104. Noting that "Americans' deficiencies in programming the VCR are so well known that they have become a staple of comedy," this article outlines how rapidly the "warfare" among manufacturers trying to outdo their competitors by loading their products with options escalates to the point where users, in a state of "cognitive overload," are left with a "machine from hell" that makes them feel stupid; not exactly the best marketing strategy.
118. Joseph G. Rosa, *Colonel Colt in London: The History of Colt's London Firearms. 1851–1857* (London: Arms and Armour Press, 1976), 33.
119. *Hartford Daily Times*, June 16, 1851.
120. *Punch*, July 5, 1851, as cited in Rosa, *Colonel Colt in London*, 15.
121. *Hartford Daily Times*, September 12, 1853, September 22, 1876.
122. Sam Colt, Richmond, to John Pearson, Baltimore, January 16, 1835. In working up models, Colt instructed his gunsmith to "go to a good cabinet maker & agree with him to make a handsome case with appartments [sic] in it to receive the gun & each of the several parts,... have it lined with green baize to prevent chafing,... [and] have the corners of the box tipt [sic] with brass and a small silver orniment [sic]... in the senter [sic] of the lid." Colt instructed Pearson to have the "lid suitable for ingraving [sic] the name of the owner... [and to have] the ornament on the stock ingraved [sic]... [with] the Colts heads, in the center of which I want my name (S. Colts PR) engraved" [Sam Colt, Richmond, to John Pearson, Baltimore, January 23, 1835].
123. Sam Colt, Vienna, to Elisha Colt, Hartford, July 18, 1849, Colt Collection, CSL, as cited in R. L. Wilson, *Samuel Colt Presents* (Hartford: Wadsworth Atheneum, 1962), 236.
124. Colt PFAMC, Journal A, 1856–63 (February 20, 1861), 347, private collection, Colt Archives, Wadsworth Atheneum.
125. Ormsby produced half a dozen "cylinder scenes," beginning as early as 1839, illustrating a deer and Indian; a stage coach robbery; Texas navy's engagement with Mexico at Compache, 1843; a frontier pioneer defending his cabin; and Hays's big fight, showing western mounted horsemen chasing Indians [Wilson, *Book of Colt Firearms*, 52, 104, 107, 123, 152]. Ormsby graduated from the National Academy of Design in 1829, practiced in Albany and Lancaster, Massachusetts, before relocating in New York where he invented a ruling machine, a transfer press, and a grammagraph, a device for engraving on steel directly from medallions [Mantle Fielding, *Dictionary of American Painters, Sculptors, and Engravers* (1927; revised, Poughkeepsie: Apollo Book, 1986), 683]. A copy of his *A Description of... Bank Note Engraving* (New York: W. L. Ormsby, 1852) was in Colt's library.
126. Newspaper advertisement from 1864, cited in Wilson, *Colt Engraving*, 124.
127. *Hartford Daily Courant*, March 10, 1864; Wilson, *Colt Engraving*, 53–55.
128. Barnard, *Armsmear*, 200; Wilson, *Book of Colt Firearms*, 20, 158; Wilson, *Colt Heritage*, 56. Although models and accessory packages are not identical, this comparison was made between the .34-caliber, five-shot Paterson pistols and the .36-caliber 1851 "navy" model, or "belt pistols," Colt made throughout the 1850s and is based on advertised price lists. During the Paterson era, Colt emphasized rifles and carbines, and his first government contract for one hundred rifles, purchased during the Seminole War, were priced at ninety dollars, then the equivalent of almost seven thousand dollars. Small wonder they weren't flying off the shelves.
129. "Committee on... petition of Samuel Colt for an extension of letters patent granted... 25, February 1836," *Reports of Committees of the Senate of the United States*, 707, no. 279 (33d Congress, 1st Session, May 23, 1854), 2; Testimony on House Bill No. 59, "for the relief of Samuel Colt," *The Congressional Globe* (33d Congress, 1st Session, June 19, 1854) (Washington, D.C.: John C. Rives, 1854), 1426; Edward N. Dickerson, New York, to *New York Herald*, January 19, 1854, as cited in "Reports of... committee... to inquire.... whether money ... offered members... to extend Colt's patent," *Reports of Committees of House of Representatives*, 353 (1854) (33d Congress, 1st Session, August 3, 1854), 41.
130. Bishop notes that by 1864 Colt produced 44 different styles of pistol, in 6 patterns, 11 lengths, and 27 finishes [*A History of American Manufacturers*, 2:741]. Ten years later the *Hartford Daily Times* [January 9, 1874] asserted that "No company... has produced so many styles and varieties of cartridge pistols."
131. Wilson, *Samuel Colt Presents*, 246–49. This gift, which featured top-of-the-line engraved decoration, three pistols, a rifle, and a shoulder stock, was worth the equivalent of about nine thousand dollars, based on price list quotations for guns and engraving.
132. Donald S. Ball, "Presentation Swords," *Man at Arms*, 2, no. 2 (March–April 1980), 10–17.
133. Sam Colt, New York, to William Colt, Hartford, July 21, 1844, cited in Edwards, *The Story of Colt's Revolver*, 153.
134. *Hartford Daily Times*, August 3, 1858.
135. In 1852 Colt offered to provide the Japan Expedition with the 500 revolvers on hand after the default of a contract with the Brazilian government [Sam Colt, Hartford, to Commodore Matthew C. Perry, Washington, D.C., March 10, 1852, Colt Collection, CHS]. The *New York Daily Times* [April 7, 1855] reported the value of the consignment at $1,400, consistent with a quantity of between 50 and 75. Commodore Perry appears to have distributed the arms liberally on his journey, and it is unclear if Colt actually donated the guns or was paid something by the Navy Department [Commodore M. C. Perry, Washington, D.C., to Sam Colt, Hartford, April 25, 1857, as transcribed in Barnard, *Armsmear*, 354–55].
136. Wilson, *Samuel Colt Presents*, 283–88, 291–92. This is the most complete list of gifts presented by Colt during his lifetime. It is, however, only a partial list of the gifts he made. A systematic analysis of Colt's gift strategy has yet to be conducted. Among the well-known military recipients are Mexican War officers Maj. Benjamin McCulloch, Col. Thomas H. Seymour, Col. Charles A. May, Gen. Franklin Pierce, Gen. Zachary Taylor, Commodore John Nicholson, Gen. Thomas Jessup, Lt. Col. William S. Harney, Col. David E. Twiggs, and Col. John C. Hays,

and Civil War generals George B. McClellan, William S. Rosecrans, Francis McDowell, Joseph K. F. Mansfield, Thomas West Sherman, Thomas F. Meagher, and Andrew Porter.

Wilson, *Book of Colt Firearms*, 572. Among the well-known Ordnance Department recipients are Col. Charles A. May, Maj. George D. Ramsay, and Col. James W. Ripley. War Department and other political officials who received gift pistols from Colt include Jefferson Davis, William Faxon, John B. Floyd, and Gov. Andrew B. Moore of Alabama.

137. Dudley Selden, New York, to Sam Colt, Washington, D.C., January 6, 1839, Colt Collection, CHS.

138. Patterson, *Colt vs. the Mass. Arms Co.*, 34; Sam Colt, Hartford, to Manby, London, June 22, 1852, Colt Collection, CHS.

139. "Colt Patent, & c., & c.," *Report No. 353 of the House of Representatives* (33d Congress, 1st Session, August 3, 1854), 1.

140. Testimony on House Bill No. 59, "for the relief of Samuel Colt," *The Congressional Globe* (33d Congress, 1st Session, July 8, 1854) (Washington, D.C.: John C. Rives, 1854), 1642.

141. "The Colt Patent in Congress," *Scientific American*, August 19, 1854, 389.

142. Keating, *The Flamboyant Mr. Colt*, 49.

143. *Hartford Daily Times*, October 25, 1848.

144. *London Times*, December 28, 1854, as cited in Colt PFAMC, Scrapbooks, July 1854–November 1860, CSL, box 28.

145. Contract, Amos H. Colt, New York with William W. B. Hartley and Colt's Fire-Arms, Hartford, October 11, 1859, Colt PFAMC, Papers, CSL, box 44.

146. Colt PFAMC, Ledger, 1856–62 (May 15, 17, July 9, 16, August 30, October 11, November 30, December 6, 1860), 381, 389, 396, private collection, Colt Archives, Wadsworth Atheneum. Private armorers in the Connecticut Valley and U.S. armories in Springfield and Harpers Ferry all supplied weapons on a massive scale to the South during 1859 and 1860. President Buchanan's war secretary, John B. Floyd, a Virginian, ordered Springfield to ship 65,000 muskets to Southern arsenals in 1860 and 85,000 from Harpers Ferry in 1859 and 1860, ordered Springfield to share manufacturing data and technology with Southern ordnance experts, and then considered closing Springfield altogether [Hallahan, *Misfire*, 106–8].

147. *Hartford Daily Times*, May 16, 1860. The *Times* described the South as "preparing for the worst."

148. *New York Daily Tribune*, April 27, 1861. Thanks to Ellinor Mitchell for tracking down this citation.

149. *New York Times*, January 14, 1861.

150. *Hartford Daily Times*, April 19, 20, 22, 25, 29, 30, 1861.

151. Sam Colt to Sales Agents, January 17, 1861, as cited in the *Hartford Daily Times*, April 22, 1861. The *Hartford Times* responded (April 30, 1861) to a *New York Times* editorial that "alleged… [Colt] has advanced the price of his arms" by insisting that only the "commission heretofore allowed to New York houses has been withdrawn" and that the price would remain the same. Whether or not Colt continued to ship arms to the South after January, the company accounts record two enormous payments ($6,250 [$368,000 today] and $18,750 [$1.1 million today]) in March and April for arms received in New Orleans by Colt's old friend and Texas Ranger Gen. Benjamin McCulloch [Colt PFAMC, Journal A, 1856–64 (March 28, April 9, 1861), 353, 355, private collection, Colt Archives, Wadsworth Atheneum, 922–23].

152. *Hartford Daily Courant*, April 22, 1861. This astonishing story involved Solomon Adams, a "former inspector at the United States Armory in Springfield" and eventually the master armorer for the Confederacy, who was in town ten days after the attack on Fort Sumter. It ended with Adams evading "his pursuers… Union men of Springfield… in disguise."

153. *New York Times*, September 27, 1851, as cited in Colt PFAMC, Scrapbooks, July 1854–November 1860, CSL, box 28; *British Army Dispatch*, as cited in *Hartford Daily Times*, October 24, 1851. Colt may have been recruited by Mississippi governor John A. Quitman to help Lopez raise money and arms for an invasion of Cuba. Repelled in a first attempt in 1850, they tried again in 1851 and were captured and executed [James M. McPherson, *Battle Cry of Freedom* (New York: Oxford University Press, 1988), 104–7].

154. Elizabeth Hart Colt, Hartford, to Thomas H. Seymour, St. Petersburg, December 1, 1857, T. H. Seymour Collection, CHS; Colt PFAMC, Cash Book A, January 1856–March 1864 (November 30, December 23, 1857), 77, 81, private collection, Wadsworth Atheneum, Colt Archives, 878. J. Deane Alden, Frederick Kunkle and Mr. Brace traveled to St. Louis and Ft. Leavenworth, Kansas, on Colt business.

155. W. A. Croffut and John M. Morris, *The Military and Civil History of Connecticut during the War of 1861–65* (New York: Ledyard Bill, 1868), 73–74.

156. *Hartford Daily Times*, September 22, 1858.

157. Commission, First Connecticut Rifles, Gov. William A. Buckingham to Colonel Samuel Colt, May 16, 1861, Colt Manuscripts, Wadsworth Atheneum.

158. *Hartford Daily Times*, May 18, 1861.

159. Ibid., May 21, 1861.

160. *Hartford Daily Post*, May 21, 1861.

161. *Hartford Daily Times*, May 31, 1861.

162. Ibid., June 15, 1861; James L. Mitchell, *Colt: The Man, the Arms, the Company* (Harrisburg, Pa.: Stackpole, 1959), 12. A Colt carbine in the collection of the Museum of Connecticut History is engraved on the barrel as a gift from Colt to a member of "Company I, 12th Regiment, Colt Guard Company."

163. John D. McAulay, *Civil War Breechloading Rifles* (Lincoln, R.I.: Andrew Mowbray, Inc., 1987), 13-19.

164. Sam Colt, New York, to William Colt, Hartford, July 21, 1844, cited in Edwards, *The Story of Colt's Revolver*, 153.

COLTSVILLE

1. Grant notes that at the height of the industrial era there were 203 "villes" located within Connecticut's 169 townships, most formally chartered and named after industrialists who founded the resident industries: Hitchcocksville for chairs, Hoadleyville for clocks, Collinsville for axes and edge tools, and Cheneyville for woven silks [*Yankee Dreamers and Doers*, 25]. Although it was reported that "a petition is to be offered to the next Legislature for the change of the name of Hartford into "Col. Colt's Town," the petition was either not submitted or not passed [*Hartford Daily Times*, April 13, 1859]. However, even before breaking ground for Colt's Armory, Hartford residents recognized that "Col. Colt… [was] planning a new city" [*Hartford Daily Times*, May 18, 1854], which his friends at the *Times* dubbed "Coltsville" in 1855 [*Hartford Daily Times*, April 24, 1855].

2. The rampant colt stood on its hind legs atop the Colt Armory's Russian-style onion dome until 1988 when it was sold to an art dealer by the real estate partnership that still owns the building. In spite of rumored promises to replace the colt with a replica (still not carried out) and repeated "attempts" to place the colt with a Hartford or Connecticut-based institution, the dealer wasted little time in marketing the statue to Japanese buyers, illustrating it on the cover of a Japanese-language publication in 1989. In 1994 a committee of Hartford-based friends of the Colt Collection at the Museum of Connecticut History in Hartford launched a campaign to raise private funds to acquire the statue, which was accomplished in 1995. The rampant colt is now part of its permanent collection, a relatively happy ending to a grim chapter in the Colt saga.

3. Hartford mayor Henry C. Deming, May 1856, as cited in Barnard, *Armsmear*, 48.

4. Ibid.

5. *Hartford Daily Times*, November 16, 1853.

6. Ibid., June 24, 1874.

7. Ibid., June 19, 1877.

8. Col. Augustus G. Hazard, Enfield, to Edward Prickett, January 12, 1861, private collection, copy, Wadsworth Atheneum, Colt Archives, 812.

9. *Hartford Daily Times*, November 23, 1850.

10. Abbott, "The Armory at Springfield," 17.

11. Samuel Taylor, Diary, March 5, 1853, Hartford, Taylor Papers, CHS.

12. *Hartford Daily Times*, March 18, 1848, December 18, 1860.

13. *Hartford Daily Courant*, March 29, 1853, January 21, 1851; *Hartford Daily Times*, January 3, 1849.

14. *Hartford Daily Times*, September 14, 1849.

15. *Hartford Daily Courant*, April 29, 1871.

16. *Hartford Daily Times*, October 25, 1847. A collection of correspondence between Windsor Locks manufacturing advocate Josiah Rice and Edwin Wesson provides a fascinating glimpse of the dynamics of business recruiting at the dawn of the industrial era [Harriet Beecher Stowe Center, Hartford]. Windsor Locks had adapted its canal, built twenty years earlier for transportation purposes, to power generation and almost overnight became a booming industrial district. However, Windsor Locks never attracted the high value-added industries such as machine tool, firearms, and cutlery manufacturing, which typically preferred more diversified urban locations like Worcester, Providence, and Hartford.

17. Samuel Taylor, Diary, March 8, 1852, Hartford, np., Taylor Papers, CHS. Taylor's comment that "Saml. Colt is purchasing land in So. Meadow probably to build him a shop" suggests that business neighbors like Taylor, whose family lumber mill was one of the first manufacturing concerns located in nearby Dutch Point, were not exactly sure what was going on.

18. *Hartford Daily Times*, September 23, 1863. In February 1853 the R. G. Dun inspectors visited Hartford and noted that Porter had been "obliged to resign his Presidency [of the State Bank] partly in consequence of difficulty with 'Sam Colt'" [R. G. Dun and Co., Report, Hartford, February 1853, 15:385, 288, R. G. Dun and Co. Collections, Baker Library, Harvard University Graduate School of Business Administration].

19. Estate Inventory, Solomon Porter, 1863, Hartford Probate District Court Records, CSL.

20. R. G. Dun and Co., Report, Hartford, May 1842, 15:385,, R. G. Dun and Co. Collections, Baker Library, Harvard University Graduate School of Business Administration. When the R. G. Dun inspectors cased out Porter's situation in May 1842, Porter, then sixty years old, was described as a "shrewd sharp business man" who "inherited $15,000," married a wife who "inherited $30,000," and was "now w[orth] $125,000." If that is true, Porter managed to quadruple his fortune during the next twenty years, by riding the wave of industrialization and investing in, among other things, real estate, the patrician-controlled Hartford Carpet Company, Colt's competitor—Sharps' Rifle Company—and the Porter Manufacturing Company [Estate Inventory, Solomon Porter, 1863, Hartford Probate District Court Records, CSL].

21. Colt must have bristled when he read the poem written by his friend, Hartford's celebrated poetess Lydia Sigourney, "On the Removal of Col. Porter and His Family to Their New Mansion" [*Hartford Daily Times*, February 25, 1851].

22. *Hartford Daily Times*, September 14, 1853.

23. Ibid., September 20, 1853, December 16, 1859.

24. Solomon Porter to the Mayor and Council, as cited in *Hartford Daily Times*, August 17, 1853. Although there is not a stitch of documentation, Colt historians have suggested that Sam installed a brothel across from Porter's house and bought enough stock in Porter's bank to force him out [Edwards, *The Story of Colt's Revolver*, 315].

Commission, Gov. Thomas H. Seymour to Samuel Colt, May 2, 1850, Colt Manuscripts, Wadsworth Atheneum. It is not clear exactly what an "Aide-de-Camp" to the governor was supposed to do, something between a personal attaché and a roaming diplomatic service for the office is presumed. The other "colonel" appointed at the same time as Colt was Charles R. Ingersoll of New Haven who later (1873) became governor. In 1851 the two colonels were appointed a "committee to... invite President [Fillmore] to visit Hartford" [*Hartford Daily Times*, May 13, 1851]; later that year Colonel Colt was appointed to the "committee on inviting guests" for Hartford's Fourth of July celebration [*Hartford Daily Times*, June 4, 1851].

25. Sam Colt himself described Thomas Seymour as "the first real democrat ever elected by the people of this state" [Sam Colt, Hartford, to [Carr ?], May 12, 1852, Colt Collection, CHS].

26. *Hartford Daily Times*, April 7, 1853.

27. After Thomas Seymour's third victory, friends boasted of how Sam Colt had "*democratized* entirely the old federal State of Connecticut" [C. Vincent, New York, to Sam Colt, April 12, 1852, Colt Collection, CHS].

28. *Hartford Daily Times*, March 15, 1853.

29. Ibid., July 28, 1851.

30. Ibid., January 21, 1850, August 7, 1852, March 15, 1853. Sharps' Rifle Company, the other major arms manufacturer in Hartford, manufactured Christian Sharps's breech-loading rifle in rented facilities, beginning in mid-1851, announced plans to build in December 1852 [*Hartford Daily Times*, December 18, 1852], broke ground the following March, and occupied a new factory in the west end of the city by March 1853. Although competitors, Sharps and Colt together gained Hartford a reputation as the first-ranking firearms manufacturing city in the United States. The Armory Ball, on February 10, 1853, was a jubilant celebration of the extraordinary progress of Hartford's most prolific upstart industry. Attended by the governor, the mayor, workmen, and civic leaders, the dancing and music lasted till dawn and was one of the most memorable parties in the most expansionist decade of the city's history [*Hartford Daily Times*, February 11, 1853].

31. *Hartford Daily Times*, August 15, 1853. Colt's breach of contract suit was brought for nonfulfillment of the lease, involving a dispute over power generation. The suit was brought to a close in April 1857 [*Hartford Daily Times*, April 3, 1857].

32. Ibid., August 18, 1853.

33. Ibid., August 26, 1853.

34. Ibid., September 14, 1853. War Secretary Jefferson Davis, as auspicious an ally as Colt ever had, visited Hartford on September 5 and, at an evening banquet, turned to Sam Colt and proclaimed that "these machines, sir, which we have seen today, are wonderful.... Your Repeating Pistols... are relied upon on the borders of Texas.... Northern manufacturers... used by southern families... are... sure bonds of Union" [*Hartford Daily Times*, September 6, 1853].

35. Ibid., October 4, 6, 7, 1853.

36. Ibid., October 10, 1853.

37. Ibid.

38. Ibid.

39. "Letter from a Visitor to Hartford," ibid., November 7, 1853.

40. R. G. Dun and Co., Report, Hartford, August 14, 1849, 15:346, R. G. Dun and Co. Collections, Baker Library, Harvard University Graduate School of Business Administration.

41. Mayor Henry C. Deming as quoted in *Hartford Daily Times*, April 18, 1855.

42. Colt PFAMC, Bills and Receipts, 1854–55, np. CSL, box 1. Included among the itemized work are bills for "drawings of gable ends to Armories" (12/10/54), "making 2 designs for dome" (1/26/55), "framing plans & full size working drawing for dome" (1/28/55), "design for small Barn" (3/15/55). Colt paid Jordan $243 (today about $13,330) during the course of construction, suggesting a significant but limited role. Jordan was paid an additional $90 in July 1856, almost certainly for design work for Armsmear, and in June 1861 was paid a mere $25 for "plans for new armory" [Colt PFAMC, Cash Book A, January 1856–March 1864 (June 18, 1861), 233, private collection, Colt Archives, Wadsworth Atheneum].

43. *Hartford Daily Times*, January 29 and April 15, 1850. The "Wheeler," so noted, was Gervase Wheeler, whose *Rural Homes* (1851) and *Homes for People*, gained him a place, next to A. J. Downing, as one of the the most influential theorists of American rural and picturesque housing and gardens.

44. *Hartford Daily Courant*, July 1, 1864. Jordan operated in partnership with his brother A. Jordan, who maintained an office in Detroit, Michigan [*Detroit General Directory*, 1853, 47]. When Jordan was recorded in the 1850 census, during his first year in Hartford, he was described as twenty-five years old and from London, with a nine-month-old son born in New York, suggesting his brief residence there en route to Hartford [*Population Schedules of the 7th Census of the United States, 1850* (Washington, D.C.: National Archives, 1964), microfilm no. 41, "Hartford," 505].

45. Until the period after Colt's Patent Fire-Arms Manufacturing Co. was incorporated, corporate accounts are sparse. The bronze rampant colt statue was first mentioned at the 1856 Armory Ball where a "pedestal supporting a bronze colt" was displayed [*Hartford Daily Times*, June 4, 1856]. It appears perched on the dome in the Kellogg Brothers lithograph produced in September to commemorate the opening of Charter Oak Hall and may have been installed by the time of Colt's marriage on June 5. Curiously, among the company accounts that record Colt's personal projects, payments today equal to about $16,000 are recorded between September 1860 and June 1861 for two "zinc horses." One of these is probably the "rampant colt" that was installed in a pond behind Armsmear. There is, in fact, little evidence to suggest that Colt's Armory was tooled up to efficiently produce such work; although the statues have traditionally been credited to the factory, it seems far more likely that Colt commissioned either a German manufacturer or his business associates at the Ames Manufacturing Company in nearby Chicopee, Massachusetts, to fabricate the original horse. In 1853, in collaboration with Henry Kirke Brown, Ames became America's first large-scale manufacturer of "statuary bronze" [Michael E. Shapiro, *Bronze Casting and American Sculpture, 1850–1900* (Newark: University of Delaware Press, 1985) 50–59].

46. Colt PFAMC, Bills and Receipts, 1855, np. CSL, box 1; Colt PFAMC, Cash Book A, January 1856–March 1864 (February 28, 1861), private collection, Colt Archives, Wadsworth Atheneum.

47. Although payment records are incomplete, Pomeroy appears to have received monthly payments at least until July 1857 [Colt PFAMC, Cash Book A, January 1856–March 1864, 19–57, private collection, Colt Archives, Wadsworth Atheneum, 874–77]. Pomeroy is an elusive character who was almost certainly related to Colt's brother Christopher's wife, whose mother was a Pomeroy. In 1858 Colt presented him with a deluxe presentation revolver set—a likely going-away present. He submitted a plan featuring a "system of paths and drives" in the competition for the design of Hartford's first city park in the summer of 1858 and then vanished [*Hartford Daily Times*, July 9, 10, 1858].

48. Colt PFAMC, Bills and Receipts, 1854, np. CSL, box 1. Hicks was appointed Hartford's first city engineer in 1853 and may have been the bridge between Colt and the city in the delicately negotiated plans for expansion. Hicks was paid more than $2,500 (today about $140,000) for work he performed during the twelve to fourteen months when construction of the embankment was underway.

49. This computation is based on half a dozen payroll entries, averaging about $3,200 per month (today about $175,000), recorded in the account titled South Meadows Improvements, August 1856 to March 1857. The crews were probably larger in 1855 when construction for both the armory and embankment was underway. Although not itemized, the cost of building Armsmear and the worker housing was included under the general accounts of South Meadow Improvements [Colt PFAMC, Cash Book A, January 1856–March 1864, private collection. Colt Archives, Wadsworth Atheneum].

50. Colt PFAMC, Journal B, August 1863–June 1869 (December 31, 1864), 174, private collection, Colt Archives, Wadsworth Atheneum, 862–63.

51. *Hartford Daily Times*, May 1, 1854.

52. Ibid., May 18, 1854.

53. Sam Colt as cited in *Hartford Daily Times*, April 30, 1854, as cited in Barnard, *Armsmear*, 65–66.

54. *Hartford Daily Times*, December 5, 1855. The four-panel panoramic view of the great flood of 1854 was later installed in the observatory of Armsmear, where the Colonel and his guests retired with their cigars to look down on the South Meadows compound and marvel at the scope of his achievement. In 1856 the painter Joseph Ropes established a studio in Rome and became one of America's esteemed expatriate artists. A photograph made by Ropes at the time of the flood is in the collection of the CSL.

55. Ibid., May 3, 1854; Gov. Charles H. Pond to Sam Colt, Commission as Major Commandant, First Company Governor's Horse Guard, September 1, 1853, Colt PFAMC, "Administrative Files," CSL, box 55. Connecticut's Governor's Horse Guard was chartered in 1778 and is still in operation today. Colt's grandfather John Caldwell initiated the petition that led to its founding. Long the province of the Standing Order, it had apparently grown inactive during the 1840s, and in April 1853 Colt and his Democratic cronies, with Gov. Thomas Seymour's blessing, declared it defunct, issued new uniforms, and appropriated its name and reputation. When Whig governor Henry Dutton was elected in 1854, the Standing Order was reinstated and Colt's group reorganized as the Seymour Horse Guard [Clyde H. Bassett, *Two Hundred Years: The First Company Governor's Horse Guard, 1788–1978* (Canton, Ct., 1978), 8, 14; James L. Howard, *The Origin and Fortunes of Troop B* (Hartford: Case, Lockwood and Brainard, 1921), 55].

56. *Hartford Daily Times*, April 20, 23, 1855.

57. Ibid., May 2, 1855; Lydia Sigourney, Hartford, to Thomas H. Seymour, St. Petersburg, May 14, 1855, Seymour Papers, CHS.

58. *Hartford Daily Times*, April 23, 1855, September 17, 1869.
At the time of the April flood in 1869, it was reported that water broke through at the "same spot that gave away during the flood of 1862" in the southeast corner [*Hartford Daily Times*, April 24, 1869]. After considerable investigation, the cause was determined to be an "old goose pond... a vein of quicksand... which in dyking was not dug out" [*Hartford Daily Times*, September 17, 1869]. "Titan" was done in by a goose!
  Hartford Daily Times, February 27, 1883. Although impressed with Colt's ability to finance his expansion without borrowing money, R. G. Dun investigators expressed concern that "should a freshet wash them away he may not be able to stand the loss" [R. G. Dun and Co., Report, Hartford, February 26, 1855, 15:346, R. G. Dun and Co. Collections, Baker Library, Harvard University Graduate School of Business Administration]. Devastating floods in 1936 and 1938 again proved the South Meadows' vulnerability.

59. *Hartford Daily Times*, December 16, 1859, July 5, 1854.

60. Ibid., June 8, 1857.

61. D. M. Seymour, Hartford, to Thomas H. Seymour, St. Petersburg, February 21, 1857, Seymour Papers, CHS; Colt PFAMC, Cash Book A, January 1856–March 1864 (October 31, 1857), 71, private collection, Colt Archives, Wadsworth Atheneum, 878; *Hartford Daily Times*, July 12, 1854.

62. *Hartford Daily Times*, March 23, 1858; *Hartford Daily Courant*, October 29, 1870.

63. Colt PFAMC, Bills and Receipts, August 13, 1858, np. CSL, box 1. The bill was from E. Merritt's "account of contract for buildings."

64. *Hartford Daily Times*, March 21, 1859.

65. Barnard, *Armsmear*, 265. As early as 1853 willow was noted in *Scientific American*, February 26, 1853, 187: "There is perhaps not a place in the country where the willow could be cultivated to as good advantage as on our alluvial meadows along the Connecticut river." The writer noted that Hartford already had "a celebrated basket maker," who had been "all over Europe" and claimed "there is no place where he has ever been, where willows grow so fine and good as here." Special thanks to Dean Nelson of the Museum of Connecticut History for this reference.

66. *Hartford Daily Times*, May 5, 1860.

67. At the time of his death in 1905, Augustus Fiege's estate was valued at $32,882 (today's equivalent of $1.15 million) [Estate Inventory, Augustus Fiege, July 25, 1905, Hartford Probate District Court Records, CSL]. Although Kunkle appears to have left Hartford before his death, in 1873 he "erected a beautiful family residence on Farmington Avenue" [*Hartford Daily Times*, September 26, 1873] and also left Colt's to go into business as a merchant [R. G. Dun and Co., Report, Hartford, June 30, 1873, 15:677, R. G. Dun and Co. Collections, Baker Library, Harvard University Graduate School of Business Administration].

68. Colt PFAMC, Cash Book A, January 1856–March 1864 (December 31, 1858), 137, private collection, Colt Archives, Wadsworth Atheneum, 881. In August of the same year, Colt sent Kunkle and Christian Deyhle to New York on unspecified company business [ibid. (August 30, 1858), 119, Colt Archives, 880], and in 1859, Kunkle made at least one other trip to New York on Willow business [ibid. (April 29, 1859), 153, Colt Archives, 882].

69. Colt PFAMC, Letters and Manuscripts, 1859, CSL, as cited in Jeremy Adamson, *American Wicker: Woven Furniture from 1850 to 1930* (New York: Rizzoli and the Renwick Gallery of the Smithsonian, 1993), 29.

70. Barnard, *Armsmear*, 267–68.

71. U.S. Department of Commerce, Bureau of the Census, Eighth Census of the United States, 1860, Schedule 1, Free Inhabitants: Connecticut, Hartford County, Hartford, CHS microfilm, 124–242. Although a couple of the willow workers were born in Connecticut and most probably spoke German, among the thirty men listed as willow workers in the 1860 census all but four were born outside the United States and included immigrants from Bavaria, Prussia, Cassel, Darmstadt, and Wurtenburg in modern-day Germany, Bern and Baden, Switzerland, England, Ireland, France, Austria, and Poland.

72. *Hartford Daily Times*, May 18, 1861.

73. Colt wrote a friend in New York in 1860 that he was about to hire Leopold B. Simon to manage the Willow-Ware Factory, describing him as "the right kind of man to take charge of the business of manufacturing & the sale of our willow ware.... Kunkle [says] that Mr. Simon is... an honest jew & will prove faithful to his trust" [Sam Colt, Hartford, to W. M. B. Hartley, New York, February 20, 1860, Colt Collection, CHS]. Barnard described the Willow Factory work force as mostly German and part Irish [*Armsmear*, 268], and in 1861 the *Hartford Daily Times* (May 18, 1861), noted that "men, girls, and boys make the baskets, chairs, and fancy articles."

74. Illustrations from *Armsmear* provide abundant evidence of the farm life at Coltsville, an image confirmed by the inventory of agricultural tools and livestock recorded in Sam Colt's estate [Estate Inventory, Samuel Colt, 1862, Hartford Probate District Court Records, CSL].
  Initially, the dike, the location of Colt's Armory near Dutch Point (Hartford's first European settlement), and the windmill, which Colt built during the summer of 1858, gave the South Meadows a Dutch ambience [Colt PFAMC, Bills and Receipts, August 13, 1858, np, CSL, box 1; the bill for the windmill was from E. Merritt's "account of con-

tract for buildings"]. The discovery of artifacts (several subsequently donated to the CHS) "found South of the place where the old [Dutch] fort stood... by workmen... employed upon Col. Colt's improvement" heightened Colt's interest in projecting a sense of the South Meadows' old-world character and history [*Hartford Daily Times*, April 19, 1854]. Unfortunately, no illustrations of the windmill survive. It pumped water from its own well and was large enough to have included a small residential apartment. Complaints about its frightening horses caused even Colt's friends at the *Hartford Daily Times* to recommend its removal [April 20, 1859], and it may not have remained long past the time of Colt's death in 1862.

75. "Recollections of Judge D. A. Lyman," *Hartford Times*, February 2, 1921, 17.

    Hartford Daily Times, May 18, 1861. The *Times* noted that "three men are employed to operate the cleaning, splitting, and stripping machines... [while] forty men, girls, and boys make the baskets, chairs, and fancy articles." A rigid division of labor was applied to the weaving, assembly, and finishing of the products. In 1866 the estate of Elisha K. Root was granted a patent for "Improvements in Machinery for Shaving... Willow Switches and other thin strips of wood," which was probably in use at the factory as early as 1860 [Letters Patent No. 58958, dated October 16, 1866].

    Barnard noted that despite competition as far away as Wisconsin, Colt's, "thanks to its heavier capital, its more perfect division of labor, and its more diversified patterns, can still compete with Westerners in their own markets." He also noted that Colt's "cool and self-ventilated" furniture "has an indescribable charm for dwellers in the tropics.... Hence, a good deal... finds its way to Cuba.... The purpose of Col. Colt was to push the sale of it throughout all South America" [*Armsmear*, 271].

76. Adamson, *American Wicker*, 25–28. Wicker furniture was prominently featured in Gervase Wheeler's *Rural Homes* (1851).

77. Colt's Willow-Ware Manufacturing Co., Inventory, 1864–70, np. CSL, boxes 3–5.

78. R. G. Dun and Co., Report, Hartford, February 27, 1872, October 1872, July 13, 1874, 15:336–37, R. G. Dun and Co. Collections, Baker Library, Harvard University Graduate School of Business Administration; William Jarvis, Hartford, to William Jarvis Jr., Eyria, Ohio, November 14, 1859, Jarvis Collection, CHS.

79. Barnard, *Armsmear*, 213.

80. *Hartford Daily Times*, May 18, 1861.

81. Bishop, *A History of American Manufacturers*, 742; "Essen of America," in *The New England States* by William T. Davis (Boston: D. H. Hurd, 1897), 826.

82. Barnard, *Armsmear*, 223.

83. "Repeating Fire-Arms: A Day at the Armory," *United States Magazine*, 240.

84. *Hartford Daily Times*, March 30, 1857.

85. Ibid., July 13, September 11, 1861. The Hartford Cartridge Works, founded in 1861, was described as "making the best cartridges in the United States" and employed "from fifty to seventy hands, mostly girls... [and] made not only for Colt's, Savages and other revolving pistols, but for the Enfield, Minié, Sharps and other rifles."

86. A story recounted years later maintains that Colt turned away from Hartford's First Congregational Church—the church of his ancestors— around 1853 over a dispute involving remodeling in which he "desired everything to remain in the old style, high backed pews and all" [*Hartford Daily Times*, June 27, 1883].

87. Carl Siracusa, *A Mechanical People: Perceptions of the Industrial Order in Massachusetts, 1815-1880* (Middletown, Conn.: Wesleyan University Press, 1979), 77.

88. Sam Colt, New York, to William Colt, Hartford, July 21, 1844, cited in Edwards, *The Story of Colt's Revolver*, 153.

89. William Hamersley, cited in Alden, *Proceedings at the Dedication of Charter Oak Hall*, 20; "Repeating Fire-Arms: A Day at the Armory," *United States Magazine*, 241. Although by 1859 Colt had "planted a large number of trees on the South Meadows, around the dike and on the various streets" including "an elm tree... on every street on the meadow," the church, school, and park were amenities that Colt expected the city to help finance, and he did not live long enough to see them completed [*Hartford Daily Times*, April 29, 1859; Colt PFAMC, Cash Book A, January 1856–March 1864 (May 30, 1859), 157, private collection, Colt Archives, Wadsworth Atheneum, 882]. By 1858 the parklike grounds at Armsmear, the closest thing Hartford

had at that time to a modern, picturesque park, served a quasi-public function but was small compared with what Colt offered to help finance and create if the state agreed to build its new state house on Charter Oak Place, the grounds overlooking the armory. The new state house was not built until twenty years later and then on another side of town.

90. Barnard, *Armsmear*, 41–44.

91. Alden, *Proceedings at the Dedication of Charter Oak Hall*, 11, 41.

92. John S. Garner, *The Model Company Town: Urban Design through Private Enterprise in Nineteenth-Century New England* (Amherst: University of Massachusetts Press, 1984), 53. At one time there were public buildings in many of the hundreds of factory villages in industrial New England. Among the best known are the Hazard Memorial Hall in Peace Dale, Rhode Island, and Cheney Hall in South Manchester, Connecticut. Unlike Colt, some waited until their deaths to distribute benefactions. Colt's friend and business associate Augustus G. Hazard of Enfield left funds to build the Hazard Institute after his death in 1868. It is today the most prized historic building in the section of town bearing his name.

93. I. W. Stuart as cited in Barnard, *Armsmear*, 45–46.

94. Elizabeth Colt probably authorized the formation of an expanded collection and museum for the company at the time she added current-production model firearms to Colt's personal collection of Colt prototype and Paterson-era firearms, and historic and contemporary competitor's revolving firearms. The collection was already begun in 1861 when the company valued the inventory of the Museum Room at $3,245.75 (today about $191,000) [Colt PFAMC, Journal A, August 1863–June 1869 (December 31, 1861), 467, private collection, Colt Archives, Wadsworth Atheneum, 932]. After the armory fire in February 1864, the company charged $466.49 (today about $20,000) as a loss against "Work in Progress in Building for Museum" [Colt PFAMC, Journal B, August 1863–June 1869 (December 31, 1864), 174, private collection, Colt Archives, Wadsworth Atheneum, 862]. Colt's "museum of arms" was first mentioned in connection with his efforts to retrieve the Japanese exchange gifts brought back from Commodore Matthew Perry's Japan expedition in 1856. As Perry explained it, "Mr. Colt placed in my hands for presentation to whosoever I might select more than a thousand dollars worth of his patent pistols.... In return various presents were sent by the emperor... as return presents for the pistols" [Commodore M. C. Perry, Washington, D.C., to Hon. J. C. Dobbin, Washington, D.C., 1856, as cited in Colt PFAMC, "Incoming Correspondence," 1856, np. CSL, box 10]. Connecticut senator and political ally Isaac Toucey intervened by writing naval secretary James C. Dobbin about the "Japanese swords" and stating that they were to be "added to the number of curious arms which I understand [Colt] has already brought together... from various parts of the world," as a "museum of arms... for the amusement and benefit of those who may be [curious] in such researches" [Hon. Isaac Toucey, Washington, D.C., to Hon. J. C. Dobbin, Washington, D.C., May 10, 1856, as cited in Colt PFAMC, "Incoming Correspondence," 1856, np. CSL, box 10]. The Patent Office and Navy Department contested Colt's ownership, and the gifts were still tied up in Washington a year later when Perry wrote Colt asking if he had yet "received the Japanese swords, lances, & c. from the Patent Office" and if he had "yet commenced your museum of arms." [Commodore M. C. Perry to Sam Colt, April 25, 1857, Colt Manuscripts, Wadsworth Atheneum]. After the fire, the company posted a payment in the museum account to New York firearms dealers Schuyler, Hartley, and Graham [Colt PFAMC, Ledger, 1856–62, 229, private collection, Colt Archives, Wadsworth Atheneum]. Unfortunately the payment is neither accurately dated nor itemized. Arms by Colt's competitors were amassed in considerable quantities between 1864 and 1867, constituting part of the company collection that was eventually donated to the Museum of Connecticut History. Among the acquisitions noted in the company account were "Eureka's Pistols, $18.87" [Colt PFAMC, Cash Book B, 1864–71 (August 31, 1864), 25, private collection, Colt Archives, Wadsworth Atheneum]; "1 Smith & Wesson pistol $14.82, 1 Remington Army $10.50" [ibid., Journal B, August 1863–June 1869 (November 11, 1865), 280, private collection, Colt Archives, Wadsworth Atheneum]; "1 Needle Revolver L 2/11/9" [ibid. (April 5, 1866), 344]; and one "Connecticut Arms Co.'s pistol $10.50" [ibid. (November 15, 1867), 503]. Herbert G. Houze suggests that part of a manuscript inventory taken in 1887 actually described firearms in the museum room in 1861, citing more than a dozen additional of both

American and French manufacture ["The 1861 Inventory of the Arms and Miscellaneous Material in the Office of Colonel Samuel Colt," *Armax*, 1, no. 1 (1987), 11–19, and no. 2 (1987), 11–17].

95. *Hartford Daily Times*, May 12, 1860.

96. Barnard, *Armsmear*, 377–78.

97. Will, Samuel Colt, June 6, 1856, Hartford Probate District Court Records, CSL.

98. Income from the endowment to support variable costs such as salaries and scholarships has been estimated based on an assumption of a 7 percent annual return on capital.

99. George Wilson Pierson, *A Yale Book of Numbers: Historical Statistics of the College and University, 1701–1976* (New Haven: Yale University Press, 1983), table A–1.4. Harvard's Lawrence Scientific, founded in 1847, and the Massachusetts Institute of Technology, which opened in 1865, were New England's other technical colleges during the period [Hugh Hawkins, *Between Harvard and America: The Educational Leadership of Charles W. Eliot* (New York: Oxford University Press, 1972), 7, 35].

100. Codicil to Will, Samuel Colt, February 2, 1859, Hartford Probate District Court Records, CSL. In a fit of pique, Colt canceled bequests to his stepmother, brother James, and heirs of his brother Christopher in 1858, and then in 1859 revoked his bequest for the "school of practical mechanics and engineers."

101. *Hartford Daily Courant*, October 27, 1870; *Hartford Daily Times*, September 7, 1871. A short history of the school was compiled when it moved to a new building provided by Elizabeth Colt "for school purposes… on Charter Oak St…. Wyllys and Van Block" in 1871. After Colt's death his "nephew" Samuel C. Colt operated Charter Oak Hall, taking in $375 annual rent from the South School District and handling the chapel used by the armory parish until Elizabeth Colt built the Church of the Good Shepherd and a space used by the South Meadows Sons of Temperance [Colt PFAMC, Journal B, August 1863–June 1869 (August 29, December 31, 1866), 382, 424, private collection, Colt Archives, Wadsworth Atheneum, 868].

102. Armory Store Receipt, Colt PFAMC, Bills and Receipts, 1859, np. CSL, box 3; Cooley's grocery is cited in *Hartford Daily Times*, November 29, 1858. It is not clear if this is one store in the same location in different years or two different stores. The "Willow Shop in Hall" was first cited in 1860 [Colt PFAMC, Journal A, 1856–63 (December 31, 1860), 336, private collection, Colt Archives, Wadsworth Atheneum, 920].

103. A presentation pistol was made up for Henry Dickinson, one of the founders of the Sons of Temperance, in February 1862 [Wilson, *Samuel Colt Presents*, 95]. At the beginning of 1859, St. John's Episcopal Church in Hartford (where the Colts were communicants) "started a Sunday School in a room in Charter Oak Hall, under the direction of the Reverend E. A. Washburn ["An Historical Sketch of the Formation and Organization of the Parish of the Good Shepherd," Church of the Good Shepherd, Vestry Records, 1859–1900, 1, microfilm, Episcopal Church Diocese Records, Hartford].

104. *Hartford Daily Times*, March 3, 1856; Alden, *Proceedings at the Dedication of Charter Oak Hall*, 7. Colt's Armory Band gave its first public appearance at the dedication of Charter Oak Hall on May 6, 1856. Among the most memorable of their many touring performances was their appearance at the "Triennial fireman's parade in New York" in October 1859. In the summer of 1860, they accompanied Colt and his family and friends on a "fishing and sailing cruise," which stopped at Newport, Watch Hill, Nantucket, New Bedford, Edgartown, and Martha's Vineyard. The band performed at the Centennial of the Battle of Bunker Hill (1875), the United States Centennial in Philadelphia [Hartford Daily Times, September 8, 1876], at a cavalcade of the National Grays in Brooklyn in 1860 [*Hartford Daily Times*, October, 19, 1860], the bicentennial of "old Hadley," Colt's ancestral town in 1859 [*Hartford Daily Times*, May 19, 1859], and at the "monster concert" at Rocky Point, which featured the American Band of Providence, the Brigade Band of Boston, and several civic and company bands [*Hartford Daily Times*, September 3, 1877].

105. *Hartford Daily Times*, March 1, 3, 1856. The instruments, which cost $1,600, were manufactured by Graves and Co. of Boston and included 3 sopranos horns, 3 repianos horns, 3 elb coronets, 2 tenor horns, 2 baritone horns, 6 brass instruments, and 1 trumpet.

   In 1856 Colt's Berlin agent C. F. Wappenhaus was paid $23.52 (equal to about $1,300) for "music" [Colt PFAMC, Journal A, 1856–63

(December 31, 1856), 63, private collection, Colt Archives, Wadsworth Atheneum, 914].

   Hartford Daily Times, March 3, 1856, September 8, 1876. A short history at the time of the band's twentieth anniversary claimed that the band was founded in November 1855 and gave the date of its first performance as May 1856. The uniform was made by Henry Schultze, Hartford's first German tailor. Curiously, three years later, on the eve of the Civil War, the uniform was redesigned and Americanized to more closely resemble the uniforms that would be worn by the Union troops. The new uniform featured a "coat of dark blue, double breasted with standing collar ornamented with real gold lace," buttons with "the State arms embossed," dark blue pants with gold stripes, a "repertoire for music, like a small cartridge box… suspended from a cross belt," and a "blue United States pattern cap" with gold epaulets, the whole costing about $45 each (today's equivalent of $2,880) [*Hartford Daily Times*, April 13, 1859].

   Hartford Daily Times, March 19, 1859.

106. *Dwight's Journal of Music* (Boston, April 16, 1853), 9, 13, as cited in John Newsom, "The American Brass Band Movement," *The Yankee Brass Band: Music from Mid-Nineteenth Century America* (New York: New World Records, 1981), an essay accompanying an album by the American Brass Quintet Brass Band, excerpted from an article that appeared in the spring 1979 edition of the *Quarterly Journal of the Library of Congress*.

107. As early as 1858, the band gave concerts for the benefit of the poor [*Hartford Daily Times*, April 19, 1858] and, despite Colt's initial injunction that they play only with his permission, quickly became independent, performing in a St. Patrick's day procession [*Hartford Daily Courant*, March 17, 1864], for a visit by President Grant [ibid., July 1, 1870]; for a German meeting in Allyn Hall [ibid., September 7, 1870]; and at the fireman's parade in Springfield [ibid., October 29, 1870], among others.

108. Adkins arrived in Hartford at the age of twenty-six [*Hartford Daily Times*, January 13, 1859]; he was described as a musician from England in the 1860 census [*Population Schedules of the 8th Census of the United States, 1860* (Washington, D.C.: National Archives, 1964), microfilm no. 41, "Hartford," 128]. Unfortunately, none of the manuscript scores used by Colt's Armory Band are known to have survived. Under its second leader (1856–59), John King, the band arranged and performed "Col. Colt's Grand March," "The Hylas Quick Step" [*Hartford Daily Times*, December 22, 1856], and "Amelia's Polka" [ibid., July 2, 1858].

   Hartford Daily Times, June 18, 1861; *Hartford Daily Courant*, July 8, 1872.

109. *Hartford Daily Times*, March 26, 1859.

110. Colt PFAMC, Ledger, 1856–62, 576, Colt Archives, Wadsworth Atheneum, 910.

111. *Hartford Daily Times*, April 4, 1856; Sam Colt, Steamer Baltic, to Milton Joslin, Hartford, June 17, 1856, Colt Collection, CHS.

112. *Hartford Daily Times*, January 14, 1860.

113. Ibid., October 23, 1858; Bill of Sale, Young, Smith and Co., New York, September 18, 1858, Colt PFAMC, "Bills and Receipts," 1858–64, np. CSL, box 3.

114. *Hartford Daily Times*, October 19, 1860.

115. Ibid., June 15, 1858.

116. Ibid., October 16, 1860.

117. *Hartford Daily Courant*, March 19, 1864.

118. Ibid., July 1, 1870; *Hartford Daily Times*, February 4, 1876. When the National Base Ball League was formed in 1876, Hartford's Morgan Bulkeley was its first president, and Hartford sponsored one of the original teams that played on a field located in Coltsville at the present site of the Colt Memorial House. Documentation is sparse, but it is likely that Hartford industries organized the first community leagues and that the field adopted by Hartford's first professional team was created for the use of Colt's employees.

119. *Hartford Daily Courant*, January 18, 1870.

120. *Hartford Daily Times*, March 23, 1858; U.S. Department of Commerce, Bureau of the Census, Eighth Census of the United States, 1860, Schedule 1, Free Inhabitants: Connecticut, Hartford County, Hartford, CHS microfilm, 126. The painting, dated about 1856, is in the collection of the Museum of Connecticut History.

121. In 1858 $400 was paid to G. W. Davis for "Photographs of Machinery" [Colt PFAMC, Cash Book A, January 1856–March 1864 (December 20, 1858), 135, private collection, Colt Archives, Wadsworth Atheneum],

and "Ambrotypes of men & arms," which Colt commissioned in 1860 [ibid. (June 30, 1860), 189]. In 1863 Richard DeLamater was described as "formerly photographic artist to Col. Colt" [*Hartford Daily Times*, November 30, 1863]. At the time of his death, Colt owned a "Large New Camera" valued at $125, and a "Photographic tent and fixtures" [Estate Inventory, Samuel Colt, 1862, Hartford Probate District Court Records, CSL].

Halfrecht, described as a former gunsmith to the Duke of Coburg, was an expert at "fancy carving" [*Hartford Daily Courant*, March 10, 1864]. In 1856 Joseph Buckardt, a Swiss immigrant and wood-carver at Colt's Armory, produced an "ingenious picture of the Charter Oak" that attracted comment [*Hartford Daily Times*, November 17, 1856]. Colt's ivory-carver, Christian Deyhle, produced Colt's Charter Oak "lamp shades" [*Hartford Daily Times*, December 21, 1861]. Deyhle appears to have carried on a side business as an artist-sculptor. *Hartford Daily Courant*, March 10, 1864.

122. For an institution of such prominence, Germania House is bewilderingly elusive. It appears in the earliest view of Colt's Armory and was probably built in 1856. During the 1870s it was run by Alexander Birkholz, a Prussian mold maker who worked at Colt's beginning in 1856 and resided in worker housing on Van Block Avenue. By 1862 he was operating the boardinghouse and beer hall in Potsdam Village, but aside from notices of its charitable functions during the 1870s, it is not clear what purpose was served by this large and prominent house [*Hartford Daily Times*, January 3, 1870].

123. U.S. Department of Commerce, Bureau of the Census, Seventh Census of the United States, 1850, Schedule 1, Free Inhabitants: Connecticut, Hartford County, Hartford, CHS microfilm; ibid., Eighth Census of the United States, 1860, Schedule 1, Free Inhabitants: Connecticut, Hartford County, Hartford, CHS microfilm.

124. Ibid., Eighth Census of the United States, 1860, Schedule 1, Free Inhabitants: Connecticut, Hartford County, Hartford, CHS microfilm, 60; Estate Inventory, Fidel Bubser, 1876, Hartford Probate District Court Records, CSL; *Geer's City Directory*, 1979, 233.

125. Samuel Taylor, Diary, Hartford, February 15, 1853, Taylor Collection, CHS; *Hartford Daily Times*, July 31, 1855.

126. *Hartford Daily Times*, July 22, 1856.

127. Ibid., August 18, 1857, December 10, 1859. The first issue of *Hartforder Zeitung* was published in April 1858; the only surviving copy is in the collection of CHS.

128. *Hartford Daily Times*, August 16, 1871. In attendance were the New York Rifle Corps, Hartford's Franklin Rifle Club, Springfield's Wesson Rifle Club, and the Boston Germania Rifle Club. Described as the "third annual" in 1871, the Connecticut Schutzenbunde featured target shooting, gymnastics performances, music, and food. It was an enlarged version of the "annual" community "pic-nics," organized at least as early as 1862, usually to raise money for such charitable causes as the German-English school and featuring performances by Colt's Armory Band [*Hartford Daily Times*, July 17, 1862; *Hartford Daily Courant*, June 7, 1864]. The German newspaper was launched on an experimental basis in 1857 and soon failed [*Hartford Daily Times*, August 18, 1857].

129. *Hartford Daily Times*, July 5, 1854. In 1858 Colt sponsored a fireworks display that lasted two hours with a picnic for the "residents on Col. Colt's South Meadows" [*Hartford Daily Times*, July 7, 1858], and in 1860 he spent the equivalent of nine thousand dollars for another grand fireworks display [Colt PFAMC, Journal A, 1856–63 (August 31, 1860), 313, private collection, Colt Archives, Wadsworth Atheneum].

130. *Hartford Daily Times*, June 28, July 5, 1856.

131. Ibid., August 18, 1853.

132. Ibid., February 11, 1853.

133. Zonderman notes that "armories and machine shops were… known for good working relationships between managers and workers… [places where the] social distance… [was] not always so great" [*Aspirations and Anxieties*, 99]. Simeon North was apparently known for making the rounds, greeting workers by name, asking about their families, and occasionally inviting them into his home.

134. *Hartford Daily Times*, July 7, 1858, September 22, 1858.

135. William Jarvis, Hartford, to William Jarvis Jr., Eyria, Ohio, January 7, 1860, Jarvis Collection, CHS; *Hartford Daily Times*, January 6, 1860.

136. Abbott, "The Armory at Springfield," 17.

137. Barnard, *Armsmear*, 214–15, 238.

138. Ibid., 238.

139. Jabez Alvord, London, to George Alvord, Winsted, Connecticut, January 23, 1854, December 18, 1852, April 10, 1853, Alvord Papers, CSL. Somewhere in Germany there probably lies a collection of family letters written to relatives back home by one of Colt's German workmen. Such a discovery would be a bonanza for scholarship on this chapter of U.S. industrial history.

140. *Hartford Daily Courant*, January 4, 1864.

141. Zonderman, *Aspirations and Anxieties*, 288, 293.

142. Barnard, *Armsmear*, 218, 238. Not surprisingly, accurate compensation information is scarce, but in a fascinating, discrete, and ultimately convincing inquiry, Henry Barnard interviewed one of the filers at Colt's Armory on his piecework production. Although Barnard discretely avoided tallying up the figures, $3.75 per day is how it breaks down.

143. Pay Schedule, Colt's Cartridge Works, ca. 1857, private collection, Colt Archives, Wadsworth Atheneum, 2084; Colt PFAMC, Payroll Records, 1860, np. CSL, box 64.

144. Colt PFAMC, Payroll Records, 1860, np. CSL, box 64. This is the only payroll document known from Colt's Armory before the Civil War.

145. Deyrup, *Arms Makers*, 217. According to payroll records at the Springfield Armory, filers, oilers, drillers, and other semiskilled workmen averaged thirty to fifty dollars per month in 1850 [United States Springfield Armory, Payroll, July 1850, Springfield Armory Historic Park Manuscripts].

146. Siracusa, *Mechanical People*, 90–108; *Hartford Daily Courant*, March 2, 1870. The *Hartford Daily Times* [July 25, 1877] reported that railroad laborers were "reduced to $.77/day," at the height of the depression in the 1870s, or today's equivalent of $11,540 per year.

147. Marcus Cunliffe, *Chattel Slavery and Wage Slavery: The Anglo-American Context, 1830–60* (Athens: University of Georgia Press, 1979), 24.

148. William J. Grayson, *The Hireling and the Slave* (Charleston, S.C.: McCarter, 1856), v–ix.

149. Cited in Christopher Lasch, *The True and Only Heaven: Progress and Its Critics* (New York: W. W. Norton, 1991), 137, 162.

150. Barnard, *Armsmear*, 240–41.

151. "Repeating Fire-Arms: A Day at the Armory," *United States Magazine*, 235.

152. *Hartford Daily Times*, July 13, 1863.

153. Eric Tucker, *Administering Danger in the Workplace: The Law and Politics of Occupational Health and Safety Regulation in Ontario, 1850–1914* (Toronto: University of Toronto Press, 1990), 21–29; Zonderman, *Aspirations and Anxieties*, 52.

154. *Hartford Daily Courant*, March 17, 1864, April 2, 1864.

155. Ibid., November 16, 1864.

Having primarily surveyed the proindustrial *Hartford Daily Times* and having noticed a decisive increase in reports of industrial accidents in its competitor the *Hartford Daily Courant*, I am unlikely to have located more than a fraction of the industrial accidents that took place in the region during Colt's era. Nonetheless, deadly explosions were noted at Hazard's powder works in 1851 [*Hartford Daily Times*, April 11, 1851], 1865 [*Hartford Daily Courant*, March 21, 1865], and 1866 when "four men were blown to fragments" [*Hartford Daily Times*, May 14, 1866].

156. *Hartford Daily Times*, March 2, 1861; *Hartford Daily Courant*, July 4, November 11, September 14, 1864. The *Courant* claimed that "accidents of this kind are quite frequent in workshops hereabouts." No attempt has ever been made to fully assess the extent of workplace dangers in the Connecticut Valley's high-tech industries.

157. *Hartford Daily Times*, March 3, 1870.

158. Ibid., October 15, 1861; *Hartford Daily Courant*, September 14, 1864.

159. *Hartford Daily Courant*, November 8, 1864, April 28, 1857, *Hartford Daily Times*, July 21, 1874.

160. *Hartford Daily Courant*, February 6, 1864.

161. *Hartford Daily Times*, April 13, 1859.

162. The *Hartford Daily Times* and its publisher Alfred E. Burr remained unrepentant and unreconstructed throughout the Civil War. Analysts for R. G. Dun noted in 1863 that the *Times* was still successful in a business sense but was run by a "disloyal 'copperhead' of the worst kind," a "'Copperhead' of the Vallandigham school" [R. G. Dun and Co., Report, Hartford, August 1851, May 31, 1863, November 5, 1863, 15:367, R. G. Dun and Co. Collections, Baker Library, Harvard University Graduate School of Business Administration]. Clement Vallandigham was the embodiment of the lunatic fringe of northern

Democrats. His prolific political career culminated with his appointment as secretary of the National Democratic Committee in 1860, in which he argued a policy of noninterference, suppression of abolitionists, and a return to the Jeffersonian doctrine of states' rights. He became the nation's most impassioned orator among the antiwar, pro-Union, Peace Democrats and in 1863 was arrested for treason and banished to the Confederacy by President Lincoln [*Dictionary of American Biography*, 19:143–45].

163. *Hartford Daily Times*, September 6, 1849.

164. Printed Broadside, March 1860, Colt PFAMC, Administration Scrapbooks, 1850–60, np. CSL, box 56,

165. *Hartford Daily Times*, April 2, 1860.

166. Printed Broadside, March 1860, Colt PFAMC, Administration Scrapbooks, 1850–60, np. CSL, box 56,

167. *Hartford Daily Times*, April 13, 1859.

168. Payroll notebook, February 15, 1860, Colt PFAMC, Administrative File, np. CSL, box 64. Aside from its political value, this document is the most comprehensive list of workmen and work categories that exists for Colt's Armory during Colt's lifetime. Workmen are designated as "Democrats," "Republicans," "don't vote," and "discharged." Although a majority of those discharged were designated as "Republicans," the discharged list includes Democrats, and almost half of Colt's department heads were Republicans.

169. Long after the skirmishes of 1853 and 1854, Colt and Hartford's Common Council battled over his plans for Coltsville, and it was the defeat of his attempts to get the city to accede to his plans for roads and bridges that finally caused Colt to revoke the bequest for a technical college. Today, you would almost need a tax attorney to sort through all the accusations and assess who was most at fault. Colt claimed to pay twice the taxes of the next largest taxpayer, to have endured endless nit-picking about rights of way, the layout of streets, and the requirement that he put a railing around the dike [*Hartford Daily Times*, May 6, 1856, June 8, 1857]. In 1858 the city ordered Colt to remove a "Bridge at the foot of Commerce St... [that he] erected, at his own expense to accommodate teamsters and workmen during the erection of his large buildings." Claiming that "his workmen... taxpayers, need it," Colt resisted, while his friends at the *Times* accused the city of "envy or sheer ugliness," noting that he was then paying "one-eighth of our entire city taxes" [*Hartford Daily Times*, May 28, 1858]. The final straw was apparently in 1861, when city authorities insisted that he run new streets in the same direction as the city's [*Hartford Daily Times*, May 18, 1861].

170. On the eve of the 1860 Democratic National Convention—held in, of all places, Charleston, South Carolina—Colt wrote to J. C. Calhoun in Bridgeport, Connecticut, about an impending convention of manufacturers and his wish for laws that "will put an end to this infernal slavery agitation.... It is our duty as employers to make the stumack *[sic]* of these Black Republican Devils feel the vacuum which can only be filled by singular probation placed upon [employing them].... If you will pen a resolution which will bind us to discharge... every Black Republican disunionist & never employ any... until the slavery question is forever put at rest & the rights of the South secured permanently ... I will offer it... with the other manufacturers... who shall attend the convention [in Meriden]" [Sam Colt, Hartford, to J. C. Calhoun, Bridgeport, January 11, 1860, Colt Papers, CHS]. The next day Colt wrote his friend Augustus Hazard about the manufacturer's convention saying that "if [these] gentlemen mean anything more than child's play by [their] meeting... [and adopt] such a course as will [touch] the [belly] of the employee then I say let us go there & take an active part.... [It] might have good effect upon the delegates to the Charlestown Convention" [Sam Colt, Hartford, to Augustus Hazard, Enfield, January 12, 1860, Colt Papers, CHS]. For what it is worth, both letters show Colt still operating within the law, not above it.

171. *Hartford Daily Times*, November 8, 1860. Special thanks to Judy Johnson at the CHS for these facts and figures.

172. Bowman, *Masters and Lords*, 169, 172.

173. Edgar Thompson, *Plantation Societies, Race Relations, and the South: The Regimentation of Populations* (Durham, N.C.: Duke University Press, 1975), 22–23.

174. Richard Slotkin, *The Fatal Environment: The Myth of the Frontier in the Age of Industrialization, 1800–1890* (New York: Atheneum, 1985), 140–46.

CHARTER OAK

1. Barnard, *Armsmear*, 70.

2. Rohan, *Yankee Arms Maker*, 182. Quote from a letter from Sam Colt to his father about 1849.

3. *Hartford Daily Times*, April 20, 1868.

4. *Register of the Connecticut Society of the Colonial Dames of America* (Hartford: Case, Lockwood and Brainard, 1907), 34.

5. George Leon Walker, *History of the First Church in Hartford* (Hartford: Brown and Gross, 1884), 103.

6. In 1992 Connecticut Public Television and the Connecticut Humanities Council teamed up to produce a documentary film titled "Between Boston and New York" about Connecticut's problem finding its identity.

7. Rev. Increase N. Tarbox, "The Charter of 1662," *The Memorial History of Hartford County, 1633–1884*, ed. J. Hammond Trumbull, 2 vols. (Boston: Edward L. Osgood, 1886), 1:59.

8. Sherman W. Adams, "The Andros Government, the Charter, and the Charter Oak," *History of Hartford County*, 1:63.

9. *The Story of the Charter Oak*, compiled under the direction of Marshall Jewell (Hartford: Case, Lockwood and Brainard, 1883), 35.

10. Barnard cites a letter written to him in 1855 by a Rhode Island woman recalling a pilgrimage to the oak in 1791 [*Armsmear*, 26–31].

11. Elizabeth M. Kornhauser, *Ralph Earl: The Face of the Young Republic* (Hartford: Wadsworth Atheneum, 1991), 55. The oak appears in a second Hartford portrait by Earl of Samuel Talcott, but as a feature of a landscape in which it had long served as a dominant landmark. Thanks to Elizabeth Kornhauser for providing this information.

12. John Warner Barber, *Connecticut Historical Collections* (New Haven: Durrier and Peck, 1838), 44, cites an account from about 1795.

13. Richard Saunders, *Daniel Wadsworth: Patron of the Arts* (Hartford: Wadsworth Atheneum, 1981), 24, 62. Years later, Wadsworth would encourage the young Frederic Church to sketch and paint the oak. Church sketched the oak in 1846 [collection of Olana State Historic Site, Hudson, New York] and produced a painting of it in 1847 [collection of the Hartford Steam Boiler Inspection and Insurance Co]).

14. The painting by artist George Francis is in the collection of the CHS and is discussed in Saunders, *Daniel Wadsworth*, 62. The frontispiece to Samuel Peters, *A History of Connecticut* (New Haven: D. Clark, 1829) contains the earliest popular-press illustration of the Charter Oak. A more sickly looking likeness appears in Barber's *Connecticut Historical Collections*, 43. When the same author published *Historical, Poetical, and Pictorial American Scenes; Principally Moral and Religious* (New Haven: J. H. Bradley, 1851), he mentioned the oak but failed to provide a vignette or illustration, even though its mythology and symbolism were perfectly compatible with the subject of the book. The omission is best explained by the fact that the oak myth was just beginning to take off while the book was being compiled in 1849. In 1846 Daniel Wadsworth commissioned Frederic Church to paint the great iconic image of "westering puritans" titled *Hooker and Company Journeying through the Wilderness from Plymouth to Hartford in 1636*. It is believed that Church consciously incorporated the Charter Oak as one of the trees in his idealized wilderness landscape. Less ambiguous are prints and sheet music by Nathaniel Currier and Kellogg Bros. from the late 1840s [Wadsworth Atheneum]. At the dedication of the CHS room at the Wadsworth Atheneum, the Charter Oak and its connection with the Wadsworth family was the theme of the dedication address [Thomas Day, *A Historical Discourse Delivered before the Connecticut Historical Society* (Hartford: December 2, 1843)].

15. Spencer, *Louis Kossuth and Young America*, 13.

16. *Hartford Daily Times*, January 2, 1850.

17. Alden, *Proceedings at the Dedication of Charter Oak Hall*, 38.

18. *Hartford Daily Courant*, October 4, 1861.

19. David Mogen, Mark Busby, and Paul Bryant, *The Frontier Experience and the American Dream* (College Station: Texas A&M University Press, 1989), 23–24.

20. Barnard, *Armsmear*, 45.

21. Alden, *Proceedings at the Dedication of Charter Oak Hall*, 7, 10–11.

22. Ibid., 13, 17.

23. Ibid., 17.

24. Ibid., 38, 39, 41.

25. *Hartford Daily Times*, August 21, 1856.

26. Ibid.; *Hartford Daily Courant*, August 22, 1856.

27. Sigourney, *Fall of the Charter Oak*.

28. *The Charter Oak: Its History and Its Fall* (Hartford: A. Collins, August, 1856), 10–12; *Hartford Daily Courant*, August 22, 1856.

29. Curtis Guild, "The Famous Charter Oak Tree," appeared in the *Hartford Daily Times*, August 26, 1856.
30. George H. Clark, "Petition to Isaac Stuart," *Hartford Daily Courant*, August 26, 1856.
31. *Hartford Daily Times*, September 22, 1854; Barnard, *Armsmear*, 35.
32. *Hartford Daily Times*, August 8, 1856.
33. D. M. Seymour, Hartford, to Thomas H. Seymour, St. Petersburg, Russia, February 21, 1857, Thomas H. Seymour Papers, 1857–59, CHS. In February 1857 Isaac Stuart packed up his belongings, vacated his cherished home on Wyllis Hill, and took a room at Hartford's United States Hotel. At the time of his death four years later the room was crammed floor to ceiling with an assortment of chairs, busts of poets, authors, and military heroes, numerous framed paintings, engravings, and photographs of the Charter Oak, a personal seal made from a block of the oak, portraits of Sam Colt and Mayor Henry Deming, a Charter Oak crucifix, a shell collection, a reading stand, and shelves crowded to capacity with books. Of basic necessities there was little more than a bed, a coal stove and bin, one of Colt's revolvers, a writing desk, and the most expensive item in the room, a pianoforte and stool made from the Charter Oak [Estate Inventory, Isaac Stuart, taken February 19, 1862, Hartford Probate District Court Records, CSL].
34. Robert F. Trent, "The Charter Oak Artifacts," *Connecticut Historical Society Bulletin*, 49, no. 3 (summer 1984), 125–57.
35. Ibid.; *Hartford Daily Times*, April 16, 1858. Although it is reported that John Most made three of these pianos, he may have manufactured more. Only one survives; it is in the collection of the Deep River, Connecticut, Historical Society.
36. Currier's print, titled *Charter Oak! Charter Oak Ancient and Fair!* accompanies Lydia Sigourney's ballad by the same title [Wadsworth Atheneum, no. 44.361]. Also that year, Hartford's Kellogg Bros. produced a lithograph simply titled *The Charter Oak* and jointly sponsored with Isaac Stuart [Wadsworth Atheneum, no. 23.69].
37. *Hartford Daily Times*, October 2, 1856.
    Brownell's portrait of 1856 was rendered as a steel engraving the next year and became the iconic image after which most popular representations were subsequently modeled. The painting, which was donated to the Wadsworth Atheneum in 1898, was showcased and described in *The Story of the Charter Oak*, compiled by Jewell.
38. *Hartford Daily Times*, November 17, 1856.
39. *Hartford Daily Courant*, April 28, 1857.
40. Bernard Pares, *A History of Russia* (New York: Alfred A. Knopf, 1953), 70–73. The city of Novgorod, one of the most ancient in Russia, was historically also one of its most independent and jealous of its liberty, associations Colt would have recognized and admired.
41. *Hartford Daily Courant*, April 28, 1857. Exactly when the Colts took possession of the chair is not certain, but it was on public display in July [*Hartford Daily Times*, July 6, 1857].
42. *Hartford Daily Times*, April 30, 1857.
43. I. S. Stuart to Samuel Jarvis Colt, April 25, 1857, Charter Oak Cradle, object file, no. 05.1580, Wadsworth Atheneum; reprinted in Barnard, *Armsmear*, 32–34.
44. *Hartford Daily Times*, June 22, 1857.
45. Ibid., March 11, 1858; March 10, 1858, Colt PFAMC, Cash Book A, January 1856–March 1864, 93, private collection, Colt Archives, Wadsworth Atheneum; Barnard, *Armsmear*, 132–35.
46. Daniel Wadsworth and the Standing Order were the first to appropriate the oak's symbolism, but by the 1850s, a decade in which the Democrats dominated city politics, it had clearly been appropriated by the party in power. Differing interpretations of "liberty" and the antislavery movement clearly fueled the Charter Oak frenzy. When the oak craze began in the mid-1840s, Whig politicians held offices as mayor and governor. Hartford's leading antislavery newspaper during this period was called *Charter Oak* [Trumbull, *History of Hartford County*, 1:385, 609].
47. *Hartford Daily Times*, July 7, 1857.
48. Barnard, Armsmear, 41.
49. *Hartford Daily Times*, September 1, 1858; Barnard, *Armsmear*, 39–40. Outraged at the desecration of the site of the oak, Henry Barnard described how "reckless speculation" and the "hungary greed of gain… wrenched out of the soil by machinery, the very roots" of the oak "whose slowly decaying fibers should have been left to consecrate forever the soil, which, for untold centuries, had nourished its majestic trunk…. Instead of a noble monument, with a suitable

inscription to perpetuate the… tree… a slab hardly sufficient to mark the grave of an infant was inserted in the road-bed…. All this was done in open day, with a cloud of witnesses looking sadly on; and so well satisfied were these 'architects of ruin'… that they perpetuated both the act and themselves in a photographic picture of the scene!" Equally incensed was George H. Clark, an accomplished Hartford poet who wrote these lines of blustering indignation in "The Oak": "Yes, blot the last sad vestige out— / Burn all the useless wood; / Root up the stump, that none may know / Where the dead monarch stood /… You spurn the loved memento now, / Forget the tyrant's yoke, / And lend oblivion aid to gorge / Our cherished Charter Oak /… Let crowds unconscious tread the soil / By Wadsworth sanctified; / Let Mammon bring, to crown the hill, / His retinue of pride; /… So may the muse of coming time / Indignant speak of them, / Who Freedom's brightest jewel rent / From her proud diadem; / And lash with contemptuous scorn / The men who gave the stroke, / That desecrates the place where stood / Our brave old Charter Oak."
50. *Hartford Daily Times*, September 22, 1857.
51. March 30 and May 31, 1858, Colt PFAMC, Cash Book A, January 1856–March 1864, 95, 105, private collection, Colt Archives, Wadsworth Atheneum.
52. *Hartford Daily Times*, November 24, 1858, March 3, 1859. This chair was custom-made by the Hartford cabinetmaking firm of Robbins and Winship and was completed in March 1859. The Know-Nothing movement, so-named because its adherents reportedly denied membership when it began as a secret society, elected governors in eight states at the height of its popularity. Anti-immigration, the Know Nothings rallied under the slogan "America for Americans." The party also championed antislavery, temperance, and worker's rights, and was understandably popular in industrial Connecticut.
53. *Hartford Daily Times*, December 21, 1861; Wilson, *Samuel Colt Presents*, 139, 163, 192, 208.
54. *Geer's Hartford City Directory, 1856* (Hartford: Elihu Geer, 1856); *Population Schedules of the 8th Census of the United States*, 1860, Hartford District, State of Connecticut, 122 (Deyhle lived in worker housing at 74 Van Block St.); Colt PFAMC, Payroll Records, February 15, 1860, np., CSL, box no. 64. Deyhle and Colt's first and most-trusted German contractor, Frederick Kunkle, traveled to New York on company business—very likely involving lobbying and politics [August 30, 1858, Colt PFAMC, Journal A, January 1856–March 1863, 119].
55. June 14, 1860–December 16, 1861, Colt PFAMC, Journal A, January 1856–March 1863, 303, 329, 334, 347, 351, 357, 368, 380, 389, 398, 410, 423, 435, 447, private collection, Colt Archives, Wadsworth Atheneum. Christian Deyhle's name appears in newspaper accounts of the lamp shades and in census records as Deykle; the company accounts do not credit him personally, but he signed one of the shades and is credited as "the artist" in the newspapers. Deyhle's specialty was ivory carving. Colt employed German wood-carvers, such as Abraham Skaats, who may have collaborated on the shades.
56. *Hartford Daily Times*, December 21, 1861; Barnard, *Armsmear*, 129. Curiously, only "American liberty" is cited in the various accounts of the lamp shades.
57. Isaac Stuart was the premier propagandist of Connecticut's Revolutionary history during Colt's time. His public lectures on topics such as Nathan Hale, the martyr spy of Connecticut, and his positions on the question of who commanded at Bunker Hill mark him as a fierce and jealous defender of Connecticut's honor and reputation [*Hartford Daily Times*, January 10, 18, 1851].
58. I. W. Stuart, Hartford, to Sam Colt, Cuba, March 15, 1861, as cited in Barnard, *Armsmear*, 361–63; *Hartford Daily Courant*, February 23, 1866. Four years after Colt's death, members of the Putnam Phalanx presented their major commandant, Timothy M. Allyn, with a Charter Oak cane; the description of it is so similar to the one given Colt it is quite possibly the same cane, which was left with the Phalanx at Colt's death.

ARMSMEAR

1. Barnard, *Armsmear*, 56–57.
2. "Map of South Meadow, Hartford… Showing Lands Purchased by Col. Colt," in Barnard, *Armsmear*, 58–59; Second Ecclesiastical Society to Samuel Colt, August 30, 1852, Hartford Land Records, Hartford City Hall. The very land on which Armsmear stands was controversial in that the vestry of the Second Church had to petition the legislature to break a bequest before they could sell the land. Heirs of the donors later contested the sale.

3. *Hartford Daily Times*, December 5, 1855. In 1876 it was reported that "this great work" was executed "from May, 1855, to January, 1862" [Mary Elizabeth Wilson Sherwood, "The Homes of America: Part 7—Armsmear," *Art Journal*, November 1876, 322].

4. William Jarvis, Middletown, to William Jarvis Jr., Eyria, Ohio, June 19, 1857, Jarvis Collection, CHS.

5. Armsmear was in the midst of being reoriented to face the east, with a major addition underway, when Sam Colt died in January 1862. Hartford architect W. T. Hallett was paid almost three times as much for "Drawing working plans and changes... for dwelling," as had been paid Octavius Jordon in 1856 [Bill, W. T. Hallett, Hartford, to Sam Colt, Hartford, November 1, 1861, Colt Papers, CHS]. The changes, described as creating a south front, were to be made by adding a "wing extending... forty-six feet to the east, with a large circular... Tower observatory... to be some seventeen feet in diameter... surmounted with an Oriental Dome in iron of open work and gilded. The Conservatory on the front... to be removed to the south side of the building and its place filled with a beautiful arcaded Terrace in stone" [*Hartford Daily Times*, November 7, 1861]. The changes were apparently still underway in May when Elizabeth Colt's father described "breaking through a wall of the mansion to connect new part," extending the hall, and redoing the stairway [William Jarvis, Hartford, to William Jarvis Jr., Eyria, Ohio, May 11, 1862, Jarvis Collection, CHS].

6. Pomeroy received a monthly salary equal today to about $100,000 per year, from mid-1856, when the only surviving accounts begin, until July 1857 when Armsmear was occupied [Colt PFAMC, Cash Book A, January 1856–March 1864, 19–57, private collection, Colt Archives, Wadsworth Atheneum, 874–77].

   Colt PFAMC, Bills and Receipts, 1854–55, np. CSL, box 1. Between December 1854 and March 1855 Jordan was paid for architectural services connected with building the armory. Jordan was paid an additional ninety dollars (today about five thousand dollars) in July 1856 for a limited range of architectural services, almost certainly connected with building Armsmear [Colt PFAMC, Cash Book A, January 1856–March 1864 (July 30, 1856), 17, private collection, Colt Archives, Wadsworth Atheneum, 874].

7. John Charlton, *Osborne House* (London: Her Majesty's Stationery Office, 1960), 3–9, 23. Special thanks to Hartford architect Jared Edwards for this very convincing analogy.

8. Andrew Jackson Downing, *A Treatise on the Theory and Practice of Landscape Gardening, Adapted to North America* (1850; reprint, New York: Dumbarton Oaks, 1991), 385–88.

9. Samuel Sloan, *The Model Architect* (Philadelphia: E. S. Jones, 1852), 12.

10. Barnard, *Armsmear*, 99.

11. Sherwood, "The Homes of America," 321–25.

12. Clive Aslet, *The American Country House* (New Haven: Yale University Press, 1990), 39.

13. Isabella Hooker, Hartford, to John Hooker, Geneva, May 29–June 1, 1857, Harriet Beecher Stowe Center, Manuscript Collection.

14. Sherwood, "The Homes of America," 321–25.

15. Colt PFAMC, Cash Book A, January 1856–March 1864 (December 10, 1856), 27, private collection, Colt Archives, Wadsworth Atheneum, 875.

16. Within months of its patent, Colt adopted "Patented Mastic Roofing Compound," patented by C. R. Mills of Detroit. (3/3/1857, U.S. patent no. 16,739) [Mills Patent Roofing Bill, J. G. Fitch to Sam Colt, September 4, 1857, Colt PFAMC, Bills and Receipts, CSL, box 2].

17. Bill, Noel and Sautter, New York, to Sam Colt, Hartford, November 5, 1857, Colt PFAMC, Bills and Receipts, CSL, box 2.

18. Bill of Sale, Ringuet-LePrince, Paris, to Sam Colt, November 7, 1856, as cited in Phillip Johnston, "Dialogues between Designer and Client: Furnishings Proposed by Leon Marcotte to Samuel Colt in the 1850s," *Winterthur Portfolio*, 19, no. 1 (spring 1984), 260–61.

19. Quotation is from untitled newspaper account, December 1867, Mary Morris, Social Scrapbook, CHS, vol. 2, 1–3.

   Colt PFAMC, Journal B, August 1863–June 1869 (December 30, 1865), 310, private collection, Colt Archives, Wadsworth Atheneum, 867. In 1865 Armsmear consumed 113,500 cubic feet of gas costing $457.40.

20. William Jarvis, Hartford, to William Jarvis Jr., Eyria, Ohio, April 1862, Jarvis Collection, CHS.

21. Johnston, "Dialogues between Designer and Client," 257, 270, 262–63, 265, 273. Marcotte's second proposal estimated the cost of furnishings at $11,019. Although Colt latter quibbled with the bill, this is probably close to the amount he finally paid.

22. Bill of Sale, Ringuet-LePrince, Paris, to Sam Colt, November 7, 1856, as cited in Johnston, "Dialogues between Designer and Client," 260–61.

23. Sam Colt, Moscow, to Leon Marcotte, New York, September 16, 1856, as cited in Johnston, "Dialogues between Designer and Client," 260–61.

   Elizabeth H. Colt, New York, to Mrs. C. L. F. Robinson, London, April 16, 1905, private collection, Colt Archives, Wadsworth Atheneum, 1441.

24. G. F. Wappenhaus, Berlin, to Sam Colt, London, January 5, February 25, 1857, Colt Collection, CHS. The lithophanes were manufactured by the Royal Porcelain factory in Berlin, which at the same time was commissioned to produce a lithophane portrait of Sam Colt, one of the most prized trophies among collectors of Coltiana.

25. Barnard, *Armsmear*, 74; "Armsmear Is Much Altered," probably *Hartford Daily Courant*, ca. 1909, Scrapbook, Hartford, np. Taylor Papers, CHS. Although more than fifty of the panes are preserved in the Colt Collection of the Wadsworth Atheneum, this contemporary newspaper account noted that the "best ones" were "removed by... Colt's heirs."

26. John L. O'Sullivan, Lisbon, Portugal, to Sam Colt, Hartford, October 26, 1861, in Colt PFAMC, "General Correspondence," 1861, np. CSL, box 10. O'Sullivan, an acquaintance of Colt's and the editor of the *Democratic Review*, coined the phrase Manifest Destiny, was a U.S. ambassador to Portugal, and ended up as a pro-South northerner in exile in Europe during the Civil War.

27. Colt PFAMC, Cash Book A, January 1856–March 1864 (December 17, 1856, January 9, 1857, May 4, 1857), 27, 47, private collection, Colt Archives, Wadsworth Atheneum, 875, 877. Payments are cited to A. T. Stewart and Co. for $439 and $637. William L. Wright and the British painter Henry J. Huxham were the principal painting contractors.

28. Estate Inventory, Samuel Colt, May 20, 1862, Hartford Probate District Court Records, CSL. Although most of the books cited in the inventory were in the house at the time of Colt's death, several were added by his wife, Elizabeth, the winter before the inventory was taken. The bulk of her library is retained as part of the collection at Armsmear, and several of the unitemized books in Colt's office at the armory are preserved in the Colt Collection at the CSL.

29. "Armsmear Is Much Altered," probably *Hartford Daily Courant*, ca. 1909, Scrapbook, Hartford, np. Taylor Papers, CHS.

30. Ibid. The pictures included a "large steel engraving," two views of Colt's Armory commissioned in 1857, one "large painting" of an unidentified subject, and the nine paintings of revolving rifles in action that Colt commissioned from the artist-explorer George Catlin.

   Nothing speaks more tantalizingly of character and aspirations than the selection of pictures in Colt's office. Colt's deep sentimentality is most revealingly suggested by this poignant panorama of hope. Here at a glance we see images of the great national experiment launched in faith, the patriotic yearning after the glory of the Revolution, an image of Christ weeping in the advent of last judgment, and a portrait of Rev. Joel Hawes, the son of a farmer and blacksmith, whose pastorate of Hartford's First Church (1818–62) spanned almost Colt's entire life. Colt was a communicant in Hawes's church into his adult life, and his grandfather Caldwell had been one of its most prominent and active parishioners [George Leon Walker, *History of the First Church in Hartford* (Hartford: Brown and Gross, 1884), 367–402]. Hawes's *Lectures to Young Men*, which sold several hundred thousand copies in numerous printings in the thirty years after it was compiled in 1828, was the kind of popular blueprint for Protestant morality and enterprise that resonates with maxims and principles like those Colt tried, and often succeeded, in making his own.

31. *Hartford Daily Times*, December 5, 1855. Colt commissioned the paintings, which were based on photographs and sketches made by the artist Joseph Ropes from the tower of Hartford's First Church during the Great Flood of 1854. A newspaper account reporting on the "panorama... intended to be placed in the sides of the cupola of Col. Colt's new house" is the earliest reference to the house, just then under construction.

32. William Jarvis, Hartford, to William Jarvis Jr., Eyria, Ohio, July 4, 1867, Jarvis Collection, CHS.

33. C. S. Mason to Milton Joslin, November 1855, Colt PFAMC, Bills and Receipts, CSL, box 2. Colt purchased "chrysanthemums, azaleas, acacias,... forsythias, begonias, plumbago, generia," and other annuals and perennials.

34. Sherwood, "The Homes of America," 321–25. In January 1856 Colt bought "Chinese primulas, rhyncospermus, kinidia, tournafortia, azaleas, forsythia, Large Begonia" [C. S. Mason to Milton Joslin, January 22, 1856, Colt PFAMC, Bills and Receipts, CSL, box 2] and in July posted his first payments to James Stubbins [Colt PFAMC, Cash Book A, January 1856–March 1864 (July 3, 1856), np., private collection, Colt Archives, Wadsworth Atheneum, F–9], described in the 1860 census as a forty-five-year-old native of Britain and a "gardener."

35. Robert Morris Copeland and Henry W. S. Cleveland, "A Few Words on the Central Park" (Boston, 1856), Winterthur Museum Library. Copeland was described as living in Lexington, Massachusetts, and Cleveland in Salem. The firm advertised their availability to "Furnish Plans for the laying out and improvement of Cemeteries, Public Squares, Pleasure Grounds, Farms and Gardens."

36. Charles E. Beveridge and David Schuyler, eds., "Creating Central Park, 1857–1861," in *The Papers of Frederick Law Olmsted*, ed. Charles Capen McLaughlin (Baltimore: Johns Hopkins University Press, 1983), 310, 316–37. In a bid for public recognition in his new home, Cleaveland published *The Public Grounds of Chicago: How to Give Them Character and Expression* (Chicago: C. D. Lakey, 1869). In 1856 Cleaveland collaborated on a book titled *Village and Farm Cottages: The Requirements of American Village Homes*, and in 1859 Copeland's *Country Life: A Handbook of Agriculture, Horticulture, and Landscape Gardening* featured plans "For Laying Out a Country Place of sixty acres" with hedges, fruit trees, grape arbors, strawberry patches, a kitchen garden, grapery, garden house, and picturesque walks, that is strongly reminiscent of the firm's designs for Colt [Henry W. Cleaveland, William Backus, and Samuel D. Backus, *Village and Farm Cottages: The Requirements of American Village Homes* (New York, 1856); Robert Morris Copeland, *Country Life: A Handbook of Agriculture, Horticulture, and Landscape Gardening* (Boston: John Jewett, 1859)].

37. Charles McIntosh, *The Book of the Garden* (London: William Blackwood and Sons, 1855), plates 2, 4, 21, 713. Although Elizabeth Colt's library contains a two-volume edition of this book published in 1865, the title is also listed in the 1862 inventory of Sam Colt's estate, together with A. J. Downing's *Cottage Residences* (New York: Wiley and Putnam, 1844), which survives in the Colt library at Armsmear.

38. Copeland and Cleveland, Boston, to Sam Colt, Hartford, February 13, 1856, Colt PFAMC, Bills and Receipts, CSL, box 1. In 1856 the first draft of Colt's will left to "Col. Luther Sargeant, now in charge of Armory in London—if he is in my employ when I die… a lot of land No. 29 in square C on condition that he build a good house on the same and occupy the same within three years of my death." Similar terms were provided for Elisha K. Root ("No. 31 in square C"), Colt's secretary, and J. Deane Alden ("No. 30 in square C"), and his chief financial officer, Milton Joslin ("No. 28 in square C"). By 1858 a codicil to Colt's will indicates that he had already abandoned this plan [Will, Samuel Colt, June 6, 1856, January 12, 1858, Hartford Probate District Court Records, CSL]. These lots are clearly delineated in an 1856 map of the homelot of Armsmear that predates its completion [Barnard, *Armsmear*].

39. Elihu Geer, *Geer's Hartford City Directory for 1860–61* (Hartford: E. Geer, 1860), 133, 150, 168, 213, 217.

40. *Hartford Daily Courant*, June 7, 1870.

41. Barnard, *Armsmear*, 96; Marcus Smith, "Map of the City of Hartford" (New York: Marcus Smith, 1850), CHS.

42. Sherwood, "The Homes of America," 323.

43. Ibid., 325; Barnard, *Armsmear*, 94, 107–8.

44. Sherwood, "The Homes of America," 325; Barnard, *Armsmear*, 94.

45. *Hartford Daily Courant*, February 15, 1864.

46. *Hartford Daily Times*, February 27, 1863.

47. "Artesian Wells for All Situations," *Scientific American*, March 7, 1857, 205. Special thanks to Dean Nelson of the Museum of Connecticut History for bringing this article to my attention. Timing, subject, and source suggest its probable role in inspiring Colt's experiment. Technology is as subject to fads and fashions as any other field, and apparently the late 1850s witnessed a vogue for artesian wells. When the work began at Colt's the *Hartford Daily Times* [January 22, 1862] noted that an "Artesian Well in Grenelle in Paris" was commenced in 1833 and that in Bavaria one commenced in 1850 reached 1,878 1/2 feet before "water burst up" in a "fifty-eight-foot jet."

48. *Hartford Daily Times*, February 13, 1868.

49. Ibid., January 22, March 18, 1862.

50. Ibid., January 22, February 20, March 18, 1862.

51. Ibid., June 10, 1868. Nothing in the life of Armsmear was reported more assiduously by the Hartford newspapers than Sam's and Elizabeth's search for the holy grail of cheap, warm, subterranean water. After noting Elizabeth's resuming the search in 1868 [*Hartford Daily Times*, February 13, 1868], the *Times* monitored her progress, from 700 feet to 956 [3/7/68] to 1,000 [4/28/68] and eventually to 1,400 feet [8/18/68]. After giving up, she resumed drilling in 1874, with the *Times* dutifully announcing on May 5 that at about 1,500 feet she had reached "water… clear and sparkling." Why she bothered remains a mystery.

52. Batterson's Monumental Works, Bill of Sale, May 30, 1861, Colt Correspondence, CHS; Batterson's Monumental Works, Bill of Sale, June 1, 1861, Colt Correspondence, CHS. Neither the Diana nor the Apollo survive, but in the spring of 1861 Colt was charged $2,189 (today, about $60,000) for supplying and installing the figures. Although Batterson was renowned for employing gifted staff sculptors, he had also been to Italy the year before on business involving Colt and probably procured these reproductions there.

53. Barnard, *Armsmear*, 104. August Kiss (1802–65) was a sculptor of the German school who attended the Academy of Berlin and specialized in subjects of mythic and romantic symbolism, including hunt scenes and a figural group titled *Faith, Love, and Hope* [E. Benezit, *Dictionnaire Critique et Documentaire des Peintres, Sculpteurs, Dessinateurs et Graveurs* (Paris: Librairie Grund, 1976), 6:229]. Kiss's *Amazon* appears to have been a popular culture sensation. It inspired a New York composer, Horatio Hewitt, to write "The Amazon Polka" and illustrates the cover of Hewitt's published sheet music.

54. Colt, "Memoir," *Armsmear*, 295, 110.

55. Barnard, *Armsmear*, 109. Although only one of the tablets is documented [Batterson's Monumental Works, Bill of Sale, November 1860, Colt Correspondence, CHS], Batterson had been personally involved with Colt in preserving and acquiring for Hartford the final work of the sculptor Edward Bartholomew, and he was one of only two monument makers in Hartford with the capability of producing the marble components of the "grove of graves." Elizabeth completed the site, burying Sam and their second daughter, Harriet, there and hiring Theo. M. Lincoln and Co. to design and install an ornamental railing [Colt PFAMC, Cash Book A, January 1856–March 1864 (July 21, 1862), 287, private collection, Colt Archives, Wadsworth Atheneum, 884]. Lincoln was a Hartford-based manufacturer of "every description of Plain and Ornamental Iron railing" located at 53 Arch Street.

56. Estate Inventory, Samuel Colt, May 20, 1862, Hartford Probate District Court Records, CSL. One of the most inscrutable details of Colt's inventory is the appraisal of his beloved horse Shamrock, appraised "dead" at his East Hartford farm at five dollars.

57. Colt PFAMC, Cash Book A, January 1856–March 1864 (August 30, October 29, November 27, 1858), 121, 129, 133, private collection, Colt Archives, Wadsworth Atheneum, 880, 881; Bill, John Welles to Mr. Hall, September 2, August 31, 1857, Colt PFAMC, Bills and Receipts, CSL, box 2.

58. William Jarvis, Hartford, to William Jarvis Jr., Eyria, Ohio, November 30, 1865, Jarvis Collection, CHS. Elizabeth's father reported the "death of three of our swans," almost like a member of the family. The upper lake was eventually filled with them, and their significance to the family was such that Elizabeth hired a photographer to make souvenir stereo photographs of their nest, which also appears as an illustration in Barnard's *Armsmear*, 98.

59. Colt PFAMC, Cash Book A, January 1856–March 1864 (October 31, 1857), 71, private collection, Colt Archives, Wadsworth Atheneum, 878.

60. Barnard, *Armsmear*, 82; *Hartford Daily Times*, April 24, 1869, April 8, 1879, April 5, 1888. The accounts of "Mrs. Colt's deer" bounding off "through the meadow, over the dike, and finally up to the avenue… on the side walk… [and] all over the south end of the city" and of the time when a "pet deer which escaped from Mrs. Colt's," was "killed by a hunter," in the "interest of humanity," after being discovered on the "river… struggling to land from a piece of floating ice" with its leg broken are the most colorful of many published accounts. Deer also got away in 1865 and 1866.

61. *Population Schedules of the 8th Census of the United States, 1860* (Washington, D.C.: National Archives, 1964), microfilm no. 41, "Hartford," 221–22. At the time of the census, Stubbins was forty-five years old and living in the gardener's house on the grounds of

Armsmear; Maltman was thirty. Both were described as of English birth. With the outbreak of the Civil War, Stubbins resigned and returned to England [*Hartford Daily Times*, September 10. 1861]. Elizabeth's father alluded to the "great scampering here of those... liable to be drafted. Nearly all of the men employed on Elizabeth's grounds and in her green-houses disappeared... bound for England or Ireland" [William Jarvis, Hartford, to William Jarvis Jr., Eyria, Ohio, August 12, 1862, Jarvis Collection, CHS]. These facts are interesting for documenting the international character of the Colts' garden work force.

62. Barnard, *Armsmear*, 84.

63. C. S. Mason to Milton Joslin, January 22, 1856, Colt PFAMC, Bills and Receipts, CSL, box 2; *Hartford Daily Times*, July 18, 1862.

64. Barnard, *Armsmear*, 84.

65. Estate Inventory, Samuel Colt, May 20, 1862, Hartford Probate District Court Records, CSL.

66. *Hartford Daily Times*, August 4, 1858. Colt's pineapples were exhibited in the exhibition of 1858.

67. Barnard, *Armsmear*, 83–84, 86; Sherwood, "The Homes of America," 323; *Hartford Daily Times*, April 25, 1861.

68. *Hartford Daily Times*, June 23, 1859, July 18, 1862; review of the annual horticultural exhibition.

69. *Hartford Daily Times*, July 19, 1862, March 30, 1861.

70. William Jarvis, Hartford, to William Jarvis Jr., Eyria, Ohio, April 21, 1862, Jarvis Collection, CHS.

71. Estate Inventory, Samuel Colt, May 20, 1862, Hartford Probate District Court Records, CSL.

72. *Hartford Daily Times*, August 3, 1866. In 1870 the Courant acknowledged the public's "debt" to "Mrs. Colt... for the display of... the rarest productions of her beautiful greenhouse" [*Hartford Daily Courant*, September 29, 1870]. Although a horticultural committee of the Hartford Agricultural Society was formed as early as 1851, it was apparently not until 1863 that the Horticultural Society held its "first exhibition... at the rooms of the society" in the Atheneum Building [*Hartford Daily Courant*, June 25, 1863].

73. Sherwood, "The Homes of America," 321–25.

74. *Hartford Daily Times*, May 11, 1868.

75. *Hartford Daily Courant*, April 18, 1870; *Hartford Daily Times*, December 24, 1869.

76. *Hartford Daily Times*, March 26, 1879, December 29, 1868.

77. Colt PFAMC, Cash Book A, January 1856–March 1864 (June 2, 1862), 281, private collection, Colt Archives, Wadsworth Atheneum, 884.

78. Richard H. Jarvis, Hartford, to William Jarvis Jr., Eyria, Ohio, November 20, 1879, Jarvis Collection, CHS.

79. *Hartford Daily Times*, December 31, 1883.

80. Sherwood, "The Homes of America," 323.

81. Barnard, *Armsmear*, 90–91.

82. Sherwood, "The Homes of America," 321–25.

83. Barnard, *Armsmear*, 84.

84. Edward Tuckerman Potter, Ledger, ca. 1868–75 (probably 1871), 105, Potter Papers, Avery Architectural Library, Columbia University. Potter visited Armsmear, probably on business related to the remodeling in February 1871, when Elizabeth Colt's father described Potter's father, an Episcopal bishop, as "my old friend and classmate" [William Jarvis, Hartford, to William Jarvis Jr., Eyria, Ohio, February 24, 1871, Jarvis Collection, CHS].

The themes of the sculptural busts incorporated into the capitals of the pilasters are the musician, sculptor, hunter, miner, explorer, chemist, physicist, communication, statesman, the muse of arts, printing, medicine, theology, a knight or warrior, painter, playwright, fisherman, sailor, female agrarian, male agrarian, and the four continents, with only Europe posed in modern garb.

85. Since about 1910 Armsmear has functioned as a retirement community for Christian women. The house has been extensively altered to accommodate its institutional mission, and Elizabeth Colt's library is one of the few public rooms that has remained somewhat as it was. It now functions as a reading and reception room for the residents. Almost all of the four hundred–plus books in the collection are inscribed with the owner's name and date of purchase. Many were gifts. Only a few belonged to Sam Colt, and more than half the titles cited in his estate inventory are no longer in the collection. Perhaps half the collection was acquired in 1862, suggesting that the formation of a book collection was a project as much as a process and one that may have served an emotionally therapeutic function for a woman who lost three children and a husband in three and a half years.

86. These eighteenth-century style vases, allegedly exhibited in Paris in 1867, are signed "Bertren" who supposedly decorated them in 1789. In spite of the fact that Elizabeth Colt paid $3,500 for the pair (today's equivalent of more than $100,000), they do not predate 1865 and, depending on how represented, were possibly fakes.

87. Efforts to learn more about the artist and subject matter come up empty. A marble figure of the same image at the National Historic Park in Jamestown, Virginia, suggests the subject may be Pocahontas. In any event, it is a remarkable work of art, which would be all the more interesting if we knew where and when she bought it and what it meant to her; it is situated in the most prominent location in the most public room at Armsmear.

88. Untitled newspaper account, December 1867, Mary Morris, Social Scrapbook, CHS, vol. 2, 1–3.

89. Sherwood, "The Homes of America," 321–25.

90. Untitled newspaper account, December 1867, Mary Morris, Social Scrapbook, CHS, vol. 2, 1–3.

91. William Jarvis, Hartford, to William Jarvis Jr., Eyria, Ohio, December 9 and 11, 1867, Jarvis Collection, CHS.

92. *Hartford Daily Times*, December 7, 1867.

93. Sherwood, "The Homes of America," 324.

94. *Hartford Daily Times*, January 8, 10, 17, 1870. Slaviansky's orchestra was the sensation of the social season that winter. He and his orchestra first performed at Allyn Hall on January 4. They returned for a second engagement sponsored by the Young Men's Institute on the twelfth and performed a third time in Hartford to a "full house" at Allyn Hall on the eighteenth. Slaviansky was described as a "tenor... one of the sweetest ever heard in Hartford," and his orchestra performed a repertoire consisting of Russian national hymns, polkas, the "Volga Sailor's Song," the "Ivan Song," and various Russian, Czech, and Polish folk songs [*Hartford Daily Courant*, January 5, 13, 15, 19, 1870].

95. *Hartford Daily Times*, February 4, 1870.

96. Ibid., November 24, 1879.

97. *Hartford Daily Courant*, November 25, 1879; the *Hartford Connecticut Post* November 29, 1879, made a point of noting that "singing quadrilles... introduced... at Newport... [the] past summer... [were here] rendered in Hartford for the first time."

98. *Hartford Connecticut Post*, November 29, 1879. The Samuel Clemenses, former governor Joseph R. Hawley, Charles Dudley Warner, and Gen. William B. Franklin were the glitterati of the Hartford contingent of a guest list estimated at nine hundred [Olivia Langdon Clemens, Hartford, to Olivia Lewis Langdon, Elmira, New York, November 30, 1879, University of California, Berkeley].

99. *Hartford Daily Times*, April 9, 1888.

100. William Jarvis, Hartford, to William Jarvis Jr., Eyria, Ohio, July 16, 1862, November 10, 1868, Elizabeth Miller Jarvis, Hartford, to William Jarvis Jr., Eyria, Ohio, August 7, 1873, September 6, 1873, September 4, 1874, October 10, 1876, July 6, 1878, March 3, 1879, April 23, 1879, April 24, 1880, November 30, 1880, Jarvis Collection, CHS.

101. Samuel Clemens, Paris, to Elizabeth Hart Colt, Hartford, February 15, 1895, Mark Twain Papers, Watkinson Library, Trinity College, Hartford.

102. Elizabeth Miller Jarvis, Hartford, to William Jarvis Jr., Eyria, Ohio, January 3, April 24, 1880, Jarvis Collection, CHS.

103. George W. Bragdon, "'Dauntless' Models Recall 'Colly' Colt, City's Famed Yachtsman," *Hartford Daily Times*, May 20, 1938, 28.

104. *Hartford Daily Times*, August 25, 1879.

The discovery of what appears to be a cock-fighting ring located upstairs and out of the way in the carriage and horse barn on the grounds of Armsmear suggests that Caldwell engaged in the most violent of gambling rituals on the grounds of his mother's house. Whether the ring predates Sam's death, it certainly predates 1875 and is too formally elaborated a space to have been a casual endeavor.

105. Leverett Bradley, "Address," in *In Memoriam: Samuel Colt and Caldwell Hart Colt* by Rev. Samuel Hart (Springfield: Clifton Johnson, 1898), np.

106. Bragdon, "'Dauntless' Models Recall 'Colly' Colt," 28.

107. *Newport Mercury*, August 10, 1844, August 11, 1868.

108. *Hartford Daily Times*, August 2, 1871.

109. Ibid., January 27, 1894.

110. The *Dauntless*, originally called the *L'Hirondelle*, was first owned by S. Dexter Bradford Jr., who sold it to *New York Herald* publisher James Gordon Bennett in 1868. It won its first race against the *Vesta*

in 1866, sailing from Gaunt Head to Sandy Hook Lightship, and in 1871 won a celebrated race against the *Dreadnought* [Henry A. Mott, ed., *The Yachts and Yachtsmen of America*, 3 vols. (New York: International Yacht Publishing, 1894), 1:62, 150]. The ship was built in Mystic, Connecticut, by Forsythe and Morgan in 1865 and was designed by J. B. Van Duesen of Noank, Connecticut. It was 130 feet long with 8,024 square feet of sails. Caldwell bought it in 1881 or 1882 from John R. Walker of New York, who owned it for two years after purchasing it from Bennett [Bill of Sale, John R. Walker, New York, to Caldwell H. Colt, Hartford, July 19, 1882, private collection, Colt Archives, Wadsworth Atheneum, 1446].

111. Mott, *Yachts and Yachtsmen of America*, 1:125. The cost of the *Dauntless* was estimated at $70,000, and perhaps his four other ships combined cost that much. A crew of sixteen, each paid wages estimated at $30 per month, totals about $5,000 per year, not including supplies, new sails, and repairs. Figuring average annual expenses at about $7,500 per year for fifteen years, $140,000 invested in ships, and at least another $2,500 a year in club memberships and entertaining, Caldwell's lifetime investment in yachting may have been as high as $300,000.

112. *Hartford Daily Times*, January 23, October 11, 1884, June 30, 1886.

113. Ellsworth S. Grant, *The Colt Legacy: The Story of the Colt Armory in Hartford, 1855–1980* (Providence, R.I.: Mowbray, 1982), 52, 54–55. Almost all of the Colts' modern biographers have repeated stories about Caldwell's wild parties and womanizing, the most famous being that he was murdered by the husband of a woman he was involved with. The genealogy of these stories is vague and no documentary evidence has ever been cited in connection with them, which is not to say that Caldwell was not something of a playboy, but that the reports are almost certainly exaggerated.

114. In 1864 Elizabeth Colt invited the Hartford photographer DeLamater to photograph the aftermath of the great armory fire for sale commercially and commissioned extensive photographic work around the grounds in 1866 to aid in the preparation of woodcut illustrations for *Armsmear*. In 1867 a newspaper account of "beautiful stereoscopic views of... 'Armsmear'... by DeLamater... taken from the top of... Colt's factory... [of] Mrs. Colt's residence and grounds" indicates another round of expensive picture taking, and in 1869, after the completion of the Church of the Good Shepherd, the Hartford photographers Prescott and White produced and sold a set of stereo pictures entitled "The Memorial Church of the Good Shepherd, Erected by Mrs. Samuel Colt" [*Hartford Morning Post*, December 13, 1867]. Almost forty different subjects involving Armsmear, Colts Armory, and the Church of the Good Shepherd are known.

115. Martha J. Lamb, ed., *The Homes of America*, (New York: D. Appleton, 1879), 177–83; Sherwood, "The Homes of America," 321–25.

116. Estate Inventory, Samuel Colt, May 20, 1862, Hartford Probate District Court Records, CSL.

117. *Hartford Daily Courant*. April 26, 1906, as cited in Mary Lincoln, Scrapbook, np., Watkinson Library, Trinity College, Hartford.

118. *Hartford Daily Times*, November 11, 1881.

119. Anderson was born and died in the Hart family ancestral town of Saybrook, Connecticut, the primary source of Elizabeth Hart Colt's illustrious pedigree and personal wealth. When she died, Elizabeth's cousin Samuel Hart of Middletown was the executor of her estate. It is possible that Robinson was related to the woman Elizabeth Colt described as "gold old Rose, a cook, and slave-child in the family of Gen. William Hart" [Colt, *A Memorial of Mrs. Elizabeth Miller Jarvis*, 8]. She died in Hartford at the age of eighty-eight as a servant to Elizabeth's Aunt Hetty [William Jarvis, Hartford, to William Jarvis Jr., Eyria, Ohio, October 24, 1866, Jarvis Collection, CHS]. Special thanks to Judy Johnson at the CHS for her research on Miranda Anderson whose childhood sampler is in the society's collections.

120. William Jarvis, Hartford, to William Jarvis Jr., Eyria, Ohio, October 21, 1862, Jarvis Collection, CHS.

121. *Population Schedules of the 8th Census of the United States, 1860* (Washington, D.C.: National Archives, 1964), microfilm no. 41, "Hartford," 221–22.

122. Elizabeth H. Colt, New York, to Mrs. C. L. F. Robinson, London, April 16, 1905, private collection, Colt Archives, Wadsworth Atheneum, 1441.

FRONTIERS OF CIVIC CULTURE

1. Barnard, *Armsmear*, 394–95.

2. *Hartford Daily Times*, March 30, 1857. Jewett "spent three years on a whaling vessel" and later studied painting in Europe [Obituary, *Hartford Daily Courant*, December 28, 1864].

    Hartford Daily Times, April 11, 1859. In 1859 Fitch studied painting in Germany "under the celebrated Richard Zimmerman." Elizabeth Colt bought Fitch's view of Mount Washington about 1866.

3. Ibid., January 7, 1846. This was the first advertisement where Batterson announced he had "opened an establishment at 323 Main St. in Hartford."

4. *Hartford Daily Times*, November 21, 1849.

5. Ibid., October 13, 1853. Batterson won a silver medal for a "marble font" submitted for judgment at the first Agricultural Society exhibition after he moved to Hartford in 1846; in 1853 he took more prizes and exhibited work in half a dozen categories.

6. Ibid., August 30, September 6, 1854.

7. Wilson, *Samuel Colt Presents*, 93. Colt presented an 1855 sidehammer pocket pistol to G. Argenti, signifying their relationship.

    The monument is signed by Batterson, and Argenti clearly had a hand in finishing it, but it is not certain if the firm produced the monument or merely added such customized components as its inscriptions and the Colt coat of arms. A near-twin to the monument (Angelica Charlotte Yeatman, 1849), recently discovered in the Bellfontaine Cemetery in St. Louis is illegibly signed by a "New York" monument cutter, probably Muldoon, Bullett and Co. The date of its completion is also undetermined and could be as early as 1853 or 1854.

8. *Hartford Daily Times*, December 27, 1871. Batterson's "energy and genius" were credited with increasing the number of workmen and bringing "the money of Cincinnati, St. Louis, New York and many other places" to Hartford.

9. Payments of almost two thousand dollars are posted to J. G. Batterson's account between July 1856 and March 1858, indicating that he supplied construction material and finished elements for the armory, mansion, and worker's housing. Although not specifically itemized, the decorated mantlepieces inside Armsmear represent Batterson's high-end architectural work [Colt PFAMC, Cash Book A, January 1856–March 1864, private collection, Colt Archives, Wadsworth Atheneum].

10. Batterson's Monumental Works, Bill of Sale, November 1860, Colt Correspondence, CHS.

11. Henry Deming, Oration, *Hartford Daily Times*, November 20, 1857. "We have gathered together this evening... to pay... tribute to American Art.... There is no country where the cultivation of the Fine Arts... is more needed... and no State... where their mission to refine, to elevate, and to spiritualize is more imperiously demanded than in Connecticut.... In our haste to accumulate wealth... we grossly neglect the sentiments... to which the Fine Arts are ministering spirits.... [And so tonight we are] honored ... by a guest, who seven years ago, left his native city... alone, friendless, poor... with nothing to animate... him but an indomitable spirit.... He returns... victorious in the maturity of his powers... [and as an example of] a moral heroism in surmounting the barriers that blockade the pathway of poor and friendless genius."

12. *Hartford Daily Times*, November 19, 1857, May 31, 1858.

13. Ibid., April 11, 1861.

14. William G. Wendell, "Edward Sheffield Bartholomew, Sculptor," *Wadsworth Atheneum Bulletin* (winter 1962), 9.

15. *Hartford Daily Times*, September 15, 1858.

16. Samuel Colt to Hiram Powers, September 10, 1858, Hiram Powers Papers, Archives of American Art (microfilm, roll 1139, 1093–94). Colt wrote: "I take great pleasure in introducing... my friend J. G. Batterson Esq. who is delegated to proceed to Rome and take charge of the effects of our late friend and townsman Bartholomew. Mr. Batterson is a young gentleman.... His standing as a monumental architect—ranks among the first in this country—he having erected in the different cities of the Union, several monuments in commemoration of distinguished men."

17. *Hartford Daily Times*, April 11, 1861.

18. Ibid., December 5, 1855. The four-panel panoramic view was originally installed in the observatory at Armsmear. The pictures were based on photographs [CSL] and sketches made from the tower of Center Church during the flood of 1854.

19. Ropes first appears as an exhibitor in the annual fair of the Hartford County Agricultural Society in 1851 [*Hartford Daily Times*, October 23, 1851]. In 1852 he exhibited paintings there with such titles as *Mountain Cascade* and *Jesus and Companions* [*Hartford Daily Times*, October 1, 5, 1852]. Ropes's specialty was landscape painting and while in Hartford he taught drawing and painting in classes most likely hosted by the Hartford Arts Union [H. W. French, *Art and Artists in Connecticut* (Boston: Lee and Shepard, 1879), 79].

20. William Jarvis, Hartford, to William Jarvis Jr., Eyria, Ohio, July 18, 1859, Jarvis Collection, CHS; a bill from Wilson to Colt, November 3, 1860, for $240, Colt Papers, box 8, CHS.

21. The *United States Magazine* story "Repeating Fire-Arms: A Day at the Armory" reiterated Colt's largely unfulfilled aspirations as an art patron and prompted the magazine's founder, John L. O'Sullivan, then in the diplomatic service in Lisbon, Portugal, to write Colt offering him a "very fine collection of pictures," assembled in Europe. Recalling the "many years since we met," O'Sullivan explained how "I have understood that you use elegantly & worthily the great wealth yielded by your great invention... and that you are a liberal purchaser of works of art." The collection of predominately religious subjects and European cityscapes allegedly included works by Zurbarán, Van Dyke, Vermeer, and other prominent old masters, including "a splendid Murillo, of a virgin & child" [John L. O'Sullivan, Lisbon, Portugal, to Sam Colt, Hartford, October 26, 1861, Colt PFAMC, Incoming Correspondence, CSL, box 10]. John Lewis O'Sullivan was a colorful and ardent expansionist, immortalized for coining the phrase Manifest Destiny. He was a journalist and diplomat who lived abroad during the Civil War, voicing Southern sympathy in pamphlets and urging the British to recognize the Confederacy. He established the *United States Magazine and Democratic Review* in 1837, and in 1854 he was appointed foreign ambassador by President Pierce, serving alternately in Lisbon, London, and Paris until 1871 [*Dictionary of American Biography* (New York: Charles Scribner Sons, 1934), 14:89].

22. *Hartford Daily Post*, July 26, 1866; *Hartford Daily Courant*, June 2, 1866; *Hartford Daily Times*, May 22, 1862. Estimates of Colt's wealth ranged from "over three million" to as high as "five million." As early as 1853, reliable sources estimated his earnings at "from $200,000 to $400,000 per year" [R. G. Dun and Co., Report, Hartford, August 1853, 15:346, R. G. Dun and Co. Collections, Baker Library, Harvard University Graduate School of Business Administration]. In reconciling the accounts and earnings after his death, the company appraised Colt's salary and travel expenses (not including dividends) at today's equivalent of $300,000 per year [Colt PFAMC, Journal A, 1856–63 (December 31, 1862), 567, private collection, Colt Archives, Wadsworth Atheneum, 937]. The May 1862 appraisal of Colt's estate arrived at a total of $3,201,086 [Estate Inventory, Samuel Colt, May 20, 1862, Hartford Probate District Court Records, CSL].

23. *Hartford Daily Courant*, January 5, 1870. As late as 1870, as measured by the property taxes paid, Elizabeth Colt's personal wealth and the assets she managed for her son, Caldwell, placed them first among the city's taxpayers. In the late 1870s Maj. James Goodwin, the insurance king of Hartford, inched ahead.

24. Batterson founded the Travelers Insurance Company in 1863 and for the next twenty years presided over growing empires in insurance, statuary monuments, and building contracting. At the time of his death in 1901, the *New York Herald* remembered him as one of the nation's "greatest sons," of whom "few more remarkable ever lived in Connecticut." By the 1930s the Travelers, the first accident insurance company in America and a pioneer in the lucrative field of automobile insurance, had assets of more than $600 million and was described as "the world's largest" insurance company [Charles W. Burpee, *The Story of Connecticut* (New York: American Historical Company, 1939), 2:897–98].

25. *Hartford Daily Courant*, June 6, 1864. The Atheneum's trustees "appointed a committee to confer with James G. Batterson, respecting a collection... purchased by him in Europe for the Atheneum."

26. Ibid., June 21, 1864.

27. *Catalogue of Original Oil Paintings Both Ancient and Modern* (Hartford: Case, Lockwood, 1864).

28. Horace Bushnell, "How to be a Christian in Trade," *Sermons on Living Subjects* (New York: Scribner, Armstrong, 1873), 265.

29. George M. Fredrickson, *The Inner Civil War* (New York: Harper and Row, 1965), 98–101, 211–13.

30. *Hartford Daily Courant*, February 20, March 15, 1864.

31. Ibid., April 25, 1864. The amount raised was about $1.1 million, at a time when skilled artisans were earning $2.50 per day.

32. *Hartford Daily Times*, May 4, 1861. The Colts donated a revolver case and floral bouquets to the Ladies' Fair at Union Hall.
    Hartford Daily Times, November 28, 1863. Also featured were scenes from the coronation of Alexander II, a Hartford favorite; an artist's studio; a volunteer's departure; and a "sad and thrilling scene" titled "Recognition on the Battle Field," in which a young woman beholds the body of her dead fiancé.
    Hartford Daily Courant, March 23, April 18, April 12, May 2, 1864.

33. *Metropolitan Fair in Aid of the United States Sanitary Commission* (New York: Charles O. Jones, 1864), 4–5, 7; *Hartford Daily Courant*, April 18, 1864; *New York Times*, April 6, 1864; *Hartford Daily Times*, April 4, 29, 1864.

34. *The Spirit of the Fair* (New York: John F. Trow, 1864), 29. August Belmont opened his gallery to the public during the Metropolitan Fair of 1864.

35. *Hartford Daily Times*, December 22, 1862; Annett Blaugrund, "The Tenth Street Studio Building: A Roster, 1857–1895," *American Art Journal* (spring 1982).

36. Mrs. Samuel Colt, Hartford, to Frederic E. Church, New York, December 18, 1863, Olana State Historic Site, Hudson, New York. Special thanks to Karen Zukowski for sharing this information.

37. *Hartford Daily Courant*, February 24, 1865. Elliot, who was paid three thousand dollars for the portrait of Sam and possibly more for the portrait of Elizabeth and Caldwell, painted many of antebellum America's most celebrated personalities, including New York governor Horatio Seymour, President Zachary Taylor, George Peabody, and William Cullen Bryant.

38. William Jarvis, Hartford, to William Jarvis Jr., Eyria, Ohio, March 10, 1865, Jarvis Collection, CHS. Elizabeth's father wrote that Elliot was then "painting a full length portrait... from the several pictures, photographs, and busts taken... in his lifetime." Elizabeth is known to have shipped medals and related memorabilia to the illustrators in New York while preparing the book *Armsmear*, and it is likely that she sent small objects and pictures of the furniture and cabinets—not artist's props but real items that belonged at Armsmear—to Elliot while he was at work on Colt's portrait.

39. *Hartford Daily Courant*, February 24, 1865; William Jarvis, Hartford, to William Jarvis Jr., Eyria, Ohio, March 10, April 26, 1865, Jarvis Collection, CHS. The portrait of Sam Colt, which was probably completed by the first of March, was exhibited at Bolles and Roberts, a store in Hartford, before being delivered to Armsmear, where it was unveiled at a reception for friends and armory workmen on March 8 [*Hartford Daily Courant*, March 9, 1865].

40. *Hartford Daily Courant*, May 26, 1865.

41. Thanks to Elizabeth M. Kornhauser whose forthcoming catalog of the Wadsworth Atheneum's American painting collection will contain detailed catalog information about these pictures.

42. Mrs. Samuel Colt, Hartford, to Frederic E. Church, Hudson, New York, May 15, October 26, 1866, Olana State Historic Site, Hudson, New York. Special thanks for permission to reproduce. When she wrote Church in October she complained about having shown Elliot "the covering I have for the walls," who apparently thought the color was "a little *too green*."

43. Mrs. Samuel Colt, Hartford, to Frederic E. Church, Hudson, New York, May 15, 1866, Olana State Historic Site, Hudson, New York. Special thanks for permission to reproduce.

44. Ibid., October 26, 1866. Special thanks for permission to reproduce. This letter contains the first reference to "your picture for me," Church's *St. Thomas in the Vale, Jamaica*, also known as *Vale of St. Thomas, Jamaica*.

45. *Hartford Daily Times*, November 16, 1869. The painting is dated 1867 but, as Elizabeth Kornhauser has noted [Wadsworth Atheneum, forthcoming collection catalog], was "retouched" and exhibited at Goupil's art gallery in New York in 1870 [*Hartford Daily Times*, April 11, 1870], suggesting that, contrary to the November notice, the painting may have remained in New York until 1870.

46. "Armsmear Is Much Altered," probably *Hartford Daily Courant*, ca. 1909, Scrapbook, Hartford, np. Taylor Papers, CHS. Written half a century after the picture gallery was completed, this is the best and one of the few verbal descriptions of the space on record.

47. Sherwood, "The Homes of America," 323; Lamb, *The Homes of America*, 181.

48. "A Ramble among the Studios of New-York," *Hartford Daily Courant*, March 22, 1865.

49. *Hartford Daily Courant*, March 17, 1868.

50. William Jarvis, Hartford, to William Jarvis Jr., Eyria, Ohio, March 29, 1865, Jarvis Collection, CHS. Elizabeth's father made special note of Church's family tragedy in corresponding with his nephew.

51. Sanford Gifford, New York, to Mrs. Samuel Colt, June 5, 1866, Colt Manuscripts, Wadsworth Atheneum.

52. John F. Kensett Papers, Archives of American Art, 1867, roll no. N68–85, frame 483. Special thanks to Elizabeth Kornhauser for discovering and sharing this documentation.

53. Hermann Fuechsel, New York, to Elizabeth Hart Colt, December 8, 1866, Colt Manuscripts, Wadsworth Atheneum.

54. *Hartford Daily Times*, June 6, 1879, November 14, 1881. Ongley was one of only three Hartford painters from whom Elizabeth Colt is known to have bought pictures. Although he was described as having a studio at 281 Main St. in 1879 and 1881, he is not listed among Hartford artists in the city directory the year the picture was signed, having already relocated to Utica, New York, where he spent most of his career catering to summer travelers in the Adirondacks.

55. Special thanks to Elizabeth Kornhauser for sharing this information. The relationship between Church and Cole will be explored in considerable depth in her forthcoming catalog of the Wadsworth Atheneum's American paintings collection.

56. *Hartford Daily Times*, September 20, 1867. In 1867 Elizabeth Colt loaned paintings by Jared Flagg, Petrus van Schendel, and Louis Gallait to the first annual art exhibition at Yale's new gallery.

57. In addition to family and the American land, the prevailing themes of Elizabeth Colt's picture gallery were rural life (5), the romance of Italy (4), the private world of books and learning (4), the mystery and allure of the Middle East (3), history as theatrical tableau (5), the nobility of the poor (3), and innocent girlhood (7), all, to varying degrees, windows into aspects of her life and personal experience. Not surprisingly, most of the portrait figures are female and in several, notably Jared Flagg's *Grandfather's Pet*, van Schendel's *Candlelight*, and Anton Dieffenbach's *Dreaming*, we see the expectant gaze of the dreamer.

58. *Hartford Daily Times*, April 17, 1868.

59. *Hartford Daily Courant*, February 11, 1868.

60. Neil Harris, *The Artist in American Society* (1966; reprint, Chicago: University of Chicago Press, 1982), 285.

61. *Hartford Daily Times*, January 4, 1884. Railing against collections built "by sheer force of money," the editorial went on to describe the gallery of William H. Vanderbilt in New York as a "shoddy, show-shop of unlimited wealth" and an example of "all that money could purchase... plastered on the walls... not so much [for the] picture as... the signature."

62. In Hartford, James Batterson, having been burned in his attempt to promote "old masters," assembled a new collection in 1867 devoted almost exclusively to Europe's "modern masters" [*Hartford Daily Courant*, January 23, February 19, 1868]. The bandwagon, however, did not move in only one direction. Across the state in Norfolk, Robbins Battell, described as "constantly seeking ways in which to gratify and cultivate the finer tastes of his townspeople," built a picture gallery during the 1870s comprising a veritable atlas of northeastern scenery, from Lake George and the Catskills to the White Mountains and Mount Katahdin in Maine, with works by such notable American painters as Worthington Whittredge, John Kensett, William Hart, A. D. Shattuck, J. M. Heade, Samuel Coleman, Sanford Gifford, Asher Durand, George Innes, John Cropsey, Frederic Church, and Albert Bierstadt [*Winsted Herald*, August 11, 1882]. In St. Johnsbury, Vermont, the industrialist Horace Fairbanks built a collection during the late 1860s and early 1870s that was also predominantly American, including Albert Bierstadt's gigantic *Domes of the Yosemite*, purchased in 1872 at the firesale price of $5,100 from the estate of its first owner [Patricia C. F. Mandel, "American Paintings at the St. Johnsbury Athenaeum in Vermont," *Antiques*, April 1980, 868–79]. The Fairbanks Gallery, which is the only collection of its scope and period still intact in the United States, includes works by Samuel Colman, Jasper Cropsey, Worthington Whittredge, Jervis McEntee, Thomas Waterman Wood, Arthur Fitzwilliam Tait, John G. Brown, Sanford Gifford, Asher Durand, William and James Hart, Seymour Guy, and others.

63. *Hartford Daily Times*, April 30, 1873, May 13, 1876.

64. William Jarvis, Hartford, to William Jarvis Jr., Eyria, Ohio, July 16, August 5, 23, 30, 1868, Jarvis Collection, CHS, Jarvis Collection, CHS.

65. *Hartford Daily Times*, July 12, 1876; "musical-boxes from Switzerland... Dutch lapidaries... repoussé work of Russia and Denmark,... [the] big porcelain vases... of Japan... the colored glass of Bohemia... silverware... pottery, glass, malachite... ermines ... bear robes [from]... Sweden and Norway... Turks, Chinamen, Frenchmen,... Danes, Germans, Italians—all are here... [in] picturesque national costumes."

66. Ibid., October 17, 1877. The *Times* described "the Ceramic Rage... here in Hartford," as a "mania for collecting curious or quaint old articles," involving collectors who "scoured the country round, in search of new discoveries."

67. Austin Abbott on Americans in Europe, *Harper's New Monthly Magazine*, July 1869, 248.

68. Richard W. H. Jarvis, Hartford, to William Jarvis Jr., Eyria, Ohio, June 2, August 2, 1883, Jarvis Collection, CHS.

69. Unfortunately, the purchase records come to us secondhand from Elizabeth Colt's estate attorney Charles Gross, who unquestionably had bills of sale for many items in the collection that are now lost. He supplied information on numerous objects, as to the dealer, place, and price paid, ranging from the painting by Antonio Paoletti (05.25) "purchased of Guggenheim, Venice, April 1882 for 1000 lire," to the reproductions of *Two columns from the Roman Forum* (05.1115) bought of "B. Boschetti, Rome, February 1882 for 230 lire," to the reproductions of arm chairs in the doge's palace (05.1881–2) "purchased in Venice, April 1882 of Moisa Rietti for 700 lire." These notes provide the closest thing on record to a time line for their travels.

70. Elizabeth Colt must have appeared outlandish with her "Oriental Tea Parties," "Russian concerts," and elaborately scripted "Union Bazaars" in which "all nations will be represented," with "articles for sale... from Russia, Paris, London, China, Japan, New York, Philadelphia," etc. [*Hartford Daily Times*, June 5, 1874]. At the Union Fair of 1886, "attended by ladies in brilliant and varied Russian,... Turkish, Algerian, Armenian, Greek & Georgian costumes," Elizabeth Colt "dispensed Russian tea, in a Russian costume" [*Hartford Daily Times*, February 10, 1886; Katherine Day, Hartford, to John Day, Hartford, February 11, 1886, Katherine Day Collection, Harriet Beecher Stowe Center]. In 1879, at the Battle Flag Day processional that coincided with the dedication of Connecticut's new state house, Elizabeth Colt decorated Armsmear by "tastefully draping" Turkish carpets "over the balconies," a gesture not mimicked on any of the nearly hundred houses and shops that were decorated for the event [*Hartford Daily Times*, September 17, 1879].

71. Retailers' marks or bills of sale indicated that Elizabeth Colt made purchases from R. H. Briggs of Boston, Henry Kohn and Sons of Hartford, and Gilman Collamore and Co. of New York.

72. Unable to exactly determine the value of lire in 1882 dollars, I have based this estimate on the fact that the most readily translated value, the 35 lire Elizabeth Colt paid for the Salviati pitcher, was probably a value no more than today's equivalent of $250.

73. The manufacture of accurate copies of shields and helmets, the originals attributed to Benvenuto Cellini (1500–71) and allegedly belonging to Francis I and Henry IV of Italy, were one of the triumphs of art brought to the attention of the U.S. public at the Centennial. Although Elizabeth Colt bought these in Florence, identical objects were exhibited in Philadelphia and are illustrated in *The Masterpieces of the Centennial International Exhibition* (Philadelphia, 1876), 2, 150, 151, 261.

74. *Hartford Daily Times*, September 10, 1896.

75. Elizabeth loaned pictures to the first annual art exhibition of the new Yale Art Gallery in 1867 [*Hartford Daily Times*, September 20, 1867]; objects and pictures to Hartford's Centennial Loan Exhibition in 1875 [*Hartford Daily Times*, November 2, 1875]; her best and most valuable picture to the Exhibition for the Veteran City Guard in Hartford in 1884 [*Hartford Daily Times*, March 6, 1884]; and to the memorial exhibition of Frederic Church's work at the Metropolitan Museum of Art in New York in 1900 [Mary Morris, Obituary Scrapbook, April 7, 1900, 38, 30, 35, CHS].

BUILDING MEMORIALS

1. *Appleton's Cyclopedia of American Biography*, ed. James G. Wilson and John Fiske (New York: D. Appleton, 1887), 695.

2. Colt, "Memoir," *Armsmear*, 323.

3. *Hartford Daily Courant*, June 6, 7, November 2, 9, December 14, 1870.

4. William Jarvis, Hartford, to William Jarvis Jr., Eyria, Ohio, February 21, 1867, Jarvis Collection, CHS.

5. Lydia H. Sigourney, Sachem's Head, to Elizabeth Hart Colt, Hartford, August 11, 1864, private collection, Colt Archives, Wadsworth Atheneum, 1457. Sigourney wrote with thinly veiled disappointment of her plan "to prepare a descriptive volume, in which your noble mansion was to take the lead… printed and bound in the highest style of excellence" and confessing that "I should… not think of invading the province of Mr. Barnard."

6. Elizabeth Hart Colt, Hartford, to Nathaniel Orr, New York, January 9, May 30, August 24, November 2, December 19, 1864, July 17, 1865, January 10, 1866, January 15, 1867, Orr Collection, University of Florida.

7. Elizabeth Hart Colt, Hartford, to Nathaniel Orr, New York, November 2, 1864, Orr Collection, University of Florida. Although once it was completed she could write of being "much pleased with" the illustrations and "superb" binding and of how it was "much praised" by "all who have seen it," Elizabeth was repeatedly discouraged by delays, owing mostly to Barnard's "keeping me from year to year with promises to finish," after he assumed the presidency of St. John's College in Annapolis in the middle of the project [Elizabeth Hart Colt, Hartford, to Nathaniel Orr, New York, January 15, 1867, April 13, 1866].

8. Ibid., April 13, 1866.

9. William Jarvis, Hartford, to William Jarvis Jr., Eyria, Ohio, February 21, 1867, Jarvis Collection, CHS. Almost all the copies that turn up from time to time bear presentation greetings from Elizabeth Colt. As late as 1889 she received an "acknowledgement of the… beautiful memorial of the life and career of your husband which you kindly sent me" from President Grover Cleveland [Grover Cleveland, Washington, D.C., to Elizabeth Colt, Hartford, February 12, 1889, private collection, Colt Archives, Wadsworth Atheneum, 1446].

10. The books were bound in New York by a Scottish immigrant custom binder, William Matthews, whose work first came to the public's attention at the New York Crystal Palace in 1853 [Willman Spawn, *Book Binding in America* (Bryn Mawr, Pa.: Bryn Mawr College, 1983), 96].

11. Jarvis, Jarvis, and Wetmore, *The Jarvis Family*. Considering the costly illustrations of both her father and husband, the six-page biographical profile on Sam, and the fact that it was printed in Hartford, it is likely that Elizabeth helped finance this publication, though no such acknowledgment is made.

12. Colt, *A Memorial of Mrs. Elizabeth Miller Jarvis*.

13. Rev. Samuel Hart, *In Memoriam: Samuel Colt and Caldwell Hart Colt* (Springfield: Clifton Johnson, 1898).

14. *A Memorial of Caldwell Hart Colt* (Hartford, 1899) and *A Memorial to the Soldiers and Sailors… in the War of 1898* (Springfield: Clifton Johnson, 1902); Mrs. Joseph Rucker Lamar, *A History of the National Society of the Colonial Dames of America* (Atlanta: National Society of the Colonial Dames of America, 1934), 2, 14, 27, 33, 73–74, 104–6. Elizabeth Colt was a member of the delegation that sat on the dais with the president during the dedication ceremony.

15. *Hartford Daily Times*, August 24, 1905.

16. *Industrial America; or Manufacturers and Inventors of the United States* (New York: Atlantic Publishing and Engraving, 1876); Davis, *The New England States*, with extended profiles on both Sam and Caldwell Colt; Lamb, *The Homes of America* (1879); Sherwood, "The Homes of America."

17. In November 1864 Elizabeth complained politely to her illustrator that "Barnard… showed me a letter from you asking for $500 more," and then asked if he would be "kind enough to send me a list of the engravings completed, with the prices annexed" [Elizabeth Hart Colt, Hartford, to Nathaniel Orr, New York, November 2, 1864, Orr Collection, University of Florida]. This and the bits of documentation regarding Frederic Church's involvement in the formation of her picture gallery hint that, as much as she may have valued input from distinguished men like Barnard and Church, their role as intermediaries may have been a necessary expedient when dealing with male contractors.

18. *Hartford Daily Times*, September 19, November 13, 1865, July 7, 1866.

19. Ibid., December 10, 1866. The monument was already designed in 1864 when Elizabeth Colt sent pictures of it to the illustrators to engrave for inclusion in *Armsmear* [Elizabeth Hart Colt, Hartford, to Nathaniel Orr, New York, September 3, 1864, Orr Collection, University of Florida].

20. *Hartford Daily Courant*, June 29, 1864.

21. Ibid., January 5, 1861; Barnard, *Armsmear*, 397.

22. Batterson's Civil War monuments found a ready market in New England and New York before 1870, with documented examples in Greenfield, Deerfield, and Stockbridge, Massachusetts [*Hartford Daily Times*, October 19, 1866], East Bloomfield, New York [*Hartford Daily Times*, October 1, 1866], and East Hartford, Connecticut. While most of Batterson's early monuments are of native red sandstone or marble, the Greenfield war monument, commissioned in 1869, is of red granite finished in the same Egyptian style as the column for Colt's monument. Batterson's business in war monuments took off after 1869, when he was awarded a fifty-thousand-dollar commission for the National Monument at Gettysburg [*Hartford Daily Times*, June 30, 1869]. Special thanks to David Ransom for identifying and documenting Batterson's work in Greenfield.

23. Large public monuments and burial obelisks predating the Civil War are rare and invariably of marble or brownstone. While detailed information is lacking, the technology to readily manipulate, cut, and polish granite does not appear to have become widely available in the United States until the mid-1870s, and certainly the Colt memorial may have been the largest granite monument in the country when it was installed in 1866. Twelve years later, Batterson's monument to Maj. Gen. John E. Wool (Oakwood Cemetery, Troy, New York), costing fifty thousand dollars—and today far and away the tallest granite obelisk in the Albany area—rivals John D. Rockefeller's in Cleveland, which is likely the tallest cemetery obelisk ever erected in the United States [*Hartford Daily Times*, April 15, 1878].

24. Barnard, *Armsmear*, 398.

25. *Hartford Daily Courant*, May 12, 1866.

26. *Hartford Daily Times*, September 25, October 1, 1866. In addition to Roger's Gabriel, Batterson displayed the prototype for the figure of a standing soldier that became the most popular of his Civil War monument figures. As early as 1853 James Batterson maintained a "collection of geological curiosities… on free exhibition," which "attracted thousands of visitors" eager to see the specimens of dinosaur tracks salvaged from the Portland Quarries, and rare and exotic stone collected in his world travels. Batterson donated the collection to Hartford High School in 1883 [*Hartford Daily Times*, May 11, 1883]. It is now lost.

27. *Hartford Daily Courant*, February 6, 1864.

28. William Jarvis, Hartford, to William Jarvis Jr., Eyria, Ohio, February 12, 1860, Jarvis Collection, CHS.

29. *Hartford Daily Courant*, February 8, 1864. In the weeks following, Hartford photographers N. A. and R. A. Moore, Prescott and Gage, and R. S. DeLamater all produced souvenir photographs in stereo and sizes suitable for framing [*Hartford Daily Courant*, February, 9, 12, 15, 1864].

30. Barnard, *Armsmear*, 253, 256. Sigourney's poem was dated the actual day of the fire.

31. J. D. Butler, "Destruction of the Armory by Fire," in Barnard, *Armsmear*, 251.

32. *Hartford Daily Courant*, February 8, 1864.

33. Butler, "Destruction," 252.

34. *Hartford Daily Courant*, February 8, 1864. The "work in progress in hands of contractors" at the time of the fire was estimated at $152,393 [Colt PFAMC, Journal B, August 1863–June 1869 (December 31, 1864), 174, private collection, Colt Archives, Wadsworth Atheneum, 862]. Estimates of the buildings and machinery lost totaled $530,652, including $200,000 for the east armory and $243,693 for machinery and tools [Account of the "loss by Fire, February 5, 1864," Colt PFAMC, Journal B, August 1863–June 1869 (December 31, 1864), 174, private collection, Colt Archives, Wadsworth Atheneum, 862–63]. The building was eventually replaced at a cost of about $300,000, bringing the total loss closer to $800,000.

35. *Scientific American*, March 12, 1864, 161.

36. *Hartford Daily Courant*, February 6, 1864.

37. Ibid., February 6, 15, 1864. Both wings of the armory were insured for $660,000. The loss was estimated at 60 percent, thus rendering a claim of about $396,000.

38. William Jarvis, Hartford, to William Jarvis Jr., Eyria, Ohio, February 12, 1864, Jarvis Collection, CHS; *Hartford Daily Times*, December 13, 1865; *Hartford Daily Courant*, January 2, 1866. The armory remained in ruins until January 1866.

39. *Hartford Daily Times*, October 23, 1865; William Jarvis, Hartford, to William Jarvis Jr., Eyria, Ohio, November 14, 1865, Jarvis Collection, CHS.

40. Thanks to Sara Wermiel for sharing this information.

41. *Hartford Daily Courant*, January 10, 1868; Barnard, *Armsmear*, 210.

42. *Hartford Daily Times*, April 5, 1866.

43. *Hartford Daily Courant*, January 10, 1868. In 1866 Colt's Firearms paid $21,234 (today about $680,000) to the Phoenix Iron Co. in Pennsylvania for iron beam essential to carrying out fireproof construction [Colt PFAMC, Cash Book B, 1864–71 (October 8, 1866), 119, private collection, Colt Archives, Wadsworth Atheneum]. Sara Wermiel, an architectural historian and graduate student at MIT, has suggested that the technology of fireproof construction only became possible in the United States in the mid-1850s through the use of brick arch construction combined with rolled iron beams first developed by the Phoenix Ironworks, the biggest steel mill of its day. One of its first applications was in constructing the new Treasury Department building in Washington in the late 1850s. According to Ms. Wermiel, the next earliest industrial building of fireproof construction was the Brown and Sharpe factory built in Providence, Rhode Island, in 1870 [interview with author, June 1995].

44. *Hartford Daily Times*, November 28, 1866.

45. *Hartford Daily Courant*, January 10, 1868.

46. *Hartford Daily Times*, September 11, 1867; "An Historical Sketch of the formation and organization of the Parish of the Good Shepherd," Vestry Records, Church of the Good Shepherd, 1–3, Episcopal Church Diocese Records, Hartford.

47. Gen. William B. Franklin, Diary, 1865–67 (April 9–10, August 2, 1866), Colt PFAMC, Administrative File, CSL, box 54.

48. Edward Tuckerman Potter, Ledger, ca. 1868–75 (September 1867–February 1868), 2, 3, 7, Potter Papers, Avery Architectural Library, Columbia University; *Hartford Daily Times*, October 8, 1868. Potter designed St. John's Episcopal Church in East Hartford, Connecticut, in 1866 or early 1867 for Rev. William C. Doane, former rector of Armory Sabbath School. Doane almost certainly played a role in Potter's selection as architect.

49. Sarah Bradford Landau, "Edward T. and William A. Potter: American High Victorian Architects," Ph.D. diss., New York University, 1978, 2–3, 16, 35, 38, 154.

50. Rev. William Jarvis, Hartford, to William Jarvis Jr., Eyria, Ohio, February 24, 1871, Jarvis Collection, CHS.

51. Elizabeth Miller Hart Jarvis, Hartford, to William Jarvis Jr., Eyria, Ohio, September 9, 1867, Jarvis Collection, CHS.

52. *Hartford Daily Times*, January 23, 1867.

53. Ibid., January 30, 1869. Among the craftsmen and manufacturing companies that participated in the construction, Mr. J. Brown was credited with the masonry; Anthony Nolle, woodcarving, furniture, and pews; A. A. Prall of Prall and Drummon, furniture and pews; William Fenner, stonecutting; Moore of New York, painting; Smith and Crane, exterior stonecarving; and Mitchell Vance and Co., gas fixtures [*Hartford Daily Times*, October 8, 1868; Colt PFAMC, Journal B, August 1863–June 1869 (January 28, 30, 1869), 225, private collection, Colt Archives, Wadsworth Atheneum, 889].

54. *Hartford Daily Times*, October 8, 1884.

55. Lamb, *The Homes of America*, 183; *Hartford Daily Times*, October 8, 1868.

56. Sherwood, "The Homes of America," 323.

57. *Hartford Daily Times*, October 8, 1868, January 27, 1869.

58. Ibid., January 3, 1870.

59. Colt PFAMC, Journal B, August 1863–June 1869 (October 12, December 22, 1868, January 22, 1869), 211, 221, 225, private collection, Colt Archives, Wadsworth Atheneum, 888–89. Sharp used the commission to advertise his firm [Advertisement for "Henry E. Sharp and Son, Glass Stainers, E. 22d Street, New York," in Henry Hudson Holly's *Church Architecture* (Hartford: M. H. Mallory, 1871), 259]. Special thanks to Virginia Raguin for sharing this information. Her work on the American stained glass industry and on the uses of fine art by stained glass artists was invaluable.

60. A loose bound print, likely removed from an edition of the *Art Journal*, a British periodical represented in the Colts' library, shows the model for the Catherine wheel component of the memorial window. The marble bust, a study for the recumbent figure by Edward Bartholomew, is in the collection of the Wadsworth Atheneum (no. 05.1119). Special thanks to Virginia Raguin for sharing this informa-

tion. She kindly provided a reproduction of the print for the Colt Archive (p. 2067), but has yet to identify the exact issue of the periodical from which it was removed.

61. *Church of the Good Shepherd: A Brief Description* (Hartford: Brown and Gross, 1869), 16, 30.

62. *Hartford Daily Times*, October 8, 1868.

63. Sherwood, "The Homes of America," 323. Elizabeth Colt added a wrinkle to the story in her "Memoir" by noting that Colt's brothers "repaid him with malice and slander" [306].

64. *Hartford Daily Times*, January 29, 1869.

65. Elizabeth Hart Colt, printed form letter, March 18, 1868, private collection, Colt Archives, Wadsworth Atheneum.

66. *Hartford Daily Times*, February 10, 1875.

67. *First Annual Report of the Connecticut Branch of the Woman's Auxiliary to the Board of Missions* (Hartford: Case, Lockwood and Brainard, 1881), 3–7.

68. *Second Annual Report of the Connecticut Branch of the Woman's Auxiliary to the Board of Missions* (Hartford: Case, Lockwood and Brainard, 1882), 27.

69. *Hartford Daily Times*, November 13, 1883; Phillips Brooks, Boston, to Elizabeth Hart Colt, January 16, 1885, CHS; Cornelius Vanderbilt, New York, to Elizabeth Hart Colt, 1885, CHS; *Fifth Annual Report of the Connecticut Branch of the Women's Auxiliary to the Board of Missions* (Southport, Conn.: Church Record Association, 1885), 5.

70. William Cowper Prime, Punta Gorda, Florida, to Charles Dudley Warner, Hartford, January 23, 1894, Warner Papers, Watkinson Library, Trinity College, Hartford.

71. *Hartford Daily Times*, January 25, 1894.

72. Colt, *A Memorial of Mrs. Elizabeth Miller Jarvis*, 19.

73. Bradley, "Address."

74. Bragdon, "'Dauntless' Models Recall 'Colly' Colt," 28. Astonishingly, Elizabeth Colt also built, that year, a second Church of the Good Shepherd in Caldwell's memory, in Punta Gorda, Florida.

75. *Hartford Daily Times*, September 10, 1896; Bragdon, "'Dauntless' Models Recall 'Colly' Colt," 28. Part of the evidence for Elizabeth Colt's involvement in developing the concept and design for the building are the details taken from objects in her possession that embellish the exterior of the building. The "heads of buffalo and mountain goats," and Florida tarpon, for example, were not generic. They were based on actual mounted trophies from Caldwell's western hunting trips. The "199 stuffed birds," "Mounted Tarpon caught in Florida," and the mounted heads of "Rocky Mountain Sheep, Reindeer, American Bison, Elk, and Common Deer"—although disposed of with no remaining photographic record—were among the collections received by the Wadsworth Atheneum from Elizabeth Colt's estate in 1905.

FIRST LADY OF CONNECTICUT

1. Kathleen D. McCarthy, *Noblesse Oblige: Charity and Cultural Philanthropy in Chicago, 1849–1929* (Chicago: University of Chicago Press, 1982), 3. Other works concerning the Victorian sense of noblesse oblige include McCarthy, *Women's Culture: American Philanthropy and Art, 1830–1930* (Chicago: University of Chicago Press, 1991), Gertrude Himmelfarb, *Poverty and Compassion: The Moral Imagination of the Late Victorians* (New York: Alfred A. Knopf, 1991), and Robert H. Bremner, *The Public Good: Philanthropy and Welfare in the Civil War Era* (New York: Alfred A. Knopf, 1980).

2. For a woman who epitomized self-sovereignty and independence, Elizabeth Colt wrote derisively about the more extreme elements of the women's movement, noting in 1881 that "so many women are fretting in discontent with the place their Creator has assigned to them in the world, striving to do a man's work poorly, instead of making their own noble as He meant it should be…. [Women's work is] not inferior, because it is necessarily not the same." The challenge, which she felt that women like her mother had met with consummate skill, was making "the work given her to do… admired and respected alike by men and women" [Colt, *A Memorial of Mrs. Elizabeth Miller Jarvis*, 25–26]. Although her perspective may have been shared by most of the women of her class and generation, it was also based in her Christian resignation and faith, a faith that sustained her in times of loss and propelled her into the realm of civic leadership. Not so, Isabella Beecher Hooker, who frequently shared the stage, the speaker's podium, and the limelight with such national icons of women's rights as Lucy Stone, Elizabeth Cady Stanton, and Susan B. Anthony.

Hooker was Hartford's leading suffragist, playing a leadership role in the 1870s convention in Washington [*Hartford Daily Courant*, January 27, 1870], and chairing the first annual meeting of the Connecticut Woman's Suffrage Association, also in 1870 [*Hartford Daily Courant*, August 19, 1870].

3. *Hartford Daily Courant*, February 2, 1864.

4. William Jarvis, Hartford, to William Jarvis Jr., Eyria, Ohio, February 1, 1864, Jarvis Collection, CHS. Elizabeth may have been elected president of the society; formal records not having been found, it is difficult to tell.

5. Fredrickson, *The Inner Civil War*.

6. *Hartford Daily Times*, November 18, 1864.

7. Ibid.

8. Bushnell, "How to be a Christian in Trade," 265.

9. *Hartford Daily Times*, August 24, 1905; *In Memoriam: Mrs. Samuel Colt* (Elizabeth Hart Jarvis) (Hartford: Connecticut Society of Colonial Dames of America, 1905), np.

10. Colt, *A Memorial of Mrs. Elizabeth Miller Jarvis*, 7–8.

11. Fredrickson, *The Inner Civil War*, 98–101, 211–13.

12. *Hartford Daily Courant*, December 10, 1864, January 20, 1865.

13. Ibid., January 31, 1865. No such home was built in Hartford, though the group may have contributed to the State Soldier's Home built after the war in Darien.

14. *First Annual Report of the Union for Home Work* (Hartford, 1872), 3–5, 40–41. Sarah Cowen was its first president, with an executive committee composed of twelve women, including Elizabeth Colt, her sister-in-law, and the wives and daughters of clergymen, attorneys, Hartford's mayor, and leading industrialists. The six-man advisory committee included Elizabeth's brother-in-law, the mayor, her long-time friend Frank Cheney, and the man who would read the last rites at her funeral, Rev. Francis Goodwin.

15. *Hartford Daily Times*, March 1897, cited in *Fourteenth Annual Report of the Union for Home Work* (Hartford, 1897), 59–61; *Hartford Daily Times*, December 6, 1883.

16. Mabel Collins Donnelly, *The American Victorian Woman: The Myth and the Reality* (New York: Greenwood Press, 1986), 108. The author of this generally thoughtful and informative book erred simply in misreading the current value of Victorian money.

17. *Fourteenth Annual Report of the Union for Home Work*, 56–58.

18. *Hartford Daily Times*, April 21, 1880.

19. *Fourteenth Annual Report of the Union for Home Work*, 56–58.

20. *Hartford Daily Times*, March 1897, cited in *Fourteenth Annual Report of the Union for Home Work*, 59–61.

21. The theories and practice of the London Charity Organization Society, founded in 1869, were first adopted in the United States in Buffalo, New York, in 1877, catching on like wildfire so that by the end of the century there were at least 237 related societies in the United States. In 1878 Hartford and New Haven were among the first cities in New England to adopt the system [*Fifty Years of Family Social Work, 1877–1927* (Buffalo: Conference on Family Life in America Today, 1927), 14–21].
    Elizabeth Sluyter is quoted in *Hartford Daily Times*, October 30, 1878.

22. *Hartford Daily Times*, December 18, 1878.

23. Ibid., January 3, 1879.

24. Ibid., October 18, 1879.

25. Ibid., April 16, 1879.

26. Ibid., June 29, 1872, June 3, 1880.

27. Ibid., October 16, 1884. Built at a cost of $25,000 (today almost $1 million) it represented a lady's "fiscal campaign" of considerable ambition.

28. Ibid., June 6, 1880; *June Days: Thirty Poems by Friends of the Union for Home Work* (Hartford: Case, Lockwood and Brainard, 1880); *Bazaar Budget* (June 1–6, 1880)

29. *Hartford Daily Times*, April 12, 1875. The event was described by participants as "one of the most unique and beautiful entertainments ever given in this city."

30. Ibid.; Henry Theodore Tuckerman, *Book of the Artists* (New York: G. Putnam and Sons, 1867), 323–24; "Mr. Huntington's 'Republican Court' in the Time of Washington," *New Path*, vol. 2, November 1865, 176–78.

31. *Hartford Daily Times*, April 22, 1875.

32. Ibid., April 21, 1875; Photographic Album, Portraits from the "Martha Washington Tea Party and Reception," Colt Archives, Wadsworth Atheneum.

33. *Hartford Daily Times*, April 22, 1875.

34. *Old-Time Relics at the Martha Washington Tea* (Hartford, 1875), Colt Archives, Wadsworth Atheneum.

35. *Hartford Daily Times*, April 12, 1875.

36. Ibid., April 24, 1875. I make this claim by dividing the reported cost of a "good meal" ($.25) by the $4,100 raised.

37. Ibid., June 23, October 5, November 1, 2, 8, 1875; *Catalogue of Works of Art of the Centennial Loan Exhibition* (Hartford, 1875). The Centennial Loan Exhibition was a much bigger deal. James Batterson, no doubt inspired by what Hartford's women had accomplished in April, teamed up with the Union ladies to double the exhibition, adding American and European pictures to the collection of antiquities. It was the greatest art exhibition ever staged in Hartford up to that time. Elizabeth Colt loaned some pictures and objects, but Batterson's collections dominated the show. Although about 85 percent of the pictures exhibited were by European artists, the exhibition included works by Frederic Church, John Kensett, Benjamin Champney, Sanford Gifford, Albert Bierstadt, and Dwight Tryon. Staged at the Phoenix Bank, it included bronzes and marble statuary, Carl Conrad's bas-relief sculpture for the Founder's Memorial in Plymouth, Massachusetts, in a room "draped in maroon... [with] spears, shields, banners and eagles." Many more Colonial antiquities were exhibited, including a "sword presented by General Washington," "Gen. Wadsworth's Turkish pipe," and "a conch-shell used to call the grandfather of W. R. Cone from the field when the news of the battle of Lexington was heard." Afterward, Elizabeth Colt and the union ladies formed the Women's Centennial Association, "to make a collection" to display in Philadelphia in 1876. Connecticut's exhibits there helped trigger the Colonial Revival Movement and antiques craze that was arguably one of the most influential movements in American culture and arts during the next half century.

38. *In Memoriam: Mrs. Samuel Colt (Elizabeth Hart Jarvis)*.

39. R. G. Dun and Co., Report, Hartford, April 10, 1867, April 14, 1870, April 28, 1873, 15:298j, R. G. Dun and Co. Collections, Baker Library, Harvard University Graduate School of Business Administration.

40. George Keller, Hartford, to sister, August 20, 1889, Keller Papers, Harriet Beecher Stowe Center.

41. *Hartford Daily Courant*, January 1, 1870.

42. *Hartford Daily Times*, May 29, 1877.

43. Decorative Art Society of Hartford, Manuscript, Secretary's Reports, 1877–86, Hartford Art School, University of Hartford, 1–27.

44. *Hartford Daily Times*, February 12, October 22, 1878. Champney, a well-respected American painter and member of the National Academy, joined in October of the society's first year, remaining until 1885 when the society occupied new quarters in the Wadsworth Atheneum.

45. Ibid., December 14, 1883.

46. McCarthy, *Women's Culture*.
    The men who championed the cause of civic art in Hartford were a divided and divisive lot. The most high-minded and cosmopolitan faction was led by Charles Dudley Warner and William Cowper Prime, a New York journalist, art critic, collector, and founding professor of Princeton's Department of Art. James Batterson and the Goodwins, and through them, J. Pierpont Morgan, represented the faction that eventually gained dominance [Gregory Hedberg, "The Morgans of Hartford," in *J. Pierpont Morgan, Collector*, ed. Linda Horvitz Roth (Hartford: Wadsworth Atheneum, 1987), 11]. A third faction was represented by the city's artists, who clamored for recognition amid a rising tide of Eurocentricity that eventually caused the best of them, notably Dwight Tryon and John Fitch, to abandon Hartford for Europe and New York. Finally, there was Timothy M. Allyn, a politician, hotel baron, and self-made millionaire, who launched the opening volley in the campaign to create a public gallery and school of art.

47. *Special Acts and Resolutions of the State of Connecticut (1885–89)* (Hartford: Case, Lockwood and Brainard, 1890), 10:285. Special thanks to David Corrigan for discovering and sharing.

48. *Hartford Daily Times*, February 6, 9, 16, 1886.

49. *Wadsworth Atheneum and Morgan Memorial, 1841–1921: Its Past and Future* (Hartford: Hartford Times, 1928), 9.

50. Elizabeth H. Colt, Hartford, to Mrs. John M. Holcombe, Hartford, June 9, 1899, CHS.

51. *A Register of the Connecticut Society of the Colonial Dames* (Hartford, 1907).

52. Mary Philkotheta Root, ed., *Chapter Sketches, Connecticut Daughters of the American Revolution* (New Haven: Connecticut Chapters,

Daughters of the American Revolution, 1904), v; *In Memoriam: Mrs. Samuel Colt (Elizabeth Hart Jarvis)*. Special thanks to Barbara Ulrich for information about the Ruth Wyllys Chapter of the DAR.

53. Lamar, *A History of the National Society of the Colonial Dames of America*, 2, 14, 27, 33, 73–74, 104–6.

54. *Hartford Times*, May 11, 1896, p. 6, May 7, 1896, p. 7.

In 1924, when the nationally eminent antiques collector and champion of colonial and rural life Wallace Nutting described Hartford as the "center of the American past," he had in mind its legions of antiques collectors and colonialists [*Connecticut Beautiful* (Garden City, N.Y.: Garden City Publishing, 1923), 130]. Buildings and hereditary organizations sprouted up from the soil, fertilized by long memories, ancient fortunes, and a growing concern about a now-dominant population of immigrants, Catholics, Eastern Europeans, and "others." The long march "back" to a sense of rootedness, the core of the so-called Colonial Revival Movement, has been explored by Michael Kammen and other scholars [Kammen, *Mystic Chords of Memory: The Transformation of Tradition in American Culture* (New York: Alfred A. Knopf, 1991]. It is the story of forging, or trying to forge, a national identity through a matrix of icons, artifacts, and legends. Although widely stigmatized as elitist, nativist, racist, and exclusionary, the turn-of-the-century champions of the past were motivated by a desire to maintain the supremacy of their Anglo-Saxon culture and, more important, to create a unified center based on a common mythology and shared values.

55. *Hartford Times*, April 29, 1896.

56. Lamar, *A History of the National Society of the Colonial Dames of America*, 2, 14, 27, 33, 73–74, 104–6.

57. Ibid.

58. *Hartford Daily Times*, August 24, 1905.

59. *In Memoriam: Mrs. Samuel Colt (Elizabeth Hart Jarvis)*.

60. Elizabeth Hart Colt, Hartford, to "Lizzie" Mrs. C. L. F. Robinson, 1901, private collection, Colt Archives, Wadsworth Atheneum, 1439. Sadly, it is easy to imagine CEOs willing to have changed the name. The Colt legacy has proven a gold mine.

61. Grant, *The Colt Legacy*, 57–58.

62. Elizabeth H. Colt, Hartford, to Elizabeth Hart Beach Robinson, Newport, June 6, 13, 21, 1901, private collection, Colt Archives, Wadsworth Atheneum, 1436–40.

63. Elizabeth Hart Colt, Hartford, to "Lizzie" Mrs. C. L. F. Robinson, June 13, 1901, private collection, Colt Archives, Wadsworth Atheneum, 1437.

64. This book notwithstanding, the quantity of documentation on the life of Elizabeth Colt is razor thin. Lost from record for an entire year of her tour of Europe, it is the rare season when records show where Elizabeth Colt was, let alone what she was involved with. Probably, the two medals in the Colt Collection from the Louisiana Purchase Exhibition of 1904 were acquired there, but there is no record of her traveling that far in 1904.

65. "The Connecticut House," *World's Fair Bulletin*, 5, no. 7 (May 1904), 58.

66. "Facts about the Universal Exposition," ibid.

67. Inventory of American Sculpture, National Museum of American Art, Smithsonian Institution, 9431–56. Charles Dudley Warner, Hartford's premier champion for the cause of public culture and civic embellishment, undoubtedly played a role in encouraging Elizabeth Colt to build the Colt Memorial Statue. Warner headed a Commission on Sculpture and through his role as a travel writer, editorialist, and publisher of the *Hartford Courant* drew on an extensive web of national and regional connections to build a case for civic embellishment. In 1898 Warner led the campaign that resulted in John Massey Rhind's first Hartford commission, the Corning Fountain; Rhind later sculpted a portrait bust of Warner.

68. *Hartford Courant*, April 26, 1906.

69. Ibid., August 31, 1905, 12. Work commenced on the site in February 1905 [*Hartford Daily Times*, April 24, 1905].

70. *Hartford Courant*, August 18, 1905.

71. Charles E. Gross, cited in the *Hartford Courant*, August 24, 1905.

72. Last Will and Testament of Elizabeth H. Colt, 1905, Colt Archives, Wadsworth Atheneum, 16–22.

73. *Hartford Courant*, November 17, 1910. A newspaper account of the gift appraised the collection at $158,000 (today $5.5 million) but noted, pointedly, on behalf of the building committee "that the gift of $50,000 has been entirely expended in its erection." With no endowment and a fund that probably did not cover more than 75 percent of the building's cost, Elizabeth Colt succeeded in making the Goodwin-Morgan faction share the cost of her own memorial. Curiously, it was only after the Goodwins relinquished control over the Atheneum's board in the 1970s that the institution's stewardship of the collection nosedived, culminating, in 1982, with the dismantling of the last vestige of the Elizabeth Hart Jarvis Colt Gallery.

74. Ibid., November 17, 1910.

75. "Armsmear Is Much Altered," probably *Hartford Daily Courant*, ca. 1909, Scrapbook, Hartford, np. Taylor Papers, CHS.

76. *Hartford Daily Times*, November 16, 1910.

77. Ibid.

78. Ibid., August 24, 1905.

79. Rev. George T. Linsley, cited in *Hartford Daily Times*, August 28, 1905.

80. *Hartford Daily Courant*, April 26, 1906.